Wood Modification: Optimisation and Characterisation of Modified Timbers

Wood Modification: Optimisation and Characterisation of Modified Timbers

Guest Editors

Morwenna Spear
Miklós Bak

MDPI

Basel • Beijing • Wuhan • Barcelona • Belgrade • Novi Sad • Cluj • Manchester

Guest Editors
Morwenna Spear
The BioComposites Centre
Bangor University
Bangor
United Kingdom

Miklós Bak
Faculty of Wood Engineering
and Creative Industries
University of Sopron
Sopron
Hungary

Editorial Office
MDPI AG
Grosspeteranlage 5
4052 Basel, Switzerland

This is a reprint of the Special Issue, published open access by the journal *Forests* (ISSN 1999-4907), freely accessible at: www.mdpi.com/journal/forests/special_issues/TI028472P2.

For citation purposes, cite each article independently as indicated on the article page online and using the guide below:

Lastname, A.A.; Lastname, B.B. Article Title. *Journal Name* **Year**, *Volume Number*, Page Range.

ISBN 978-3-7258-2960-6 (Hbk)
ISBN 978-3-7258-2959-0 (PDF)
https://doi.org/10.3390/books978-3-7258-2959-0

Cover image courtesy of Morwenna Spear

Contents

About the Editors

Morwenna Spear

Morwenna Spear is a research fellow in the BioComposites Centre at Bangor University, providing research and consultancy for the timber industry. Her main areas of interest are wood modification, the chemistry and physics of wood (from nano scale to industrial processes), and adding value to underutilized woods. She also has a strong interest in the biogenic carbon storage role of timber and accounting for carbon dynamics associated with extending the service life of wood products. Morwenna is currently the Chair of the Wood Technology Group of the IOM3, a learned society promoting excellence and sustainability in materials science, where she also engages with the Construction Materials Group. She also participates in various industry- and policy-facing committees to offer technical insights and maintain awareness of industry needs and trends.

Miklós Bak

Miklós Bak obtained an M.Sc. degree in Timber Engineering in 2007, M.Sc. in engineering teacher in 2009, and Ph.D. in Materials Science and Technology in 2013 from the University of Sopron, as well as an M.Sc. in Materials Science in 2015 from the University of Miskolc. Miklós is a researcher at the Faculty of Wood Engineering and Creative Industries of the University of Sopron. Miklós has participated in 12 scientific projects, being the coordinator in 3 projects. To date, Miklós has contributed to 230 scientific publications, 167 of which are in international conferences, books, and theses, and 63 are in international journals with referees.

forests

MDPI

Editorial

Wood Modification—Trends and Combinations

Morwenna J. Spear [1,*] and Miklós Bak [2,*]

1 The BioComposites Centre, Bangor University, Deiniol Road, Bangor, Gwynedd LL57 2UW, UK
2 Faculty of Wood Engineering and Creative Industries, University of Sopron, 9400 Sopron, Hungary
* Correspondence: m.j.spear@bangor.ac.uk (M.J.S.); bak.miklos@uni-sopron.hu (M.B.)

Wood modification is a field that has enjoyed sustained interest over the past two decades, although its history can be tracked back significantly further, to the pioneering work of Alfred Stamm and co-workers at the Forest Products Laboratory in the USA in the 1930s, 1940s, and 1950s [1–4]. The steady surge in interest over the last twenty years reflects a period in which there has been a series of European conferences dedicated to this topic, starting in Ghent in 2003 [5], and migrating around Europe biennially ever since [6]. The conference series itself was initiated by the EU Thematic Network on wood modification, which started in 2000 with partners from 15 European countries [7]. In parallel, a European Thermally Modified Timber workshop was also developed, and has been run by the IHD in Dresden since 2003 [8]. This has been a driver for a great deal of innovation in Europe and, indeed, globally.

As a result, commercially available wood modification techniques such as thermal modification, acetylation, and furfurylation are well established and have been well studied [9–13]. A wealth of textbooks are available to consolidate this topic and communicate the state of the art in this field [14–16]. Many other modification systems for wood are still under development, using resins, bio-based chemicals, biopolymers, and novel combinations of heat, pressure, steam, or other active components [17,18]. Mechanical and physical modification systems have also evolved in parallel with thermal and chemical methods [19,20]. Most wood modification systems seek to enhance the dimensional stability of wood [21–23], while many also extend its service life [24] and provide durability and resistance to fungal decay [25–28] or insect and marine borer attacks [29].

In recent years, innovation in wood modification has focused on three main themes: combining modifications to increase the range of benefits offered [30–34]; seeking bio-based or green chemistry modifications [35–38]; and introducing nanotechnologies to provide innovative functional materials [17,37,39–42]. As a result, an invitation was extended to authors for papers relating to each of these aspects for a Special Issue, and the papers received met this demand well.

Five of the papers in this Special Issue address thermal modification or thermal processes [43–47]. Four of the papers address innovative chemical treatment agents [45,48–50], while another considers the effect of delignification on the densification process [51]. One paper considers the permeability gains made using microwave drying as an intended pre-treatment for impregnation-based modification systems [47].

Two papers consider wood functionalisation methods, namely, phase change materials [52] and the use of silver nanoparticles [50]. These suggest potential for use in advanced building products (thermal efficiency) and resistance to biodegradation, respectively. Clearly, there is plenty of overlap within this cohort of papers, and several papers use a combination of approaches [45,50,51], reflecting the current trends in research related to this subject.

Excitingly, two papers use machine learning, algorithm optimisation, or modelling approaches to improve prediction and processing [43,44]. This area is set to develop in parallel with advances in Industry 4.0 and as concepts become incorporated within wood modification technologies.

check for updates

Citation: Spear, M.J.; Bak, M. Wood Modification—Trends and Combinations. *Forests* **2024**, *15*, 1268. https://doi.org/10.3390/f15071268

Received: 17 June 2024
Accepted: 17 July 2024
Published: 20 July 2024

Looking at the published articles, we can say that the demand for sustainable, natural materials and technologies is steadily emerging even in wood modification procedures. Thes concepts have always been in at the heart of the basic principles of wood modification. All the scientific results published in this area ultimately contribute to the industrial application of as many promising modification processes as possible, hopefully within the shortest possible time.

We hope that you will enjoy this selection of recent research in the field of wood modification and property optimisation. With so many advances being made by such a wide group of researchers around the globe, it remains essential to tune in to what is happening in different sub-topics within the wider field of wood modification.

Funding: This research received no external funding.

Conflicts of Interest: The author declares no conflicts of interest.

References

1. Stamm, A.J.; Hansen, L.A. Minimizing Wood Shrinkage and Swelling: Effect of Heating in Various Gases. *Ind. Eng. Chem.* **1937**, *29*, 831–833. [CrossRef]
2. Stamm, A.J.; Seborg, R.M. Minimizing wood shrinkage and swelling–treating with synthetic resin-forming materials. *Ind. Eng. Chem.* **1936**, *28*, 1164–1169. [CrossRef]
3. Stamm, A.J.; Seborg, R.M. Resin-treated wood (IMPREG) Forest Products Laboratory Report 1380 (Revised 1962). Forest Products Laboratory, Forest Service, US Department of Agriculture: Madison, WI, USA, 1942.
4. Stamm, A.J.; Burr, H.K.; Kline, A.A. Staybwood—Heat-Stabilized Wood. *Ind. Eng. Chem.* **1946**, *38*, 630–634. [CrossRef]
5. Van Acker, J.; Hill, C. (Eds.) European thematic network for wood modification. In Proceedings of the First European Conference on Wood Modification, Ghent, Belgium, 3–4 April 2003; ISBN 9080656526.
6. Jones, D.; Militz, H. Overview of the Tenth European Conference on Wood Modification (ECWM10). *Wood Mater. Sci. Eng.* **2023**, *18*, 1–2. [CrossRef]
7. Jones, D.; Homan, W.; Van Acker, J. Modified Wood—Methods, testing and applications: Outcomes of the EU-Thematic Network. In Proceedings of the International Research Group on Wood Preservation 38th Annual Meeting, Brisbane, Australia, 18–23 May 2003; p. IRG/WP 03-40268.
8. Scheiding, W. (Ed.) IHD Dresden. In Proceedings of the 11th European TMT Workshop, Dresden, Germany, 2–3 March 2023; Available online: https://www.ihd-dresden.com/en/knowledge-portal/tagungsbaende/ (accessed on 6 June 2024).
9. Esteves, B.M.; Pereira, H.M. Wood Modification by Heat Treatment: A Review. *BioResources* **2008**, *4*, 370–404. [CrossRef]
10. Gérardin, P. New alternatives for wood preservation based on thermal and chemical modification of wood—A review. *Ann. For. Sci.* **2016**, *73*, 559–570. [CrossRef]
11. Mantanis, G.I. Chemical modification of wood by acetylation or furfurylation: A review of the present scaled-up technologies. *BioResources* **2017**, *12*, 4478–4489. [CrossRef]
12. Sandberg, D.; Kutnar, A.; Mantanis, G.I. Wood modification technologies—A review. *Iforest–Biogeosciences For.* **2017**, *10*, 895–908. [CrossRef]
13. Jones, D.; Sandberg, D. A review of wood modification globally–updated findings from COST FP1407. *Interdiscip. Perspect. Built Environ.* **2020**, *1*, 1. [CrossRef]
14. Hill, C.A.S. *Wood Modification–Chemical, Thermal and Other Processes*; Wiley Series in Renewable Resources; Wiley and Sons: Chichester, UK, 2006.
15. Jones, D.; Sandberg, D.; Goli, G.; Todaro, L. *Wood Modification in Europe: A State-of-the-Art about Processes, Products and Applicationsi*; Firenze University Press: Florence, Italy, 2019; online PDF; 113p, ISBN 978-88-6453-970-6.
16. Sandberg, D.; Karlsson, O.; Kutnar, A.; Jones, D. *Wood Modification Technologies. Principles, Sustainability, and the Need for Innovation*; CRC Press: Baton Rouge, FL, USA, 2021; ISBN 978-1-138-49177-9.
17. Spear, M.J.; Curling, S.F.; Dimitriou, A.; Ormondroyd, G.A. Review of functional treatments for modified wood. *Coatings* **2021**, *11*, 327. [CrossRef]
18. Zelinka, S.L.; Altgen, M.; Emmerich, L.; Guigo, N.; Keplinger, T.; Kymäläinen, M.; Thybring, E.E.; Thygesen, L.G. Review of Wood Modification and Wood Functionalization Technologies. *Forests* **2022**, *13*, 1004. [CrossRef]
19. Sandberg, D.; Navi, P. Introduction to thermo-hydro-mechanical (THM) wood processing. Växjö Universitet Skog & Trä: Växjö, Sweden, 2007.
20. Sandberg, D.; Haller, P.; Navi, P. Thermo-Hydro and Thermo-Hydro-Mechanical Wood Processing: An Opportunity for Future Environmentally Friendly Wood Products. *Wood Mater. Sci. Eng.* **2013**, *8*, 64–88. [CrossRef]
21. Hill, C.A.S.; Jones, D. The dimensional stabilisation of Corsican Pine sapwood by reaction with carboxylic acid anhydrides. *Holzforschung* **1996**, *50*, 457–462. [CrossRef]
22. Ohmae, K.; Minato, K.; Norimoto, M. The Analysis of Dimensional Changes Due to Chemical Treatments and Water Soaking for Hinoki (Chamaecyparis Obtusa) Wood. *Holzforschung* **2002**, *56*, 98–102. [CrossRef]

23. Hill, C.A.S.; Altgen, M.; Rautkari, L. Thermal modification of wood—A review: Chemical changes and hygroscopicity. *J. Mater. Sci.* **2021**, *56*, 6581–6614. [CrossRef]
24. Ormondroyd, G.; Spear, M.; Curling, S. Modified wood: Review of efficacy and service life testing. *Constr. Mater.* **2015**, *168*, 187–203. [CrossRef]
25. Weiland, J.J.; Guyonnet, R. Study of Chemical Modifications and Fungi Degradation of Thermally Modified Wood Using DRIFT Spectroscopy. *Holz Als Roh-Und Werkst.* **2003**, *61*, 216–220. [CrossRef]
26. Thybring, E.E. The decay resistance of modified wood influenced by moisture exclusion and swelling reduction. *Int. Biodeterior. Biodegrad.* **2013**, *82*, 87–95. [CrossRef]
27. Rademacher, P.; Nemeth, R.; Bak, M.; Fodor, F.; Hofmann, T.; Baar, J.; Paril, P.; Rousek, R.; Paschova, Z.; Sablik, P.; et al. European co-operation in wood research from native wood to engineered materials. Part 1: Impregnation with native impregnation agents. *Pro Ligno* **2018**, *13*, 341–350.
28. Biziks, V.; Bicke, S.; Koch, G.; Militz, H. Effect of phenol-formaldehyde (PF) resin oligomer size on the decay resistance of beech wood. *Holzforschung* **2021**, *75*, 574–583. [CrossRef]
29. Gellerich, A.; Brischke, C.; Militz, H.; Klüppel, A. Resistance of modified wood against marine borers. *Holztechnologie* **2018**, *59*, 5–11.
30. Esteves, B.; Ribeiro, F.; Cruz-Lopes, L.; Ferreira, J.; Domingos, I.; Duarte, M.; Duarte, S.; Nunes, L. Combined treatment by densification and heat treatment of maritime pine wood. *Wood Res.* **2017**, *62*, 373–388.
31. Montanari, C.; Li, Y.; Chen, H.; Yan, M.; Bergland, L.A. Transparent wood for thermal energy storage and reversible optical transmittance. *ACS Appl. Mater. Interfaces* **2019**, *11*, 20465–20472. [CrossRef] [PubMed]
32. Popescu, C.-M.; Jones, D.; Kržišnik, D.; Humar, M. Determination of the effectiveness of a combined thermal/chemical wood modification by the use of FT–IR spectroscopy and chemometric methods. *J. Mol. Struct.* **2020**, *1200*, 127133. [CrossRef]
33. Jones, D.; Kržišnik, D.; Hočevar, M.; Zagar, A.; Humar, M.; Popescu, C.-M.; Popescu, M.-C.; Brischke, C.; Nunes, L.; Curling, S.F.; et al. Evaluation of the Effect of a Combined Chemical and Thermal Modification of Wood though the Use of Bicine and Tricine. *Forests* **2022**, *13*, 834. [CrossRef]
34. Mariani, A.; Malucelli, G. Transparent Wood-Based Materials: Current State-of-the-Art and Future Perspectives. *Materials* **2022**, *15*, 9069. [CrossRef] [PubMed]
35. Grosse, C.; Noel, M.; Thevenon, M.F.; Rautkari, L.; Gerrardin, P. Influence of water and humidity on wood modification with lactic acid. *J. Renew. Mater.* **2017**, *6*, 259–269. [CrossRef]
36. Larnøy, E.; Karaca, A.; Gobakken, L.R.; Hill, C.A.S. Polyesterification of wood using sorbitol and citric acid under aqueous conditions. *Int. Wood Prod. J.* **2018**, *9*, 66–73. [CrossRef]
37. Montanari, C.; Ogawa, Y.; Olsen, P.; Berglund, L.A. High Performance, Fully Bio-Based, and Optically Transparent Wood Biocomposites. *Adv. Sci.* **2021**, *8*, 2100559. [CrossRef]
38. Kurkowiak, K.; Emmerich, L.; Millitz, H. Biological durability and wood–water interactions of sorbitol and citric acid (SorCa) modified wood. *J. Wood Sci.* **2023**, *69*, 34. [CrossRef]
39. Berglund, L.A.; Burgert, I. Bioinspired Wood Nanotechnology for Functional Materials. *Adv. Mater.* **2018**, *30*, 1704285. [CrossRef] [PubMed]
40. Bi, W.; Li, H.; Hui, D.; Gaff, M.; Lorenzo, R.; Corbi, I.; Corbi, O.; Ashraf, M. Effects of chemical modification and nanotechnology on wood properties. *Nanotechnol. Rev.* **2021**, *10*, 978–1008. [CrossRef]
41. Bak, M.; Molnár, F.; Rákosa, R.; Németh, Z.; Németh, R. Dimensional stabilization of wood by microporous silica aerogel using in-situ polymerization. *Wood Sci. Technol.* **2022**, *56*, 1353–1375. [CrossRef]
42. Bak, M.; Takács, D.; Rákosa, R.; Németh, Z.; Németh, R. One-step process for the fabrication of hydrophobic and dimensional stable wood using functionalized silica nanoparticles. *Forests* **2023**, *14*, 651. [CrossRef]
43. Zhang, R.; Zhu, Y. Predicting the Mechanical Properties of Heat-Treated Woods Using Optimization-Algorithm-Based BPNN. *Forests* **2023**, *14*, 935. [CrossRef]
44. Li, J.; Li, N.; Li, J.; Wang, W.; Wang, H. Prediction of Thermally Modified Wood Color Change after Artificial Weathering Based on IPSO-SVM Model. *Forests* **2023**, *14*, 948. [CrossRef]
45. Yang, C.; Yang, S.; Yang, H.; Puangsin, B.; Qiu, J. Improvement of the Dimensional Stability of Rubber Wood Based on the Synergies of Sucrose and Tung Oil Impregnation. *Forests* **2023**, *14*, 1831. [CrossRef]
46. Spear, M.J.; Dimitriou, A.; Marriott, R. Chemical Composition of Larch Oleoresin before and during Thermal Modification. *Forests* **2024**, *15*, 904. [CrossRef]
47. Mascarenhas, F.J.R.; Simões, R.M.d.S.; Dias, A.M.P.G.; Christoforo, A.L.; Cunha, A.E.P. Moisture and Temperature Profiles of Heartwood Pinus pinaster Ait. Wood Specimens during Microwave Drying. *Forests* **2024**, *15*, 944. [CrossRef]
48. Crisostomo, M.; Del Menezzi, C.; Militz, H.; Kurkowiak, K.; Mayer, A.; Carvalho, L.; Martins, J. Effect of Citric Acid on the Properties of Sapwood of Pinus sylvestris Submitted to Thermomechanical Treatment. *Forests* **2023**, *14*, 1839. [CrossRef]
49. Scharf, A.; Dernegård, H.; Oja, J.; Sandberg, D.; Jones, D. Wood Modification Using Imidazole and Succinimide: Effects on Dimensional Stability and Bending Properties. *Forests* **2023**, *14*, 1976. [CrossRef]
50. Spavento, E.; de Troya-Franco, M.T.; Acuña-Rello, L.; Murace, M.; Santos, S.M.; Casado-Sanz, M.; Martínez-López, R.D.; Martín-Gil, J.; Álvarez-Martínez, J.; Martín-Ramos, P. Silver Nanoparticles and Chitosan Oligomers Composites as Poplar Wood Protective Treatments against Wood-Decay Fungi and Termites. *Forests* **2023**, *14*, 2316. [CrossRef]

51. Mania, P.; Kupfernagel, C.; Curling, S. Densification of Delignified Wood: Influence of Chemical Composition on Wood Density, Compressive Strength, and Hardness of Eurasian Aspen and Scots Pine. *Forests* **2024**, *15*, 892. [CrossRef]
52. Can, A.; Lee, S.H.; Antov, P.; Ghani, M.A.A. Phase-Change-Material-Impregnated Wood for Potential Energy-Saving Building Materials. *Forests* **2023**, *14*, 514. [CrossRef]

forests

MDPI

Article

Moisture and Temperature Profiles of Heartwood *Pinus pinaster* Ait. Wood Specimens during Microwave Drying

Fernando Júnior Resende Mascarenhas [1,2,*], Rogério Manuel dos Santos Simões [3], Alfredo Manuel Pereira Geraldes Dias [1,2], André Luis Christoforo [4] and André Eduardo Palos Cunha [3]

1 University of Coimbra, ISISE, ARISE, Department of Civil Engineering, Rua Luís Reis Santos—Pólo II, University of Coimbra (UC), 3030-788 Coimbra, Portugal; alfgdias@dec.uc.pt
2 SerQ—Innovation and Competence Forest Centre, Rua J, No 9, Zona Industrial da Sertã, 6100-711 Sertã, Portugal
3 Unit of Fiber Materials and Environmental Technologies (FibEnTech-UBI), Department of Chemistry, University Beira Interior (UBI), 6201-001 Covilhã, Portugal; rmss@ubi.pt (R.M.d.S.S.); aepcunha@gmail.com (A.E.P.C.)
4 Department of Civil Engineering, Federal University of São Carlos (UFSCar), Rodovia Washington Luís (SP-310), Km 235, São Carlos 13565-905, Brazil; christoforoal@yahoo.com.br
* Correspondence: fer.jr.resende@hotmail.com

Abstract: Microwave (MW) drying of wood has gained popularity in the field of wood modification. The rise in temperature during MW drying leads to increased steam pressure, enhancing wood permeability but potentially decreasing mechanical properties. Understanding temperature and moisture behaviors during MW drying is crucial for its industrial application in wood drying. Therefore, this study aimed to characterize the temperature and moisture behaviors during MW drying of small Portuguese maritime pine (*Pinus pinaster* Aiton.) wood samples to support a wider use of this technology. The effects on water uptake and the compressive strength parallel to the grain were also investigated. The results indicated three distinct phases in the MW drying rates, with an average of 0.085% of water removed per second. Moreover, the temperature underwent three distinct stages: an initial rapid increase, a period of constant temperature, and a slight decrease until drying was complete. At the beginning of MW drying, the temperatures were below 100 °C, with average temperatures ranging from 126 to 145 °C. Specimens with lower initial moisture content had higher temperatures, and a positive correlation was found between initial moisture content and drying time. In contrast, negative correlations were found between the initial moisture content and average temperature, as well as average temperature and MW drying time. Additionally, the operating condition parameters used in MW drying of pine samples enhanced water impregnability by 65%, generating a slight reduction of 11% in compressive strength. It was also noticed that the initial moisture content did not impact MW-dried samples' water uptake or compressive strength. Finally, although small clear wood samples of maritime pine were utilized, the temperature and moisture patterns observed closely matched real-scale specimens. Thus, the findings corroborate a wide utilization of MW technology for wood drying, mainly demonstrating positive possibilities for structural-sized wood specimens.

Keywords: electromagnetic heating; initial moisture content; temperature characteristics; impregnability; compressive strength

check for updates

Citation: Mascarenhas, F.J.R.; Simões, R.M.d.S.; Dias, A.M.P.G.; Christoforo, A.L.; Cunha, A.E.P. Moisture and Temperature Profiles of Heartwood *Pinus pinaster* Ait. Wood Specimens during Microwave Drying. *Forests* **2024**, *15*, 944. https://doi.org/10.3390/f15060944

Academic Editors: Morwenna Spear and Miklós Bak

Received: 26 April 2024
Revised: 24 May 2024
Accepted: 28 May 2024
Published: 30 May 2024

1. Introduction

The limited permeability of many wood species presents challenges throughout the transformation process, including drying and impregnation. The conventional drying methods might have certain drawbacks, such as long drying times, substantial material losses, and expensive drying operations. Moreover, poor drying processes result in drying tensions and cell collapse, leading to defects and significant losses in the recovery of sawn timber. Consequently, the timber industry continuously seeks methods to enhance wood permeability, reduce internal tensions, and prevent downgrading caused by drying errors [1].

Wood drying using electromagnetic waves as an energy source has gained popularity, particularly with the use of microwave (MW) energy, which is a promising technology in the field of wood modification [2]. The various benefits of this technology include high-quality drying combined with a fast process and significant improvements in porosity and permeability, which, for example, make it possible to solve the problem of wood species that are difficult to treat (impregnate) [3].

When considering the process of drying wood, it becomes essential to understand how water molecules are present within the wood, mainly because that topic has been investigated for decades [4]. They can exist in two distinct manners within the structure of the wood. One manner is free or capillary water, found within the capillaries, void spaces, cell lumen, and cavities. The interaction between free water and the wood takes place due to the surface forces. The second manner is known as bound water, in the cell wall, which is chemically combined with the chemical constituents of the wood, such as cellulose, hemicelluloses, and lignin. This combination primarily occurs through the formation of hydrogen bonds with the hydroxyl groups [5,6]. The theoretical stage at which all free water has been completely depleted and only bound water remains is called the fiber saturation point (FSP), which is inherently difficult to directly quantify since it depends on the wood species and the method used to measure it [5].

Regarding the drying process, the bound water is more difficult to be removed [5]. One of the reasons is that "the water molecules 'bound' in the first uni-molecular layer at the surface of the material are less rotationally free than the water in capillaries and cavities" [6]. In the process of MW drying, the entirety of the sample is permeated by electromagnetic waves, which facilitates the absorption of heat throughout the volume of the wet wood specimen [6,7].

Water molecules have polar configurations, which makes them prone to interact with the MW electromagnetic field [6,8]. In the presence of an alternating external electrical field, the molecule oscillates as it endeavors to align with the prevailing direction of the field [8]. Hence, microwave heating operates by converting electromagnetic energy into thermal energy locally. As a result, rapid heating occurs across the thickness of the material, accompanied by a reduction in thermal gradients. This volumetric heating process can effectively reduce drying times and conserve energy [7]. Du et al. also claim that the energy consumption savings in the MW drying process can be up to 50% [9].

Depending on the power supplied by the MW device, the energy absorbed, and the duration of the drying process, the temperature within the wood can experience a rapid rise, leading to the generation of significant steam pressure [2,3,10]. The potential of microwave electromagnetic energy to rapidly remove moisture from wood [11] results from the formation of steam within the wood, which can displace capillary liquid water and provide a favorable driving force for mass transfer. The total pressure (mainly derived by the water vapor) difference between the different positions inside the wood and on the inside and outside generates tension that can damage the more vulnerable cell structures, such as pit membranes and ray parenchyma cells.

As a consequence, this phenomenon leads to the emergence of novel pathways through which liquids can flow within the wood. As a result, there is an overall increase in the wood's porosity (and therefore permeability). However, if the MW parameters are not properly adjusted, for instance, elevate powers, the MW drying process may also negatively affect strength and stiffness [3,12,13], which is not desirable, especially if the wood dried in the MW is to be used for structural purposes later on.

Xiao et al. [14] explain that the extent of structural impairment is ascertained by the magnitude of internal pressure within the wood element during the process of microwave drying. Therefore, the internal water pressure assumes a pivotal function in determining the extent of damage to the microstructure [14]. As the internal water pressure generated during

MW drying is directly linked to temperature, it is expected that a balance between the drying rate and wood damage should be established by controlling the internal temperature and, consequently, the pressure generated by the MW drying. In other words, knowing the MW drying rates and the profile of internal temperatures throughout the MW drying process are crucial.

In the literature, temperature measurements were made in different ways and positions inside and outside the wood samples. Due to operational issues, in some works, the temperatures of the wood specimens were analyzed outside the oven after each MW drying cycle [15–17]. Other authors [14,18,19], in turn, used a continuous measurement of the temperature inside the samples during the MW drying, which seems to be the most appropriate. In addition, most of the works presented in the literature that investigated the temperature in MW-dried samples used wood samples containing only sapwood, or both stem parts (with no differentiation between the two within the same wood specimen).

Acknowledging that the research works in the literature have demonstrated that the wood species and the stem part are important parameters that influence MW drying [2,20], there are still scarce investigations about the moisture and temperature profiles during MW drying of one of the most used wood species in the construction, civil engineering, and furniture industry in southwestern Europe, the native maritime pine (*Pinus pinaster* Ait.) [21]. Besides those different uses, pine is also recognized as an ecologically versatile wood species [22].

Hence, based on the benefits of MW drying, the importance of temperature profiles throughout the MW drying process, and the economic and environmental noteworthiness of maritime pine wood species, the objective of this study was to foster the use of MW technology to dry wood by analyzing the interior temperature profile of small clear specimens of Portuguese maritime pine containing only heartwood during the MW drying, examining the drying and temperature behavior over the MW drying time and the reduction of the amount of water in the samples, and investigating the changes in the water impregnability and compressive strength.

The studied wood samples had different initial moisture contents (IMC), and the influence of this parameter on the drying rates, temperature profile, water uptake, and compressive strength parallel to the grain was explored. In addition, the drying and temperature profiles of the small clear maritime pine wood samples studied in this work were also compared with studies that investigated MW drying with full-scale wood elements (structural-sized samples).

Although the literature indicates that sapwood and heartwood react differently when dried [23], studies have demonstrated that heartwood is more severely affected by MW drying in terms of microstructural modifications than sapwood [2]. Also, different authors have focused their studies only on one of the wood stem parts to properly understand the effects of MW energy, either in sapwood or heartwood. Thus, it was chosen to work with maritime pine samples containing only heartwood since this wood stem part is usually much less permeable than sapwood [24], and it is extremely difficult to be impregnated with preservative products [25,26]. Hansmann et al. [27] explain that in a specific species, the permeability is typically higher in sapwood compared with heartwood due to the latter's increased level of pit aspiration and encrustation with extractives.

2. Experimental

2.1. Material

A total of 84 small clear Portuguese maritime pine specimens containing only heartwood, with an average oven-dry density of 697 ± 43 kg/m^3, were prepared from commercial boards, with dimensions 10 mm × 10 mm × 200 mm (Radial × Tangential × Longitudinal), and they were MW-dried (Figure 1). Another 21 samples of the same dimensions were prepared and dried using conventional methods (without MW drying), and they were referenced as the control group. These control samples were used in the water impregnability and compressive strength tests.

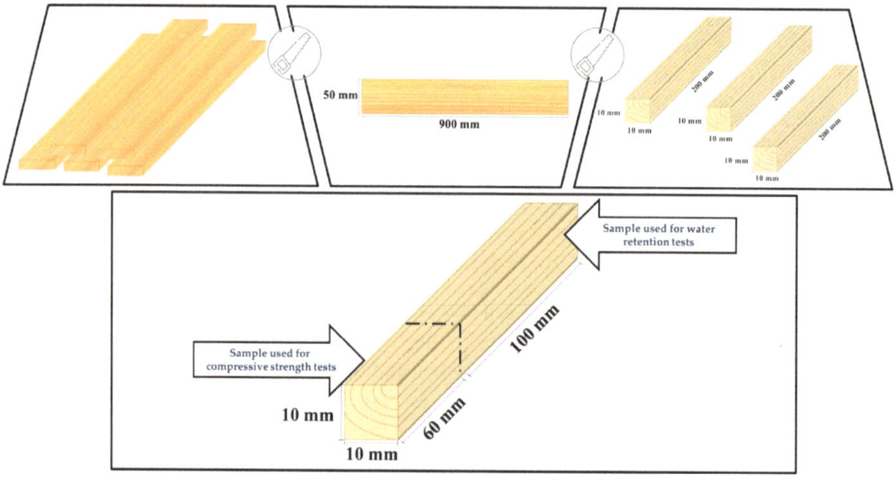

Figure 1. The preparation of wood samples. The dashed line represents the wood section in which the cut was done.

Following the same approach that Kol and Çayır [11] used in their work, 16 other extra wood samples were randomly selected from the same commercial boards and cut in the dimensions of 10 mm × 10 mm × 200 mm to determine their oven-dry mass. Using the average oven-dry mass of these samples, the IMC of the 84 samples used in the MW drying process and their moisture content (MC) evolution during the MW drying were assessed. The MC was determined according to ISO 13061-1 [28]. The wood specimens subjected to the MW drying were grouped based on their IMCs, i.e., wood samples with similar IMCs were placed in the same group, as shown in Table 1. The IMCs of the four MW-dried wood groups were statistically different.

Table 1. Samples' average IMC and quantity.

Group	IMC (%)	Number of Samples
1	63.39 (2.65) D	21
2	75.50 (5.20) C	21
3	81.12 (3.56) B	21
4	92.19 (2.70) A	21

The numbers in parentheses are the standard deviation. Values associated with different letters in Tukey tests show significant differences with a p-value of 0.05.

Furthermore, what can be observed in the literature is that in most studies involving MW drying, small clear wood samples are used [2]. Small clear samples have certain advantages, such as being easier to work with, more economically attractive in terms of material, resources, and laboratory space required for testing, and allowing different tests and analyses to be carried out before moving on to full-scale structural parts. The British Standard BS 373:1957 [29] states that the results from the tests might also be used to assess the impact of different treatments on strength, density, and other properties and aid in creating design equations for structural timbers.

2.2. Microwave Drying

MW dryings were performed using a conventional domestic MW device measuring 200 mm × 300 mm × 300 mm (inner chamber), with a frequency of 2.45 GHz, maximum output power of 800 W, and a specific applied power of 5714 kW/m^3, per MW drying cycle. A homogenous wave distribution inside the chamber was expected through a fan at the top of the MW device, through which the waves come out, which contributed to a better distribution of the waves inside the equipment's cavity. The MW oven also has a turntable, which contributes significantly to the homogeneous distribution of the waves. As presented in Table 1, the wood specimens were divided into four groups, and each group of 21 samples was divided into three sub-groups of seven samples (Figure 2). The seven samples of each sub-group were placed together in the MW to be dried simultaneously, and they were dried until they achieved an average of 12% ± 2% final moisture content.

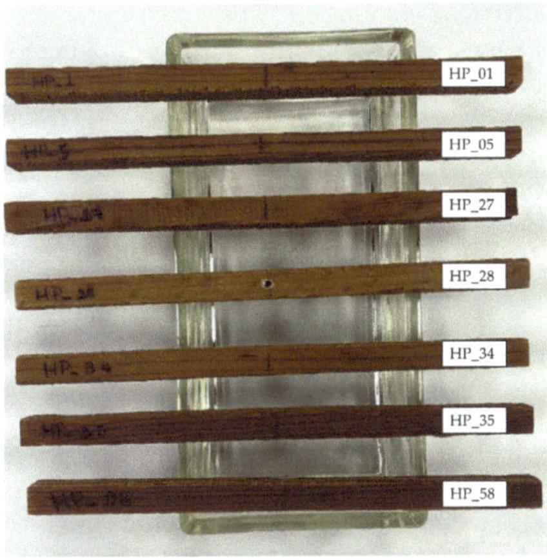

Figure 2. Wood samples dried together. "HP" in the figure stands for "heartwood pine".

The drying program used in this work was based on previous tests and also followed what Mascarenhas et al. [30] adopted in their work, with 30 s inside the MW oven (wood samples continuously exposed to MW), followed by a 30 s cooling interval (Figure 3).

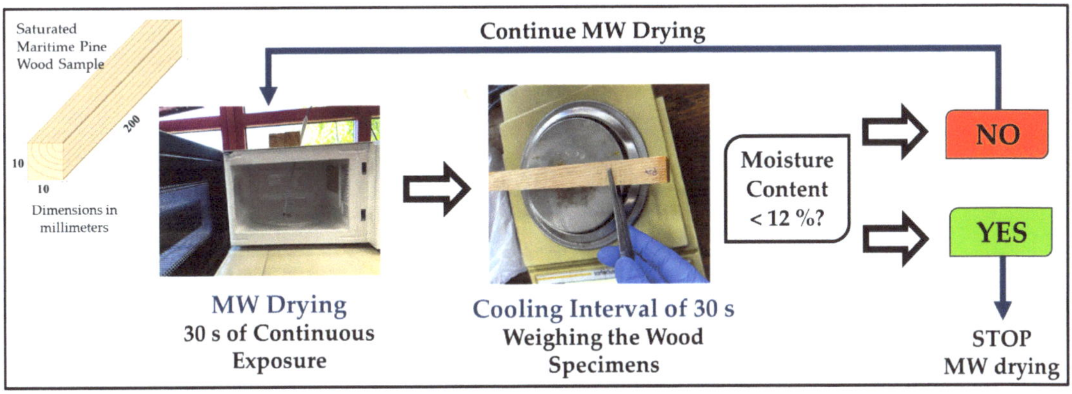

Figure 3. MW drying process program.

The MW applied energy intensity, E, was calculated using Equation (1) [11]:

$$E = \frac{P \cdot t}{V \cdot 10^6} \quad (1)$$

where E is the energy supplied during the MW process, in MJ/m^3; P is the MW power supplied (W); V is the volume of wood specimen, in m^3; and t is the drying time, in s.

Adopting the same procedure that other authors used [11,31], the weight percentage loss (*WPL*) and relative water loss (*RWL*) were calculated using Equations (2) and (3):

$$WPL = \frac{m_2 - m_1}{m_{od}} \times 100 \tag{2}$$

where m_2 is the weight of the specimen before MW drying in g; m_1 is the weight of the specimen after MW drying, in g; and m_{od} is the oven-dry weight of the specimen, in g.

$$RWL = \frac{WPL}{IMC} \tag{3}$$

The MW drying and heating rates were calculated using the following Equations (4) and (5), respectively:

$$dr = \frac{MC_t - MC_{t+\Delta t}}{\Delta t} \tag{4}$$

$$hr = \frac{T_{t+\Delta t} - T_t}{\Delta t} \tag{5}$$

where dr is the drying rate, in %/s; Δt is the time interval of drying, in s; hr is the heating rate, in °C/s; and T is the measured temperature in wood, in °C.

2.3. Temperature Measurements

A Heidolph EKT 3001 thermometer was used to measure the temperature inside the wood samples (5 mm deep at the center and middle) (Figure 4). The position was chosen based on previous works from the literature [14,18] and because the MW energy is provided in order to produce heat internally within the solid wood, which then moves to the outer surface [32]. The temperature measurements were made during the entire drying process and recorded at intervals of 5 s during each 30 s drying cycle inside the MW device. The measuring probe was removed while weighing the specimens, and a silicone seal was placed in the hole to avoid excessive evaporation.

Figure 4. Temperature measurement.

Based on the temperature measurements, the water saturation pressure was calculated according to Equation (6) [33], assuming liquid/vapor equilibrium:

$$P_s = -0.59 + 0.2846 \cdot T - 0.00867 \cdot T^2 + 0.000159 \cdot T^3 \tag{6}$$

where P_s is the water saturation pressure at a given temperature, in kPa; and T is the measured temperature in wood, in °C.

2.4. Water Uptake and Compressive Strength Parallel to the Grain Tests

To better investigate the effects of MW drying on wood samples, i.e., possible impacts on the microstructure and mechanical integrity of the wood samples, two tests were conducted to analyze the modifications: (1) the impregnability to distilled water and (2) the compressive strength parallel to the grain ($f_{c,0}$). Both tests were carried out on specimens dried with MW and those without MW (control).

The impregnability tests were performed to indirectly verify if the treatability of the heartwood of maritime pinewood samples changed. Then, wood samples measuring 10 mm × 10 mm × 100 mm were placed in an autoclave filled with distilled water, and a pressure of 600 kPa was applied for 30 min. Based on this, the water uptake, WU, was calculated using Equation (7):

$$WU = \frac{m_{ai} - m_{bi}}{m_{od}} \times 100 \quad (7)$$

where m_{ai} is the weight of the specimen after water impregnation, in g; m_{bi} is the weight of the specimen before impregnation, in g; and m_{od} is the oven-dry weight of the specimen, in g.

Besides that, tests were carried out to determine the compressive strength. Compressive strength is one of the most important mechanical properties of wood [34–38], mainly because modified wood with reduced compression strength might have restricted use [35]. Following the recommendations of Standard EN 408 [39], as performed by Hermoso and Vega [40], the tests to determine $f_{c,0}$ were carried out on test pieces measuring 10 mm × 10 mm × 60 mm. Subsequently, the results were adjusted to 12% moisture content, resulting in the compressive strength parallel to the grain at 12% moisture content ($f_{c,0,12\%}$).

2.5. Analysis of the Results

To assess the significant differences between the variables studied in this work, the classical analysis of variance (ANOVA) was employed at a significance level of 5% associated with the Tukey test. According to ANOVA and Tukey, in the event that the p-values are lower than the designated significance level (p-value < 0.05), it is plausible to assert that the analysis holds statistical significance. Moreover, when conducting a statistical analysis involving 2–9 groups with ANOVA methodology, it is necessary that the minimum sample size for each group is 15 [41]. For the current study, four groups (MW drying and temperature profiles) and five groups (water uptake and compressive strength) were utilized, with each group consisting of 21 specimens.

Pearson's correlation coefficient (r) was used to measure linear relationships between (1) MW drying time and IMC, (2) average temperature of the samples and IMC, and (3) average temperature of the samples and MW drying time. In addition, regression models, adjusted by the least squares method, were applied to determine the equations to estimate (1) MW drying time as a function of the IMC, (2) average internal temperature of the samples as a function of the IMC, and (3) average internal temperature of the samples as a function of the MW drying time. The R-squared (R^2) was used to measure the proximity of the data points to the regression line that has been fitted. All statistical analyses were conducted utilizing the Minitab 18 software (version 2017) [41].

3. Results and Discussion

3.1. MW Drying Profile

The time profile of the average MC of the four MW wood-dried groups is presented in Figure 5. Despite the difference in IMC, the MC profiles over the MW drying time were very similar. The findings using small clear maritime pine wood specimens were generally consistent with Sethy et al. [17], Xiao et al. [14], and Weng et al. [42], who utilized the MW technology to dry structural-sized (industrial scale) samples of different wood species.

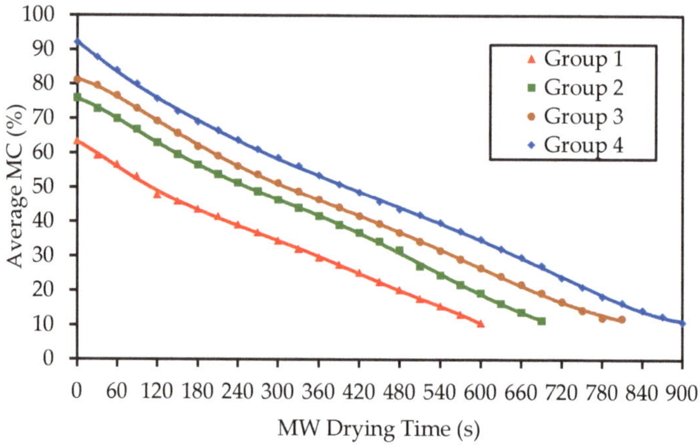

Figure 5. Average moisture content along the MW drying process.

The total MW drying time (total time the specimens were exposed to MW energy), the average initial and final moisture contents, the average MW drying rates, the average MW applied energy, the WPL, and RWL are shown in Table 2. The average MW energy supplied to the wood samples increased as the IMC increased. Additionally, higher volume of water requires more time for removal from the wood samples, directly impacting the total drying time.

Table 2. MW drying characteristics.

Group [1]	Average IMC (%)	Average Final MC (%)	Total MW Drying Time (s) [2]	Average MW Drying Rates (%/s)	Average *E* (MJ/m^3)	WPL (MJ/m^3)	RWL
1	63.39 (2.97)	10.66 (0.44)	600 ^{D}	0.082 (0.033) ^{A}	3429 (74) ^{D}	52.73 (1.14) ^{D}	83.18 (1.80) ^{A}
2	75.50 (5.20)	11.48 (0.66)	690 ^{C}	0.089 (0.023) ^{A}	3943 (276) ^{C}	64.02 (4.48) ^{C}	84.79 (5.94) ^{A}
3	81.12 (3.56)	11.46 (0.55)	810 ^{B}	0.085 (0.021) ^{A}	4629 (258) ^{B}	69.66 (3.88) ^{B}	85.87 (4.79) ^{A}
4	92.19 (2.70)	10.96 (0.57)	900 ^{A}	0.082 (0.027) ^{A}	5143 (159) ^{A}	81.23 (2.51) ^{A}	88.11 (2.72) ^{A}

[1] All four groups had 21 wood specimens each. [2] MW drying time represents the total time the samples remained inside the MW device, excluding the cooling intervals. The numbers in parentheses are the standard deviation. Values associated with different letters in Tukey tests show significant differences with a *p*-value of 0.05 within the same variable analyzed.

Despite the difference in IMC of the four groups of samples, the average drying rates for groups 1, 2, 3, and 4 were statistically equivalent (confidence interval, CI, of 95%) and equal to 0.082, 0.089, 0.085, and 0.082 percentage of removed water per second (%/s), respectively. Leiker and Adamska [43] measured MW drying rates up to 0.075%/s for *Fagus sylvatica* and *Picea abies*. Ouertani et al. [18] MW-dried *Pinus banksiana* with 12 mm thickness with MW power ranging from 300 to 1000 W, and the average MW drying rates varied from 0.017 to 0.133%/s. A linear interpolation of these results [18] for 800 W, the same used in this work, shows that the drying rate was 0.093%/s, which is within the values presented here. An analysis with further details was carried out in the following section, where the MW drying rate results were correlated with the temperature profile.

The WPL values ranged from 53 to 81%, and the RWL from 83 to 88%. Kol and Çayir [11] had WPL values ranging from 47 to 50%. Additionally, the amount of energy supplied to the maritime pine specimens to remove 1% of the water was 65.0, 61.6, 66.5, and 63.3 MJ/m^3, respectively, for wood groups 1, 2, 3, and 4. For instance, MW drying structural-sized samples of *Eucalyptus obliqua* and *Pinus radiata* heartwood, Torgovnikov and Vinden [13] indicated that 66 and 21 MJ/m^3 were applied to dry 1% of the water of each wood species, respectively, evidencing the dielectric properties of wood [44].

3.2. Temperature Profile during MW Drying

In order to analyze the evolution of the temperature inside the specimen at each 30 s MW drying cycle, the temperature was measured every 5 s (5, 10, 15, 20, 25, and 30 s). Figure 6a shows the average temperature profile over the MW drying time for group 1. In addition, the temperature profile at three different moments of the MW drying process during 30 s of continuous exposure time is presented in Figure 6b–d. Whether at the beginning, middle, or end of the drying process, the temperature peaks can be noticed at the end of each 30 s cycle. Those peaks may be responsible for causing the most significant damage to the microstructure of the wood due to the high pressures generated in the interior (as shown further ahead), which, on the other hand, leads to increased porosity and, consequently, permeability. The degree of structural alterations or even damage is typically determined by these temperature peaks, which are a function of the energy intensity of the MW drying, which, together with MW drying time, are important variables in the change in anatomical structure [2].

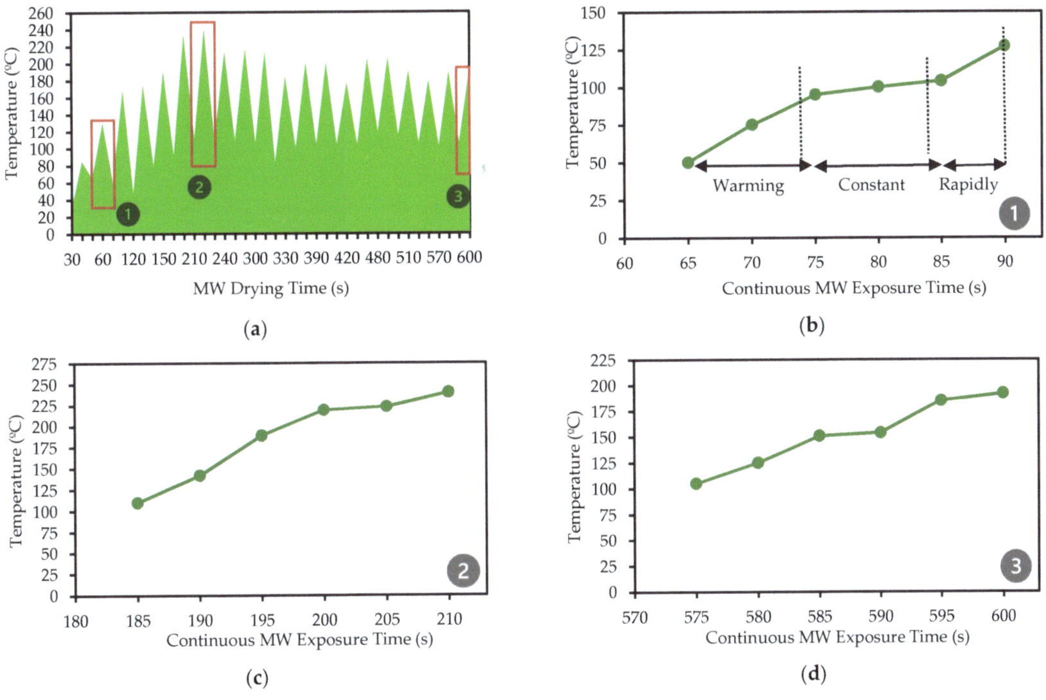

Figure 6. Temperature behavior (average) in each interval of measurement for group 1 (**a**) during the complete drying process, with a focus on the (**b**) beginning, (**c**) middle, and (**d**) ending of the drying process. Continuous MW Exposure Time represents the time the wood samples were inside the MW device.

Despite the reduced dimensions of the cross-section of the samples, these findings suggest that the cooling intervals of 30 s (outside the MW oven) were necessary and effective to avoid prolonged exposure of the wood specimens to MW energy. Hence, two main consequences could arise if the wood specimens were continuously exposed to these MW drying conditions (for too long, without the cooling intervals). First, extremely high pressure could be generated due to the extremely high temperatures, leading to excessive damage to the wood's microstructure and the appearance of cracks. When analyzing the surface of the pine wood specimens dried in the MW, no visible drying defects were observed, which is due to the aforementioned factors.

Second, the wood samples could burn, leading to the charring of the wood. A similar approach was used by other authors [45], who dried *Abies alba* L. wood samples for 30 to 60 s and made cooling intervals of 60 to 120 s. Oloyede and Groombridge [8] also explain that each consecutive exposure enables a deeper penetration level. Each specimen's weight was carefully measured with a scale and recorded for weight control purposes to follow the moisture content decrease along the drying process. Thus, a non-continuous MW drying process was carried out.

The drying time interval of 60 to 90 s, which is shown in Figure 6b, has three behavior patterns: a warming period, constant temperature around the water's normal boiling point, and a rapidly rising temperature with heating rates of 3.33, 0.90 and 4.60 °C/s, respectively for each phase. Although the drying in the MW carried out in this work was not continuous, as is more common for structural-sized pieces, the beginning of the MW drying process showed a profile similar to that found by the authors [14], using a continuous tunnel MW equipment and *Pinus sylvestris* L. var. *mongolica* Litv. samples of 480 mm (longitudinal) × 980 mm (tangential) × 930 mm (radial).

In Figure 6c, we have the average temperatures during the 30 s measured in the middle of the drying process (210 of total MW drying time), when the highest temperatures were achieved, with a heating rate of 5.20 °C/s and a drying rate of 0.10%/s. Finally, Figure 6d shows the temperatures during the final 30 s drying interval (total MW drying time equal to 600 s), with a heating rate of 3.48 °C/s and a drying rate of 0.08%/s.

The temperature profiles of the MW-dried samples for each group are shown in Figure 7. By analyzing the average inside-temperature profile along the entire MW drying process, it is possible to identify three distinct stages. Stage (1) a rapid increase in temperature (heating-up period), (2) a continuous period of relatively constant temperature, and (3) a decrease in temperature by the end of the drying process. By studying the MW drying in *Pinus sylvestris* wood samples, Hansson and Antti [46] obtained similar temperature profiles for Scots pine (*Pinus sylvestris*).

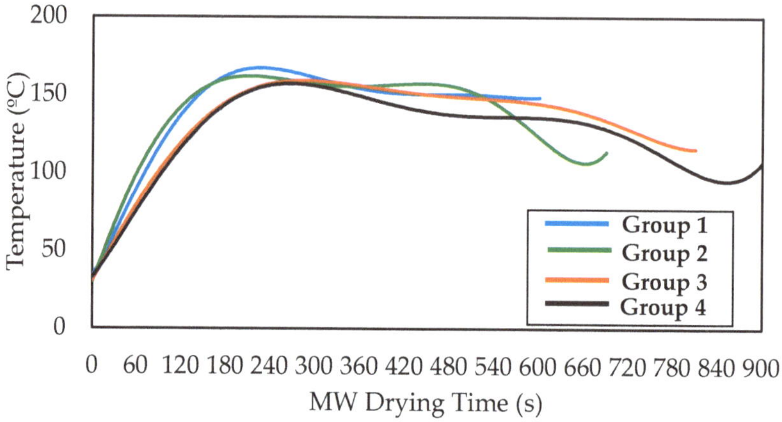

Figure 7. Average temperature over the MW drying process of groups 1, 2, 3, and 4.

Furthermore, the average temperature inside the samples remained over 100 °C for most of the MW drying time. Only at the beginning of the MW drying they were less than 100 °C, which ties well with previous studies developed by Vongpradubchai and Rattanadecho [47]. The rapid increase in temperature identified in the first couple minutes of the MW drying process demonstrates the volumetric heating in MW drying, which causes a fast transfer of energy throughout the wet wood sample. Internal evaporation and a rise in the total pressure might occur within the pores as the wet material's temperature approaches the liquid's boiling point [6,48]. In addition, the initial rise in temperature observed during the heating-up phase signifies the peak level of MW power absorption by the wood. In this stage, wood with higher moisture content absorbs a greater amount of energy compared with dry wood towards the completion of the drying [18].

The data presented in Table 3 shows that there is a significant difference in the temperature profiles of the samples with the lowest and the highest initial moisture content, particularly at the beginning of the drying process. The samples from groups 1, 2, 3, and 4 took approximately 50, 55, 80, and 105 s, respectively, to reach 100 °C for the first time. When looking at the results of different wood groups, from the smallest to the highest IMC, the average temperatures in the first 5 s within the cycle remained below 100 °C more frequently as the initial water content increased. For instance, from the temperatures measured in the first 5 s of each cycle, the percentage that was equal to or less than 100 °C represented 42% for group 1 (IMC = 63%), 61% for group 2 (IMC = 76%), 89% for group 3 (IMC = 81%), and 100% for group 4 (IMC = 92%). Xiao et al. [14] explain that under similar conditions of MW drying, a higher amount of energy was required to raise temperatures when the moisture level was elevated, resulting in a reduction of the drying rate and maximum measured temperature.

Table 3. Average measured temperatures and estimated pressure inside the wood samples.

Group	Average Temperature (°C)	Time to Achieve 100 °C (s)	Maximum Temperature (°C)	Average Pressure (kPa)	Estimated Maximum Pressure (kPa)
1	145	50 (after 1st cooling interval)	240	331	1766
2	140	55 (after 1st cooling interval)	236	291	1681
3	136	80 (after 2nd cooling interval)	232	282	1584
4	126	105 (after 3rd cooling interval)	222	215	1375

Also, based on Table 3, two contributions to these results can be drawn. First, the samples in the studied groups with higher IMCs had more free water in their cell spaces, so the MW energy was dissipated into more water. The temperature increase was slower than in the case of wood with lower IMCs. Therefore, the majority of them stayed in the liquid condition for more time; more energy was used to heat and vaporize the water, and no significant increase in temperature occurred. Second, softwoods are permeable due to pit pairs connecting tracheid lumens, allowing flow when the vapor pressure is generated within the wood above the fiber saturation point and at high temperatures [32], the wood samples from the groups with higher quantities of IMC might indicate that those samples had the interior structure more "open", with, for example, the bordered pit not aspirated, which ended up letting steam out [4,49], and, with that, the temperature increase is more moderate. The experimental data did not enable us to establish the relative importance of these two mechanisms.

The average water saturation pressure estimated based on Eq. 6 of the small clear wood samples ranged from 215 to 331 kPa (Table 3), which is within the stem pressure measured by Torgovnikov and Vinden [13], who MW-died structural-sized wood in a continuous MW device. The authors measured values up to 600 kPa and, for these values, reported severe wood microstructures and cell damage corresponding to/associated with high losses in mechanical properties.

Figure 8 shows the average drying rates of the four groups. Comparing the drying rate profiles with the corresponding temperature profile (Figure 7), an initial boost in drying rate is observed for the four groups. This boost can be due to what some authors call liquid movement [6]. Similar drying rate behaviors were observed for other wood species, *Pinus banksiana* [18], oriented strand board made of pine [9], and *Eucalyptus globulus* [30].

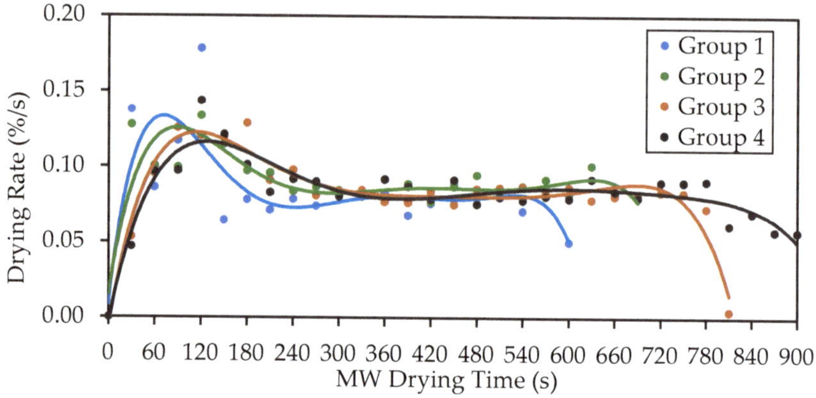

Figure 8. MW drying rate behavior over the MW drying process for the four wood groups.

When observing the drying rate behavior over the reduction of MC (Figure 9) of the non-continuous drying process in the MW of small specimens of maritime pine, it can be seen that there are three stages involving the drying rate: (1) a liquid movement period, (2) a constant drying rate, and (3) a falling drying rate period. These same stages were observed when a continuous MW drying process was carried out with large-scale samples [6]. The average drying rates during the heating-up period (which coincides with the liquid movement stage) for groups 1, 2, 3, and 4 were 0.094, 0.096, 0.091, and 0.086%/s, respectively. The heating rates in the same period were 4.08, 3.85, 3.53, and 3.52 °C/s for groups 1, 2, 3, and 4, respectively. The extremely high drying rates in the last part of the heating-up period can be related to the water's movement (expulsion) in a liquid state due to the high pressure developed inside the wood. Other authors reported this phenomenon as filtrational flow caused by the total pressure gradient [6].

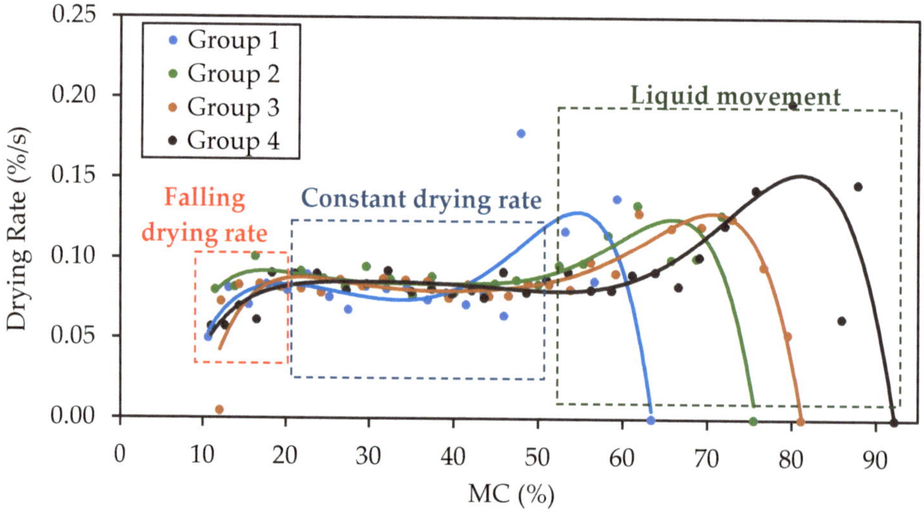

Figure 9. MW drying rates as a function of MC (%) for the four groups.

Also analyzing Figure 9, it can be seen that the constant drying rate started at different moments because of their different IMCs. In addition, the duration of the constant drying rate lasted longer in groups with higher initial water content. This might have happened because the improved moisture flow to the surface was superior "than the additional moisture evaporated at the surface due to the heat conducted from the wet" wood [6].

According to Metaxas and Meredith [6], during the stage of constant drying rate, the moisture content is quite high, and evaporation from the surface will proceed at a steady pace as long as the ambient circumstances stay constant. The authors [6] also point out that the moisture moves from the interior of the wood to the surface by capillary forces to replace the evaporated water required for the constant drying rate period.

This part of the drying process demonstrates the advantage of MW wood drying. Zielonka and Dolowy [15] explain that in traditional drying methods, the movement of water from the inner layers to the outer surface of the wood is decreased. However, because in MW drying, the energy is effectively transmitted to the water molecules present in the wood samples, it becomes feasible to sustain a suitable flow of moisture towards the surface where evaporation occurs [15,16]. In other words, the constant drying rate period lasts longer, and as the moisture content of the wood decreases below a critical point, the MW drying rate decreases progressively, limited by the decreased amount of water that migrates from the interior of the wood to its surface [6].

Thus, we have the last part of the drying process, the falling drying rate period. With the decrease in moisture levels, the evaporation rate decreased progressively, restricted by the reduction of water migration from the core of the solid to its surface. As soon as the wood sample became drier, a continuous introduction of air pockets made it difficult for the liquids to flow from the interior to the surface. Hence, the drying started to be impeded because the water in the pores' menisci dropped below the evaporating surface [6].

As the drying time increases, the free water decreases, leaving only the bound water, and removing the bound water in wood is more difficult because this residual water absorbs little energy [6,50,51] since it is strongly bound with polar groups and less rotationally free than free water [52]. In fact, when analyzing the drying rate over the moisture content reduction (Figure 9), it was noticed that from the IMC until a MC of about 24% (theoretical FSP for maritime pine [53]), the average drying rates of the four wood groups were greater than the average drying rates from that point to the ending of MW drying. Weng et al. [42] identified the same pattern in MW-dried structural-sized *Cunninghamia lanceolata* (Lamb.) Hook wood elements, where the drying rates with MC higher than the FSP were approximately 65% greater compared with when the MC is smaller than the FSP.

The temperature also started to decrease gradually by the end of the constant drying rate period and the beginning of the falling period, which is more evident in groups with higher IMC (Figure 10). One possible reason is the increase in wood permeability reached at the maximum temperature and pressure inside the structure, which facilitates the water vapor transfer to the outside and corresponding water evaporation inside, which provokes a temperature decrease inside the wood, where the sensor is located. The temperature reduction by the decrease of water vapor in contact with the temperature probe cannot be discarded. The maximum temperatures were reached at moisture contents of 41.44 (MC ratio of 0.65), 55.35 (MC ratio of 0.73), 59.16 (MC ratio of 0.73), and 66.54% (MC ratio of 0.73) for groups 1, 2, 3 and 4, respectively.

It was also observed that by the end of the MW drying process, the temperatures of the wood samples were, on average, approximately 150, 120, 115, and 105 °C for groups 1, 2, 3, and 4, respectively. Even though the temperatures were lower than the values measured in previous moments of drying, they were still high and above 100 °C, which could indicate a risk of overheating and charring the wood samples.

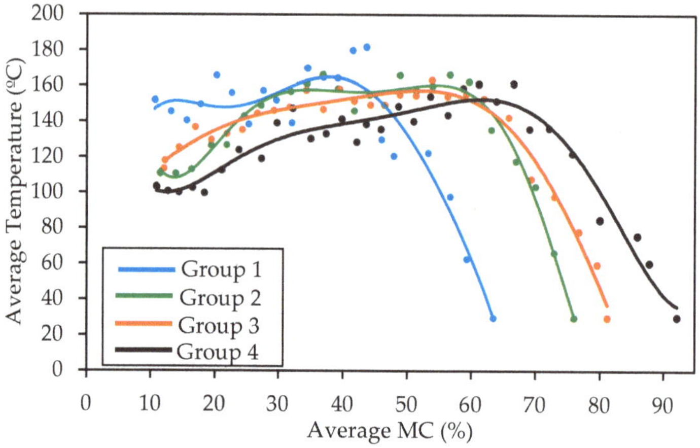

Figure 10. Average temperature behavior along with MC reduction for the four wood groups.

Finally, correlation analyses were used to corroborate what was previously presented and discussed. Figure 11 shows Pearson's correlation coefficient r and the p-values for the four wood groups. First, all the correlations were high and significant (p-value < 0.05). When analyzing the IMC and MW drying time (Figure 11a), there is a positive-strong correlation, 0.908, very close to 1. The higher the IMC, the more time is required to dry wood samples at the same power input, as expected.

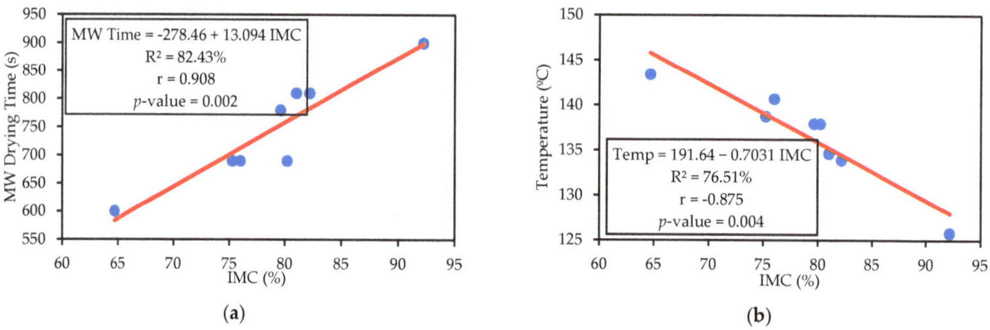

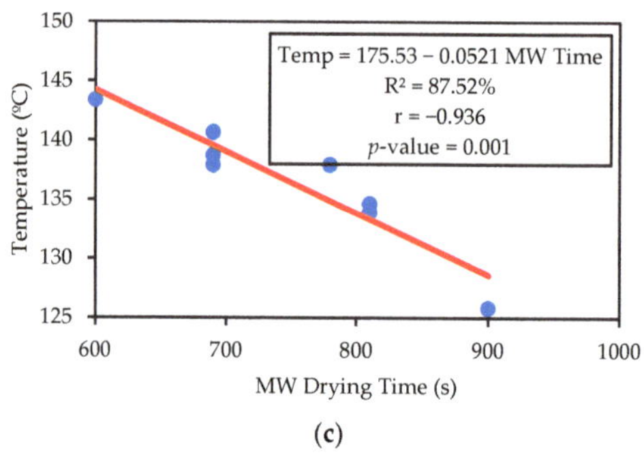

Figure 11. Correlation analysis: (**a**) MW Time and IMC, (**b**) Temperature and IMC and (**c**) Temperature and MW Time. Temp = average temperature; IMC = initial moisture content; MW Time = total MW drying time.

When analyzing the relationship between IMC and the temperature (Figure 11b), a negative-strong correlation of −0.875 indicated that the higher the IMC, the smaller the average temperature measured inside the sample. When evaluating the average temperature and MW drying time (Figure 11c), a negative-strong correlation of −0.936 indicates that the higher the MW drying time, the smaller the wood's internal average temperature. It is important to state that the MW drying time depends on the MW power and IMC of the wood samples. This is due to the fact that longer time is linked to higher IMC, which in turn leads to lower temperatures. This hypothesis is valid when MW power is a limiting factor.

In addition, by evaluating the equations presented in Figure 11, it is possible to notice that besides the elevated Pearson correlation coefficients, the high R^2 (from 75 to 87%) indicates that the three equations adequately estimate their respective variables. It is important to highlight that the idea of establishing equations to predict the total MW drying time and the average temperature has as objectives to plan better and organize the experimental drying campaigns, which can result in time saving and even material savings. Thus, it is possible to estimate both parameters before the drying process starts within the range of values analyzed here and with species with a density similar to the species studied in this work.

3.3. Evaluation of the Water Impregnability and Compressive Strength

The water uptake (WU) of the four wood groups and control samples is presented in Figure 12. The WU of MW-dried pine samples increased from 51 to 65% compared with control samples. Although the measurement of permeability was not conducted directly for the wood samples (both control and MW-dried) due to the existence of specialized techniques for such measurements, the observed enhancement in water uptake in the MW-dried samples in comparison with the control one could potentially suggest an elevation in their permeability levels.

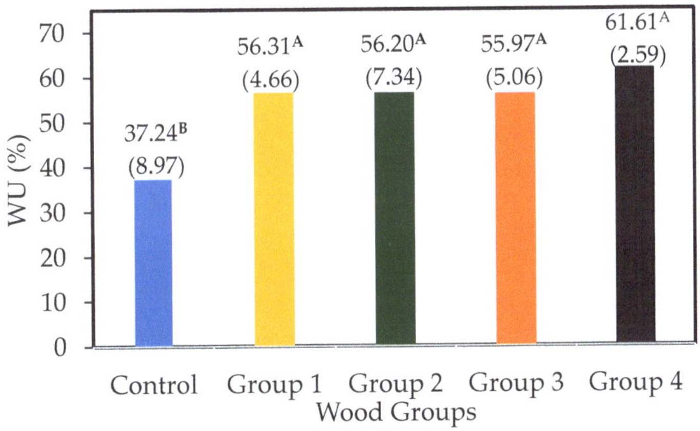

Figure 12. Average water uptake of the four wood groups. The numbers in parentheses are the standard deviation. Values associated with different letters in Tukey tests show significant differences with a *p*-value of 0.05.

He et al. [54] employed a comparable method in their study, utilizing water uptake measurements to indirectly assess the changes in the permeability of *Cunninghamia lanceolata* wood following MW drying on real-scale wood elements. Our results align with those from the authors [54] since they measured the retention of control samples of approximately 34%, and the MW-dried ones ranged from 44 to 77%.

The IMC was not a decisive factor in the WU since the four MW-dried groups had statistically equivalent WU values. Wang et al. [55] MW-died *Pinus sylvestris* var. mongolica Litv with different IMCs in real scale, and they observed that while the samples with an

IMC of 40% had no significant alteration in the cell wall morphology, and the cell structures remained unchanged, the cell wall structure of wood samples with an IMC of 20% exhibited the most severe modification.

The results showed that the peak temperatures measured on the pine specimens were probably not extreme. However, even so, they may have been high enough for the vapor pressure generated to lead to an increase in the impregnability of these samples. Hence, the rupture of weak ray cells might have created pathways that facilitate the transport of liquids, leading to enhanced permeability in the transverse direction when subjected to MW drying [13,23,42].

The compressive strength parallel to the grain was also measured (Figure 13). It is possible to conclude that despite slight reductions of 8 to 13% in the compressive strength parallel to the grain at 12% moisture content ($f_{c,0,12\%}$) of MW-dried samples compared with the control ones, the values of all wood groups were statistically equivalent. MW-drying small clear *Quercus pyrenaica* Wild wood samples, Machado [56] identified reductions in the compressive strength of 10 and 20%. Balboni et al. [57] measured reductions of approximately 11% of *Eucalyptus macrorhyncha* wood samples after MW treatment. No statistically significant reductions in the compressive strength were measured by Kol and Çayir [11] and Hermoso and Vega [40].

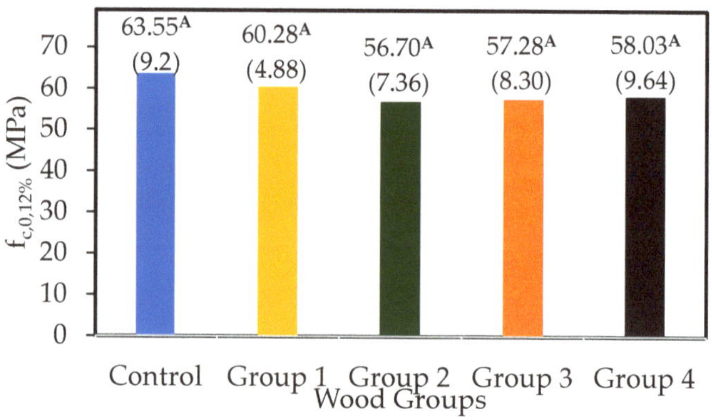

Figure 13. Average compressive strength parallel to the grain of the four wood groups. $f_{c,0,12\%}$ is the compressive strength parallel to the grain at 12% moisture content. The numbers in parentheses are the standard deviation. Values associated with the same letters in Tukey tests do not show significant differences with a *p*-value of 0.05.

Similarly to the WU results, the experimental results of $f_{c,0,12\%}$ also indicated that the IMC of the four wood groups had no statistically significant effects on this mechanical property. These findings are consistent with a previous research work developed by He et al. [54], who MW-dried real-scale wood samples whose IMCs were 40, 60, and 80%. From these results, it is important to note that both the small clean specimens dried in this work and the larger-scale specimens dried by He et al. [54] had a higher IMC than the FSP.

Torgovnikov and Vinden [13] explain that three degrees of MW modification in wood can happen and that a low level of MW modification can enhance the permeability of wood by a factor ranging from 1.1 to 1.5. However, at the same time, it does not have a statistically significant impact on the strength properties of wood, and this may have been the case with the results presented in this work. Hence, given that maritime pine is a commonly used wood species for wooden poles but whose heartwood faces challenges in being impregnated with preservative products [25,58–60], the acceptable results of the compressive strength of MW-dried wood samples and their improved impermeability due to MW drying are favorable points for MW dryings.

Thus, adopting the appropriate parameters of MW drying, such as MW power and drying time, including the MW cycle time in intermittent drying, can lead to minor impacts on the mechanical properties of wood samples since the achieved temperatures will not be so elevated and, consequently, the generated steam pressure.

Finally, even though the results presented in this research came from small specimens, the temperature profiles and moisture behavior presented in this work were similar and supported by other works that used full-scale specimens. Therefore, the outcomes obtained from MW drying small pine specimens not only allow us to infer the effectiveness and relevance of carrying out tests on small specimens, but also serve as a basis for encouraging greater use of MW technology for drying wood.

4. Conclusions

Using small clear heartwood specimens of Portuguese maritime pine to study the temperature profile under MW drying with non-continuous exposition, it was possible to draw the following conclusions:

1. Given the configuration of the MW drying process used in this work, the cooling intervals of 30 s were important to avoid excessive heating, which could result in the burning of some parts of the wood samples.
2. Three stages of temperature behavior were noticed during MW drying, and maximum temperatures were observed in the first few minutes of MW drying.
3. The drying rate comprises three distinct stages: the initial period of liquid movement, followed by a period of constant drying rate, and finally, a period of falling drying rate. A strong positive Pearson's correlation coefficient was identified between the IMC and MW drying time, and strong negative correlations were identified between IMC and average temperature and average temperature and MW drying time.
4. The parameters used in the MW drying of the Portuguese maritime pine samples associated with the cooling intervals provided moisture and temperature behaviors that were effective in increasing the water impregnability of the samples but without generating statistically significant reductions in compressive strength. Additionally, the IMC did not have a statistically significant impact on the water uptake and compressive strength of MW-dried samples.
5. Despite the fact that small clear wood samples of maritime pine were used, the temperature and moisture patterns observed in this study closely matched those of real-scale specimens, which validates the use of small pine samples for testing purposes and demonstrates the positive possibilities of using MW drying in structural-sized wood specimens. Therefore, with adequate MW parameters, it is possible to obtain the benefits of MW without deleterious effects and move from a laboratory to an industrial scale.

Finally, based on the results found here, future research could be carried out in order to investigate, through electron microscopy images, for example, possible changes in the microstructure of the wood, as well as investigations into changes in the porosity and permeability of pine samples. It is also important to indicate that due to the elevated temperatures measured, studies need to be conducted regarding the chemical changes in the MW-dried wood samples. Additionally, based on the results of this work, further tests can be carried out using small clear wood samples or real-scale wood elements.

Author Contributions: Conceptualization, F.J.R.M., R.M.d.S.S., A.M.P.G.D. and A.L.C.; methodology, F.J.R.M. and R.M.d.S.S.; software, F.J.R.M. and A.L.C.; validation, R.M.d.S.S.; formal analysis, F.J.R.M., A.M.P.G.D., A.L.C., R.M.d.S.S. and A.E.P.C.; investigation, F.J.R.M., R.M.d.S.S. and A.E.P.C.; resources, F.J.R.M. and A.M.P.G.D.; writing—original draft preparation, F.J.R.M.; writing—review and editing, F.J.R.M., R.M.d.S.S., A.M.P.G.D. and A.L.C.; supervision, A.M.P.G.D., A.L.C. and R.M.d.S.S.; project administration, A.M.P.G.D.; funding acquisition, F.J.R.M. and A.M.P.G.D. All authors have read and agreed to the published version of the manuscript.

Funding: This work is financed by national funds through the Foundation for Science and Technology—FCT, from Portugal under grant agreement 2021.07636.BD, the FCT, the Ministry of Science, Technology and Higher Education (Portuguese: Ministério da Ciência, Tecnologia e Ensino Superior or MCTES), the Programa Operacional da Região Centro (CENTRO 2020), and the European Social Fund (Portuguese: Fundo Social Europeu or FSE) attributed to the first author (https://doi.org/10.54499/2021.07636.BD). This work was partly financed by FCT/MCTES through national funds (PIDDAC) under the R&D Unit Institute for Sustainability and Innovation in Structural Engineering (ISISE), under reference UIDB/04029/2020 (doi.org/10.54499/UIDB/04029/2020), and under the Associate Laboratory Advanced Production and Intelligent Systems ARISE under reference LA/P/0112/2020.

Data Availability Statement: Data are contained within the article.

Acknowledgments: We are also thankful to the Department of Civil Engineering of the University of Coimbra, the Innovation and Competence Forest Centre—SerQ, and the Department of Chemistry, Unit of Fiber Materials and Environmental Technologies (FibEnTech-UBI), University Beira Interior, for all support provided in the development of this work.

Conflicts of Interest: The authors declare no conflicts of interest.

References

1. Torgovnikov, G.; Vinden, P. MW and RF Applications Case Histories. In Proceedings of the Fourth World Congress on Microwave and Radio Frequency Application, Austin, TX, USA, 7–12 November 2004; pp. 77–78.
2. Mascarenhas, F.J.R.; Dias, A.M.P.G.; Christoforo, A.L. State of the Art of Microwave Treatment of Wood: Literature Review. *Forests* **2021**, *12*, 745. [CrossRef]
3. Torgovnikov, G.; Vinden, P. Microwave Wood Modification Technology and Its Applications. *For. Prod. J.* **2010**, *60*, 173–182. [CrossRef]
4. Engelund, E.T.; Thygesen, L.G.; Svensson, S.; Hill, C.A.S. A Critical Discussion of the Physics of Wood-Water Interactions. *Wood Sci. Technol.* **2013**, *47*, 141–161. [CrossRef]
5. *Forest Products Laboratory Wood Handbook: Wood as an Engineering Material*; General Technical Report FPL-GTR-282; USDA: Washington, DC, USA, 2021; ISBN 1892529025.
6. Metaxas, A.C.; Meredith, R.J. *Industrial Microwave Heating*, 3rd ed.; Short Run Press Ltd.: Exeter, UK, 2008; Volume 24, ISBN 9780906048894.
7. Sahin, H.; Ay, N. Dielectric Properties of Hardwood Species at Microwave Frequencies. *J. Wood Sci.* **2004**, *50*, 375–380. [CrossRef]
8. Oloyede, A.; Groombridge, P. The Influence of Microwave Heating on the Mechanical Properties of Wood. *J. Mater. Process. Technol.* **2000**, *100*, 67–73. [CrossRef]
9. Du, G.; Wang, S.; Cai, Z. Microwave Drying of Wood Strands. *Dry. Technol.* **2005**, *23*, 2421–2436. [CrossRef]
10. Terziev, N.; Daniel, G.; Torgovnikov, G.; Vinden, P. Effect of Microwave Treatment on the Wood Structure of Norway Spruce and Radiata Pine. *Bioresources* **2020**, *15*, 5616–5626. [CrossRef]
11. Kol, H.Ş.; Çayır, B. The Effects of Increasing Preservative Uptake by Microwave Pre-Treatment on the Microstructure and Mechanical Properties of Oriental Spruce Wood. *Wood Mater. Sci. Eng.* **2022**, *18*, 732–738. [CrossRef]
12. Kol, H.Ş.; Çayır, B. Increasing the Impregnability of Oriental Spruce Wood via Microwave Pretreatment. *Bioresources* **2021**, *16*, 2513–2523.
13. Torgovnikov, G.; Vinden, P. High-Intensity Microwave Wood Modification for Increasing Permeability. *For. Prod. J.* **2009**, *59*, 84–92.
14. Xiao, H.; Lin, L.; Fu, F. Temperature Characteristics of Wood during Microwave Treatments. *J. Res.* **2018**, *29*, 1815–1820. [CrossRef]
15. Zielonka, P.; Dolowy, K. Microwave Drying of Spruce: Moisture Content, Temperature, and Heat Energy Distribution. *For. Prod. J.* **1998**, *48*, 77–80.
16. Zielonka, P.; Gierlik, E. Temperature Distribution during Conventional and Microwave Wood Heating. *Holz Als Roh Und Werkst.* **1999**, *57*, 247–249. [CrossRef]
17. Sethy, A.K.; Torgovnikov, G.; Vinden, P.; Przewloka, S. Moisture Conditioning of Wood Using a Continuous Microwave Dryer. *Dry. Technol.* **2016**, *34*, 318–323. [CrossRef]
18. Ouertani, S.; Koubaa, A.; Azzouz, S.; Bahar, R.; Hassini, L.; Belghith, A. Microwave Drying Kinetics of Jack Pine Wood: Determination of Phytosanitary Efficacy, Energy Consumption, and Mechanical Properties. *Eur. J. Wood Wood Prod.* **2018**, *76*, 1101–1111. [CrossRef]
19. Huang, R.; Wu, Y.; Zhao, Y.; Lu, J.; Jiang, J.; Chen, Z. Factors Affecting the Temperature Increasing Rate in Wood during Radio-Frequency Heating. *Dry. Technol.* **2013**, *31*, 246–252. [CrossRef]
20. Ouertani, S.; Hassini, L.; Azzouz, S.; Torres, S.S.; Belghith, A.; Koubaa, A. Modeling of Combined Microwave and Convective Drying of Wood: Prediction of Mechanical Behavior via Internal Gas Pressure. *Dry. Technol.* **2015**, *33*, 1234–1242. [CrossRef]
21. Alonso-Esteban, J.I.; Carocho, M.; Barros, D.; Velho, M.V.; Heleno, S.; Barros, L. Chemical Composition and Industrial Applications of Maritime Pine (*Pinus pinaster* Ait.) Bark and Other Non-Wood Parts. *Rev. Environ. Sci. Bio/Technol.* **2022**, *21*, 583–633. [CrossRef]

22. Meredieu, C.; Brunet, Y.; Danjon, F.; Défossez, P.; Dupont, S.; Kamimura, K.; Gardiner, B. *Pinus pinaster* in Europe: Distribution, Habitat, Usage and Threat. In *European Atlas of Forest Tree Species*; San-Miguel-Ayanz, J., de Rigo, D., Caudullo, G., Houston Durrant, T., Mauri, A., Eds.; Publications Office of the European Union: Luxembourg, 2016; pp. 128–129.
23. Ganguly, S.; Balzano, A.; Petrič, M.; Kržišnik, D.; Tripathi, S. Effects of Different Energy Intensities of Microwave Treatment on Heartwood and Sapwood Microstructures in Norway Spruce. *Forests* **2021**, *12*, 598. [CrossRef]
24. Leggate, W.; Redman, A.; Wood, J.; Bailleres, H.; Lee, D.J. Radial Permeability of the Hybrid Pine (*Pinus elliottii* × *Pinus caribaea*) in Australia. *Bioresources* **2019**, *14*, 4358–4372. [CrossRef]
25. *CEN EN 350*; Durability of Wood and Wood-Based Products—Testing and Classification of the Durability to Biological Agents of Wood and Wood-Based Materials. iTeh Inc.: Brussels, Belgium, 2016.
26. Sedighi-Gilani, M.; Vontobel, P.; Lehmann, E.; Carmeliet, J.; Derome, D. Liquid Uptake in Scots Pine Sapwood and Hardwood Visualized and Quantified by Neutron Radiography. *Mater. Struct./Mater. Constr.* **2014**, *47*, 1083–1096. [CrossRef]
27. Hansmann, C.; Gindl, W.; Wimmer, R.; Teischinger, A. Permeability of Wood—A Review. *Drev. Vysk./Wood Res.* **2002**, *47*, 1–16.
28. *ISO 13061-1*; Physical and Mechanical Properties of Wood—Test Methods for Small Clear Wood Specimens—Part 1: Determination of Moisture Content for Physical and Mechanical Tests. ISO: Geneva, Switzerland, 2014.
29. *BS 373:1957*; Methods of Testing Small Clear Specimens of Timber. BSI: London, UK, 1999.
30. Mascarenhas, F.J.R.; Dias, A.M.P.G.; Christoforo, A.L.; dos Santos Simões, R.M.; Cunha, A.E.P. Effect of Microwave Treatment on Drying and Water Impregnability *Pinus pinaster* and *Eucalyptus globulus*. *Maderas Cienc. Tecnol.* **2023**, *25*, 1–14. [CrossRef]
31. Al-Khuseibi, M.K.; Sablani, S.S.; Perera, C.O. Comparison of Water Blanching and High Hydrostatic Pressure Effects on Drying Kinetics and Quality of Potato. *Dry. Technol.* **2005**, *23*, 2449–2461. [CrossRef]
32. Niemz, P.; Teischinger, A.; Sandberg, D. (Eds.) *Handbook of Wood Science and Technology*; Springer: Cham, Switzerland, 2023.
33. He, Z.; Qian, J.; Qu, L.; Wang, Z.; Yi, S. Simulation of Moisture Transfer during Wood Vacuum Drying. *Results Phys.* **2019**, *12*, 1299–1303. [CrossRef]
34. Benabou, L. Predictions of Compressive Strength and Kink Band Orientation for Wood Species. *Mech. Mater.* **2010**, *42*, 335–343. [CrossRef]
35. Šefc, B.; Trajković, J.; Sinković, T.; Hasan, M.; Ištok, I. Compression Strength of Fir and Beech Wood Modified by Citric Acid. *Drv. Ind.* **2012**, *63*, 45–50. [CrossRef]
36. Aicher, S.; Stapf, G. Compressive Strength Parallel to the Fiber of Spruce with High Moisture Content. *Eur. J. Wood Wood Prod.* **2016**, *74*, 527–542. [CrossRef]
37. Zhao, C.; Liu, D.; Zhang, C.; Li, Y.; Wang, Y. Influence of Specimen Size on the Compressive Strength of Wood. *Buildings* **2024**, *14*, 1156. [CrossRef]
38. *ABNT NBR 7190-1*; Projeto de Estruturas de Madeira Parte 1: Critérios de Dimensionamento. ABNT: Rio de Janeiro, Brazil, 2022.
39. *CEN EN 408*; Timber Structures—Structural Timber and Glued Laminated Timber—Determination of Some Physical and Mechanical Properties. iTeh: Brussels, Belgium, 2012.
40. Hermoso, E.; Vega, A. Effect of Microwave Treatment on the Impregnability and Mechanical Properties of *Eucalyptus globulus* Wood. *Maderas Cienc. Tecnol.* **2016**, *18*, 55–64. [CrossRef]
41. *Software Minitab*, version 18; Minitab: State College, PA, USA, 2018.
42. Weng, X.; Zhou, Y.; Fu, Z.; Gao, X.; Zhou, F.; Jiang, J. Effects of Microwave Pretreatment on Drying of 50 mm-Thickness Chinese Fir Lumber. *J. Wood Sci.* **2021**, *67*, 13. [CrossRef]
43. Leiker, M.; Adamska, M.A. Energy Efficiency and Drying Rates during Vacuum Microwave Drying of Wood. *Holz Als Roh Und Werkst.* **2004**, *62*, 203–208. [CrossRef]
44. Vinden, P.; Torgovnikov, G.; Hann, J. Microwave Modification of Radiata Pine Railway Sleepers for Preservative Treatment. *Eur. J. Wood Wood Prod.* **2011**, *69*, 271–279. [CrossRef]
45. Dashti, H.; Tarmian, A.; Faezipour, M.; Hedjazi, S.; Shahverdi, M. Mass Transfer Through Microwave-Treated Fir Wood (*Abies alba* L.): A Gymnosperm Species with Torus Margo Pit Membrane. *Dry. Technol.* **2013**, *31*, 359–364. [CrossRef]
46. Hansson, L.; Antti, L. Modeling Microwave Heating and Moisture Redistribution in Wood. *Dry. Technol.* **2008**, *26*, 552–559. [CrossRef]
47. Vongpradubchai, S.; Rattanadecho, P. The Microwave Processing of Wood Using a Continuous Microwave Belt Drier. *Chem. Eng. Process. Process Intensif.* **2009**, *48*, 997–1003. [CrossRef]
48. Zhang, Y.; Jia, K.; Cai, L.; Shi, S.Q. Acceleration of Moisture Migration in Larch Wood Through Microwave Pre-Treatments. *Dry. Technol.* **2013**, *31*, 666–671. [CrossRef]
49. Petty, J.A.; Preston, R.D. Permeability and Structure of the Wood of Sitka Spruce. *Proc. R. Soc. B Biol. Sci.* **1970**, *175*, 149–166. [CrossRef]
50. Koubaa, A.; Perré, P.; Hutcheon, R.; Lessard, J. Complex Dielectric Properties of the Sapwood of Aspen, White Birch, Yellow Birch, and Sugar Maple. *Dry. Technol.* **2008**, *26*, 568–578. [CrossRef]
51. Antti, A.L.; Perre, P. A Microwave Applicator for on Line Wood Drying: Temperature and Moisture Distribution in Wood. *Wood Sci. Technol.* **1999**, *33*, 123–138. [CrossRef]
52. Henry, F.; Gaudillat, M.; Costa, L.C.; Lakkis, F. Free and/or Bound Water by Dielectric Measurements. *Food Chem.* **2003**, *82*, 29–34. [CrossRef]

53. Laboratório Nacional de Engenharia Civil (LNEC). *Ficha M9—Madeira Para Construção: Humidade Da Madeira [Wood for Construction: Wood Moisture]*; LNEC: Lisbon, Portugal, 1997.
54. He, S.; Lin, L.; Fu, F.; Zhou, Y.; Fan, M. Microwave Treatment for Enhancing the Liquid Permeability of Chinese Fir. *Bioresources* **2014**, *9*, 1914–1923. [CrossRef]
55. Wang, Z.; Xu, E.; Fu, F.; Lin, L.; Yi, S. Industrial Crops & Products Characterization of Wood Cell Walls Treated by High-Intensity Microwaves: Effects on Physicochemical Structures and Micromechanical Properties. *Ind. Crop. Prod.* **2022**, *187*, 115341. [CrossRef]
56. Machado, J.S. Effect of Microwave Treatment on Oak Compression Strength. *Silva Lusit* **2006**, *14*, 51–58.
57. Balboni, B.M.; Ozarska, B.; Garcia, J.N.; Torgovnikov, G. Microwave Treatment of *Eucalyptus macrorhyncha* Timber for Reducing Drying Defects and Its Impact on Physical and Mechanical Wood Properties. *Eur. J. Wood Wood Prod.* **2018**, *76*, 861–870. [CrossRef]
58. Dias, A.M.A.; Santos, P.G.G.; Dias, A.M.P.G.; Silvestre, J.D.; de Brito, J. Life Cycle Assessment of a Preservative Treated Wooden Deck. *Wood Mater. Sci. Eng.* **2022**, *17*, 502–512. [CrossRef]
59. Morgado, T.F.M.; Rodrigues, J.; Machado, J.S.; Dias, A.M.P.G.; Cruz, H. Bending and Compression Strength of Portuguese Maritime Pine Small-Diameter Poles. *For. Prod. J.* **2009**, *59*, 23–28.
60. Morgado, T.F.M.; Dias, A.M.P.G.; Machado, J.S.; Negrão, J.H.; Marques, A.F.S. Grading of Portuguese Maritime Pine Small-Diameter Roundwood. *J. Mater. Civ. Eng.* **2017**, *29*, 04016209. [CrossRef]

forests

MDPI

Article

Chemical Composition of Larch Oleoresin before and during Thermal Modification

Morwenna J. Spear *, Athanasios Dimitriou and Ray Marriott

The BioComposites Centre, Bangor University, Deiniol Road, Bangor LL57 2UW, UK; a.dimitriou@bangor.ac.uk (A.D.); ray.marriott@bridgefarmbioscience.co.uk (R.M.)
* Correspondence: m.j.spear@bangor.ac.uk

Abstract: Larch is a strong timber, which grows rapidly in the UK climate, but can contain abundant resin pockets. To address the resin exudation issue, a mild thermal modification process has been developed, promoting the curing of the resin. This paper reports a series of studies which characterised the chemical profile of larch oleoresin before and after the mild thermal treatment, explaining the changes which occur when resin is dried. Further experiments were used to simulate specific points in time during the mild treatment process. The non-polar components of the fresh (untreated) and treated larch oleoresin were profiled using gas chromatography mass spectrometry (GC-MS). Fresh larch oleoresin was also subjected to isothermal experiments at different temperatures in a thermogravimetric analyser–differential scanning calorimeter (TGA/DSC), followed by re-analysing the resin composition. This demonstrated the loss of monoterpenes at temperatures of 120 °C and above, with complete loss by isothermal conditions of 150 °C and 60 min. The partial loss of sesquiterpene alkanes and alkenes were also observed at all temperatures, although completeness of this loss was achieved at isothermal temperatures of 150 °C and above. The diterpene composition was seen to change for isothermal experiments conducted at 150 °C and above, with a dehydration of terpenols to form the equivalent terpene alkenes. The observed physical changes in the TGA/DSC experiment were in good agreement with observations of the oleoresin sampled from thermally modified larch planks.

Keywords: terpenes; larch; thermal modification; gas chromatography; mass spectrometry

Citation: Spear, M.J.; Dimitriou, A.; Marriott, R. Chemical Composition of Larch Oleoresin before and during Thermal Modification. *Forests* **2024**, *15*, 904. https://doi.org/10.3390/f15060904

Academic Editors: Christian Brischke and Milan Gaff

Received: 27 March 2024
Revised: 25 April 2024
Accepted: 16 May 2024
Published: 23 May 2024

1. Introduction

Japanese larch (*Larix kaempferi* Lamb.) is a strong durable timber, which grows rapidly in the UK climate, yet has a higher density, and higher tensile and bending strength than Sitka spruce and Scots pine. While slow-grown European larch and Siberian larch from Alpine or Scandinavian forests is imported to the UK for use in cladding and joinery, the faster-grown material from the UK has often occupied the garden fencing and rough-sawn timber market due to perceived properties relating to its machinability and fast growth rate. In addition, fast-grown larch is prone to resin pockets, which may lead to issues in processing. The outbreak of *Phytopthera ramorum*, which has led to the widespread felling of larch in the UK is also related to the high resin content in the timber [1].

Thermal modification of UK-grown larch timber has shown potential to reduce issues in machining [2], and to 'cure' the resin, leaving dry deposits on the timber surface or within resin pockets, which are relatively easily machined and leave less residue on the planer blades [3]. The mild thermal modification conditions employed mean that only a limited onset of chemical changes to the structural wood components (cellulose, hemicellulose, and lignin) are observed [4]. In this paper, the chemical changes which give rise to the 'drying' process that the larch oleoresin undergoes is considered.

It is well known that the kiln-drying of softwoods, and therefore also the thermal modification of softwoods, results in the liberation of terpenes and other volatile organic compounds (VOCs) at elevated temperatures [5–9]. It is increasingly recognised that the control of terpene emissions from timber-drying kilns and other drying operations requires consideration [10]. Studies by Bannerjee and co-workers have shown that the evolution of monoterpenes versus sesquiterpenes is influenced by the moisture content of and the quantity of moisture being driven off from the wood [11,12]. During the thermal modification of wood, various extractive components of wood are redistributed, and the terpene content of softwoods following thermal modification, such as in the Thermowood process, is reduced while other volatile compounds may have increased [8,13–15]. In thermal modification at high temperatures (usually 180 °C and above), such as in the Thermowood process, degradation of hemicellulose contributes to the formation of new volatile compounds, and the study of this process is frequently reported [4,8,15,16]. The mild modification system described above uses temperatures closer to 170 °C, so it does not result in the same degree of conversion of hemicellulose [3,16], and negligible change in lignin would be expected at this temperature [4]. The mild modification may therefore demonstrate changes in terpene profile, which are intermediate between those reported for high-temperature kiln-drying and full thermal modification processes. As a result, the chemical composition of oleoresin and extractives in mild thermally modified wood are likely to reveal interesting trends with potential relevance to kiln-drying and modification systems.

Oleoresin is present in several species of coniferous timber, and is principally distributed in resin canals that run longitudinally and radially throughout the wood. Its production can be enhanced as a result of injury, disease or insect attack, and in these cases, resin pockets may increase in prominence. Conifer resins are known to contain a range of terpenes [17] and have been widely used in the art industry for many years, both as turpentine and as colophony or rosin, the higher-molecular-weight solid which remains after the distillation of the oleoresin. As a result, some of the data relating to pine or larch resin composition have been determined to allow for the identification of rosin in lacquers and other artefacts [18,19]. Venice turpentine was formed from European larch oleoresin, while many rosins were sourced from pines. The term 'rosin' will be used in this paper to refer to dried resin, from which the more volatile compounds have evaporated, while resin will be used to refer to fresh or pliable oleoresin from wood.

Monoterpenes, such as the pinenes, comprise two isoprene units and a structure containing ten carbon atoms, while sesquiterpenes contain three isoprene units and so are based on a fifteen-carbon skeleton (Figure 1). Many diterpenes, for example, are based on a pimarane skeleton and 20 carbons, with three six carbon rings and various substituents. Migration of the single double bonds around the ring structures during formation permits the biosynthesis of many closely related but different molecules. Terpenoid molecules may have a large number of skeletal forms—at least 38 are known for the monoterpenes, increasing to over 200 for the sesquiterpenes [20]. The terpenes and terpenoids may have alkane, alkene, aldehyde, ester, alcohol or carboxylic acid functionalities. Two examples of monoterpenes, sesquiterpenes, and diterpenoids are shown in Figure 1. The commonly occurring α- and β-pinenes are monoterpenes, with the formula $C_{10}H_{16}$ due to the two-ringed structure and alkene functionality.

The natural larch oleoresin contains various mono-, sesqui-, and di-terpenoids, as well as some alkanes and fatty acids [17,21]. The larch monoterpenes tend to be alkene in structure, e.g., α- and β-pinenes and Δ-3-carene; similarly, the sesquiterpenes present tend to be alkenes, e.g., germacrene D and longifolene. The diterpenes in larch resin include alkenes, alcohols, and aldehyde functional groups, as well as the resin acids such as abietic acid, dehydroabietic acid, pimaric acid, and palustric acid [17,22]. The larch diterpenes also contain the characteristic larixyl acetate and its free diol larixol, or torulosol acetate and epi-torulosol, depending on species [18,23]. The larixol and its acetate occur in *Larix decidua* and *Larix gmelini*, while epi-torulosol and its acetate are detected in the species where larixol is absent, such as *Larix kaempferi*. Hybrid larch (*Larix X. eurolepis* Henry) contains components of both types as an intermediate between its parent species. *Larix kaempferi* is reported to have a similar composition to *Larix laricina* ((DuRoi) K.Koch.) (tamarak) and the other North American species, but a higher quantity of thunbergol [18].

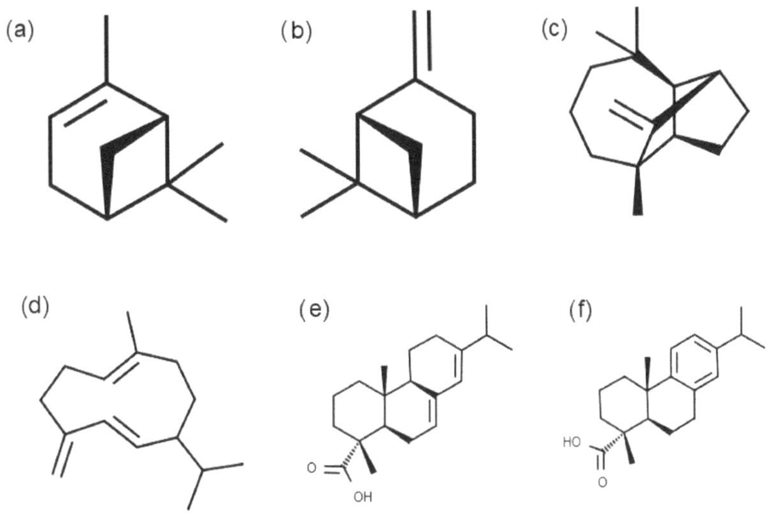

Figure 1. Structures of monoterpenes, sesquiterpenes, and diterpenes, including (**a**) α-pinene (**b**) β-pinene, (**c**) longifolene, (**d**) germacrene D, (**e**) abietic acid, and (**f**) dehydroabietate.

This study reports a set of chemical profiling experiments that compare the larch resin before and after thermal modification under different conditions. The aim of this study was to explain the changes which occur when resin dries during the thermal modification process. The composition of oleoresin from fresh and thermally modified larch timber was observed using gas chromatography–mass spectrometry (GC-MS). In a second study, DSC experiments were conducted to expose fresh oleoresin to different isothermal temperature sequences in a thermogravimetric analyser–differential scanning calorimeter (TGA/DSC). The composition of the material generated by TGA/DSC experiments was also evaluated.

2. Materials and Methods

2.1. Thermal Modification

Japanese larch (*Larix kaempferi*) harvested in Wales, UK, was thermally treated in a pilot-scale kiln by Coed Cymru according to a three-stage thermal treatment procedure. This used a ramp phase, a high-temperature hold phase, and a conditioning phase during which the temperature of the kiln decreases. A pre-treatment step was used to initiate drying—this is referred to as Day 1 in later sections—and had a target temperature of 120 °C. During the high-temperature phase, the maximum kiln temperature was 190 °C, but the internal temperature of the timber lagged below this. Thermocouple data confirmed the wood core to have been above 170 °C for between 2 h and 4.5 h [2]. The mild thermal treatment was achieved with 2–3 h above 170 °C, while the moderate thermal treatment required longer duration with the wood core at a temperature of 170 °C or above. The treatment was developed to ensure a process which enhanced machinability, rather than targeting dimensional stability or durability aspects, as are commonly the fixed objective of commercial systems; the treatment details therefore differ from the Thermowood or Retiwood processes, which are discussed elsewhere [2,24].

The treated planks were 110 mm by 30 mm in cross-section, and 90 cm in length, cut from longer planks which had been air-dried outdoors under cover. The moisture content of the planks prior to treatment was 18.04%, determined by the removal of a 2 cm moisture content block from within the length of each longer plank while preparing the

90 cm length planks for treatment (average of 30 samples). A further three moisture content samples were taken from selected planks and used for moisture content determination after treatment. These moisture content samples were also used to evaluate the weight change due to the thermal treatment process, based on calculated oven dry weight per plank before and after treatment. The apparent weight loss was 2.2 to 2.6%.

Planks were stacked in a pilot-scale kiln, in a stack that was seven planks wide by seven planks tall, and good air circulation was forced by the use of a baffle above the stack to direct the heated air through the stack. Thermocouples were drilled into five planks within the stack at the centre of the top and bottom layers of planks, and three in the central layer (left, centre, and right). Two additional thermocouples were stapled onto the surface of the planks on the left and right of the kiln stack, providing paired internal and surface temperature data for the planks nearest to and furthest from the hot air infeed on the right-hand side of the oven.

2.2. Gas Chromatography–Mass Spectrometry of Oleoresin

Resin was sampled from resin pockets identified in untreated and treated planks. These resin pockets were exposed in the sawn face of the untreated planks, and as the planks were recently cut, the resin was a viscous liquid which could be sampled using a spatula for the analysis. In the case of thermally modified planks, resin pockets were exposed by cross-cutting for sampling.

Approx. 10 mg of resin was dissolved in the selected solvent for gas chromatography–mass spectrometry analysis (GC-MS). Heptane was used as solvent for the first run, as this is suitable for a broad range of non-polar molecules present in larch resin. The dissolution was repeated using dichloromethane and with methanol to increase the polar fractions represented in the chromatograms; however, these changes were relatively small.

GC-MS was carried out on a Perkin Elmer Clarus 600 chromatograph fitted with a VF5 column. The initial oven temperature was 60 °C, held for one minute prior to increasing the oven temperature at a rate of 6 °C/min to 300 °C, where the temperature was held for 10 min. Mass spectra were analysed using TurboMass 6.10 software and compared with reference spectra from the NIST 2011 and Adams spectral libraries and selected relevant publications (e.g., [25]) to verify the identification of the dominant components within the chromatograph.

2.3. Fourier Transform Infrared (FTIR) Spectroscopy

Resin from the untreated and treated planks was also observed using FTIR spectroscopy in attenuated total reflectance (ATR) mode with a Gladi-ATR (PIKE Technologies, Madison, WI, USA) and Thermo 8000 FTIR spectrophotometer (Thermo Fisher Scientific, Waltham, MA, USA). The oleoresin was smeared directly on the diamond crystal, and the ATR surface was cleaned using the solvent between each sample of resin. Wood and dried rosin were sampled from resin pockets in the treated wood using a sharp blade to provide a flat surface, which was pressed directly against the diamond crystal, allowing the ATR spectrum to be recorded.

2.4. Temperature Simulations Using TGA/DSC

Small samples of untreated resin were heated under controlled conditions using a Mettler Toledo TGA/DSC 1 STAR^e System, under constant flow of nitrogen gas (50 cm^3/min). The resin samples were placed in open 40 μL aluminium crucibles to allow for evaporation to occur during heating. A ramp rate of 10 °C/min was used to heat the resin to the selected isothermal temperature, which was held for a 60-minute hold period before cooling. Four different profiles were investigated, with isothermal temperatures at 120 °C, 150 °C, 170 °C, and 190 °C. These isotherms were selected to reflect significant temperature steps within the thermal modification process. Resin samples from these experiments were also subjected to GC-MS analysis.

3. Results

3.1. FTIR Spectra

Samples of fresh larch resin were observed, and the typical spectrum is presented in Figure 2, alongside the spectrum for the larch earlywood. The oleoresin is a mixture of many terpenes, but it is clear that the resin spectrum contains compounds containing a hydrocarbon structure with some sp^3 CH units in addition to alkane and alkene structures (absorptions at 2843 to 2925 cm^{-1}). There are also carboxylic acids and alcohols present, as evidenced by strong absorptions at 1690 and 1638 cm^{-1} [26]; however, these occur within a broad peak (1730–1550 cm^{-1}) where many absorptions overlap, possibly relating to other carbonyl absorptions such as aldehydes or alcohol absorptions, which occur within a similar range.

Many traits within the spectrum for the oleoresin were seen to correlate with the spectra for α- and β-pinenes, such as the small absorption at 3078 cm^{-1}, the dominant 2929 cm^{-1} peak, and the pair of alkane bend absorptions at 1450 and 1373 cm^{-1}; while other aspects, e.g., the 1690 cm^{-1} and 1638 cm^{-1} carboxylic acid absorptions and 1448, 1366, 1262, 1240 and 1149 cm^{-1} fingerprint absorptions, correlated with the spectra for diterpene-based plant rosins such as pine rosin, sandarac, and copal [19].

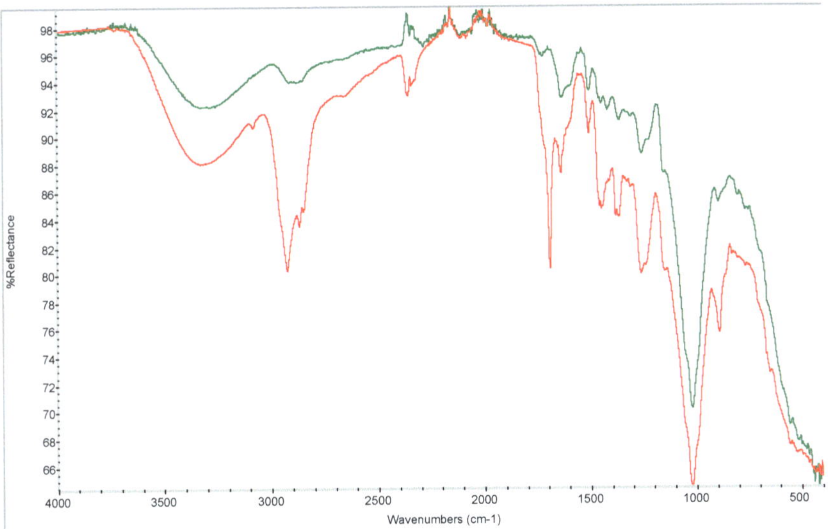

Figure 2. FTIR-ATR spectra for fresh resin from a resin pocket (red line) and larch wood (green line) before modification.

Rosin was sampled from within dried resin pockets found within treated planks for FTIR-ATR spectral determination (Figure 3). This revealed a reduction in several strong absorptions previously present in the fresh resin. For example, the strong CH stretches at 2925 and 2863 cm^{-1} were reduced, and peaks within the carbonyl and alcohol region decreased (1750–1550 cm^{-1}), indicating the loss or conversion of resin components. The relative proportion of hydroxyl appeared to increase, as the broad absorption centred on 3300 cm^{-1} relating to OH stretches became more prominent. This would be consistent with the loss of alkene dominated structures such as the monoterpenes, which are expected to evaporate in large quantities from the oleoresin.

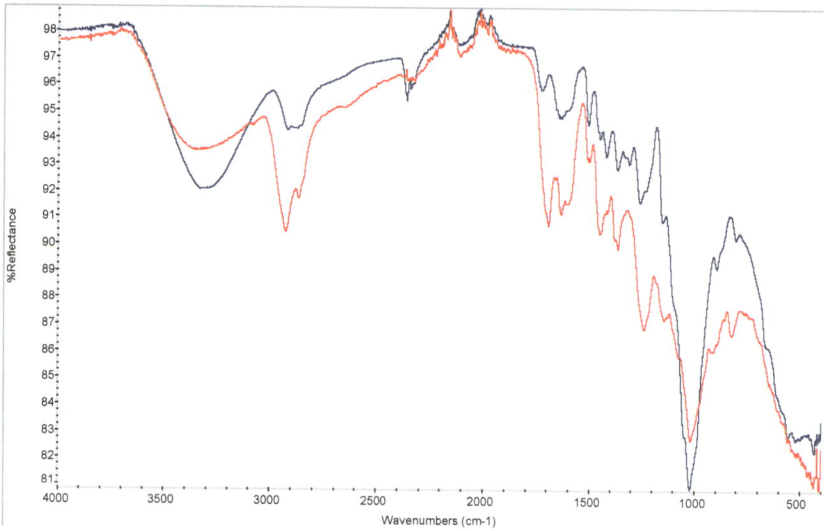

Figure 3. A sample of rosin from a mild thermally treated plank (blue line), compared with unheated (fresh) resin (red line).

3.2. *GC-MS for Oleoresins from Untreated and Thermally Treated Wood*

The composition of fresh larch resin, and samples of larch resin and rosin collected from planks subjected to mild thermal modification, are presented in Figures 4 and 5, as well as Table 1.

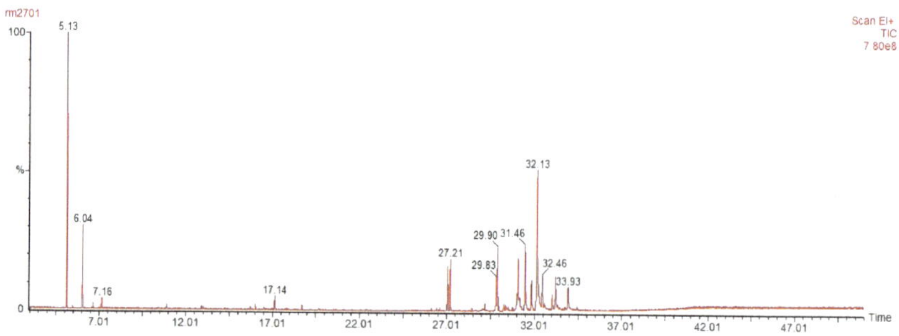

Figure 4. GC-MS chromatogram for oleoresin sampled from an untreated plank.

The components observed in the chromatogram for the unmodified larch resin (Figure 4) were broadly in agreement with the profiles observed by previous authors [18]. The main three monoterpenes identified were α-pinene (retention time = 5.13 min), β-pinene (rt = 6.04 min), and Δ-3-carene (rt = 7.16 min), all of which were seen before the 10 min elution time. The main sesquiterpenes were longifolene, γ-elemene, germacrene D, β-cadinene, and germacrene B. These were seen between 15 and 19 min, and all were present in relatively low abundance. The diterpenes were identified at higher retention times, above 25 min; this region contained a mixture of many compounds, including diterpenes (thunbergene), diterpene alcohols (e.g., epimanool, isopimarol, torulosol), acetates of some diterpene alcohols (e.g., torulosol acetate), and diterpene resin acids (palustric acid, abietic acid, neoabietic acid).

Table 1. Oleoresin composition for freshly cut larch and for larch after mild thermal modification process.

Compound	Larch Oleoresin	Oleoresin from Mild Thermally Modified Wood
α-pinene	dominant	medium
β-pinene	large	small
camphene	minor	absent
Δ-3-carene	medium	absent
d-limonene	minor	absent
l-α-terpineol	minor	minor
α-terpineol acetate	minor	absent
longifolene	minor	absent
γ-elemene	minor	minor
one of the germacrenes	minor	absent
germacrene D	small	absent
β-cadinene (or δ-cadinene)	minor	absent
germacrene B	minor	absent
thunbergol	medium	absent
epi-manool	medium	present—strong
isopimara 7,15-dien-3-one	small	minor
isopimara 7,15-dienol?	medium	minor
a compound peak mixture	medium	one component still present
larixol	medium	absent
torulosol	small	strong
dehydroabietol or neoabietinol	small	minor
palustric acid	large	absent
torulosol acetate	small	medium—large
isopimaric acid (?)	small	minor
abietic acid	small	minor
neoabietic acid	small	absent
thunbergene	not present	present—medium
pimaradiene	not present	present—small
benzenedicarboxylic acid, ester	not present	(33.76)
unidentified	not present	(35.29)
squalene ($C_{30}H_{50}$)	not present	present—dominant peak (37.08)

The typical profile of terpenes changed in the thermally treated material, with α-pinene content falling to 6%–11% from 21% in untreated material, and β-pinene content falling to 0.6%–1.1% from 7% (Table 2, Figure 5). The proportion of sesquiterpenes also decreased, while the proportion of all diterpene compounds increased, as these form the greatest component of the residue. These changes would reflect the expected evaporation of mono- and sesquiterpenes due to low boiling points. Differences were also observed between the dry rosin and the liquid resin from further inside the same treated plank; however, the reduction in α- and β-pinene content was very small (Figure 5 middle and lower traces, Table 2).

The components present in the diterpene fraction of the resin were not static and appear to have been altered during heating. There was evidence of a thermal conversion of the thunbergol and isopimaradienol into thunbergene and pimaradiene, with a reduction in the peaks at 27.18 and 30.02 min, and the creation of new peaks at 25.27 and 25.50 in the oleoresin sampled from treated wood (Figure 5). This dehydration reaction has been observed in laboratory studies on pure terpenes [27] and in high-temperature reactions during the distillation of tall oil rosins [22], but has not been previously reported in wood modification. The yield of thunbergene (25.27 min) was greater than that of pimaradiene (25.50 min) in the liquid fraction of the sampled material (lower trace of Figure 5), which corresponds with the near disappearance of the thunbergol peak (expected at 27.18 min) from the chromatograph, while the neighbouring epimanool peak (27.33 min) was still present. The newly formed pimaradiene is believed to have been formed from the isopimaradienol; there was a strong reduction in this peak, at 29.90 min in the upper trace, and minor in the lower trace (30.02 min).

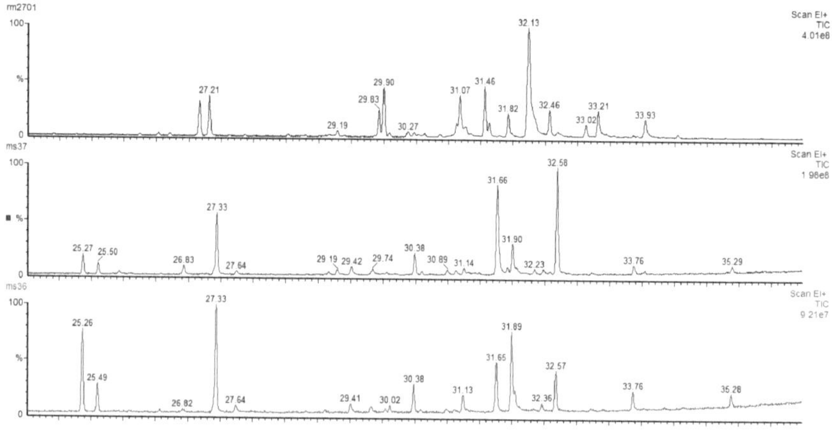

Figure 5. Gas chromatograph for untreated resin (**upper trace**), and resin sampled from low-temperature thermally modified wood (**middle and lower traces**). Middle trace is dry rosin from the surface of the treated wood and lower trace is liquid resin samples from within the resin pocket of the same plank (TM CC moderate, Table 2). The upper trace was run on a slightly longer column resulting in an approximately 12 s offset compared to the treated wood. The epimanool peak in the middle trace is at 27.33, and in upper trace is at 27.21.

Table 2. Proportion of α- and β-pinenes relative to total terpene content. Sesquiterpenes and diterpenes are presented as an aggregate figure due to variability of relative proportions of compounds within these groups.

	α-Pinene	β-Pinene	Sum of Sesquiterpenes	Sum of Diterpenes
Untreated CC	20.67%	6.66%	2.37%	70.30%
TM CC mild	6.41%	1.14%	2.51%	89.9%
TM CC moderate, liquid resin	10.71%	0.67%	0.66%	87.96%
TM CC moderate, dry resin	9.48%	0.60%	0.98%	88.94%
Untreated CM	35.19%	2.94%	3.04%	58.84%
TM high CMT	31.7%	3.64%	4.52%	60.14%

The identification of torulosol and torulosol acetate was not possible using the NIST database; however, the structure of torulosol was inferred from the fragmentation pattern of an unidentifiable component (eluted at 31.53 min) with reference to [28]. Additional samples of larch from other planks used by the project team from other sources revealed that the 31.53 min peak was dominant in some, while the larixol peak (31.46 min) was dominant in others. It has been reported that the dominance of torulosol relates to Japanese larch, which is the focus of this study, but that in hybrid larch (which is also grown in Wales) both torulosol and larixol may be identified [18]. Rosin from hybrid larch which had been thermally modified was also evaluated using GC-MS; however, for simplicity, the results will not be presented here. The majority of changes seen in the oleoresin of Japanese larch after thermal modification were also observed in oleoresin from hybrid larch when exposed to the same thermal treatment process.

3.3. TGA/DSC Temperature Simulations

The weight change of the resin samples at each stage in the thermal profile was calculated from the TGA/DSC outputs. The experiments were run on four different untreated resins from different larch planks, and consistent results were seen for all resin sources. In all cases, weight was lost during the ramp stage, and this increased as the isotherm temperature increased—relating to the additional ramp time. The quantity of resin weight lost during the isotherm period increased with an increasing set point, as would be expected (Table 3). The total weight loss doubled on increasing the isotherm temperature from 150 °C to 170 °C (16% and 38% weight loss, respectively), with a further increase in loss seen on increasing the isotherm temperature to 190 °C (59% weight loss). None of the isotherms reached a stable weight during the 60-minute hold period, although the 120 °C isotherm showed the greatest sign of approaching a plateau at an 11% weight loss.

Table 3. Weight loss on the ramp and the isotherm within DSC runs (1 h at 120 °C, 150 °C, 170 °C or 190 °C); --- indicates experiment at this isotherm was not conducted for this resin sample.

		120 °C Isotherm	150 °C Isotherm	170 °C Isotherm	190 °C Isotherm
Untreated Resin A	ramp	-	−2.3270%	−3.6629%	-
	isotherm	-	−12.5246%	−23.1080%	-
Untreated resin CC	ramp	−1.0444%	−2.0191%	−7.1936%	−10.5407%
	isotherm	−9.5621%	−14.8054%	−31.0611%	−48.5842%
Untreated resin D1	ramp	−3.0689%	−3.1069%	−7.3155%	−11.9764%
	isotherm	−8.2138%	−15.5338%	−27.6076%	−60.5482%
Untreated resin D2	ramp	−2.6315%	−4.8903%	−6.4693%	-
	isotherm	−8.1963%	−11.3445%	−28.0898%	-

It was observed that weight loss followed a gradient, relating to temperature and duration, or ramp rate (Figure 6), with a corresponding intake of energy, but no distinct endotherms relating to single components were seen. This implies that evaporation of the resin components occurs as a mixture rather than by an evolution of distinct components or fractions at their boiling points during the temperature profile. Neither was there much indication of a separate evolution of water vapour at 100 °C. The strong interaction between monoterpenes and moisture is known [11,12], and an azeotropic behaviour of mixtures of terpenes with water can be expected.

The residue which remained in the crucibles at the end of each TGA/DSC run was prepared for GC-MS analysis (Figure 7). In the case of the 170 °C isotherm, which will be presented later, the weight of residue remaining was very small, and peaks were identified by elution time, and not mass spectrum. Further reduction in residue weight for the 190 °C sample led to a chromatograph in which the compounds were approaching the detection limit, which prevented analysis.

When larch resin A was heated to 120 °C using the TGA/DSC to simulate a Day 1 heating cycle, the GC/MS trace showed a clear reduction in monoterpenes (Figure 7). The β-pinene and Δ-3-carene peaks (6.04 and 7.16 min) were not present, and only a very small α-pinene peak (5.13 min) remained. Similarly, the sesquiterpene peaks normally observed at 10 to 21 min were greatly reduced, although trace amounts of the main sesquiterpenes were observed. The diterpenes were still present; for example, both thunbergol and epimanool (27.06 min, 27.21 min) are visible, and palustric acid appears the dominant component at 32.17 min. The great reduction in α-pinene peak and disappearance of the β-pinene and Δ-3-carene peaks during the 120 °C isotherm experiment clearly supports the loss of these monoterpenes well below their boiling points (at 155, 163, and 170 °C, respectively) probably within an azeotropic mixture with water vapour.

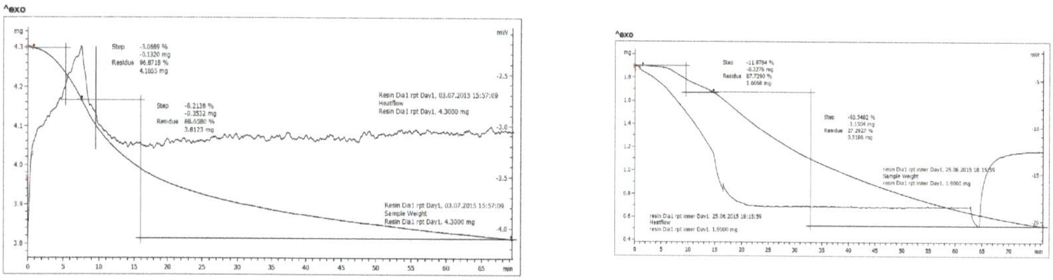

(**a**) (**b**)

Figure 6. (**a**) TGA-DSC output for Day 1 run, showing 3% weight loss on the ramp and 8% loss on the isotherm. (**b**) TGA-DSC output for 190 °C isotherm run, showing 12% weight loss on the ramp and 60% weight loss on the isotherm.

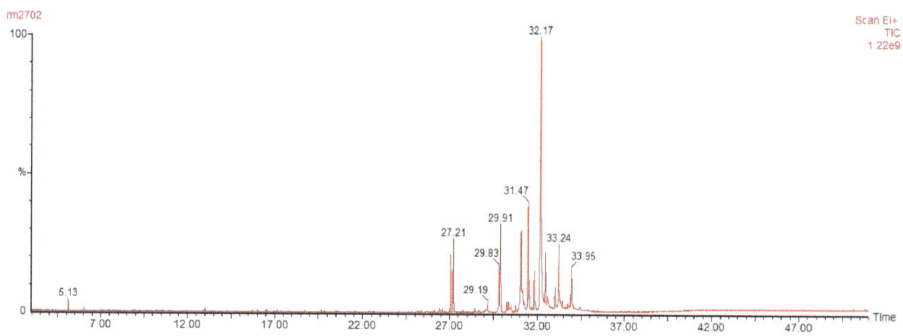

Figure 7. Gas chromatograph of resin A after heating on the TGA/DSC to 120 °C.

Samples of the same resin were exposed to isothermal temperatures of 150, 170, and 190 °C (Figure 8). For these samples, both the monoterpenes and sesquiterpenes were completely absent. In the 150 °C sample, there was a change in the pair of peaks relating to thunbergol and epimanool, with the thunbergol (27.16 min) being greatly reduced, while epi-manool remained a strong peak. This is in line with the reduction in thunbergol reported for thermally modified wood oleoresin (Table 1). Small new peaks had appeared at 25.25 and 25.46 min, relating to thunbergene and pimaradiene; these had also been seen in the samples from thermal modification experiments (Figure 5). Other minor peaks in the sesquiterpene region were also observed, including 25.80, 26.00, and 26.27 min. These were too weak to allow for identification; however, the mass spectra contained characteristic fragments for pimaratriene or androstandiene structures. This would be consistent with their development due to thermally activated dehydration reactions of the equivalent diterpene alcohol. Within the same region, there was also a somewhat stronger peak at 26.82 min. The mass spectrum of this compound revealed a phenyl ester structure, possibly characteristic of lignin degradation or of lignans within the resin. Callus resins (formed in response to injury) are typically rich in lignans and lignin esters [21], so a trace of lignin may have been present. Another peak at 27.63 min (with no good library matches) appeared to have a furan structure, and may relate to the degradation of polysaccharide (if present as contamination, e.g., traces of wood dust) occurring at elevated temperatures.

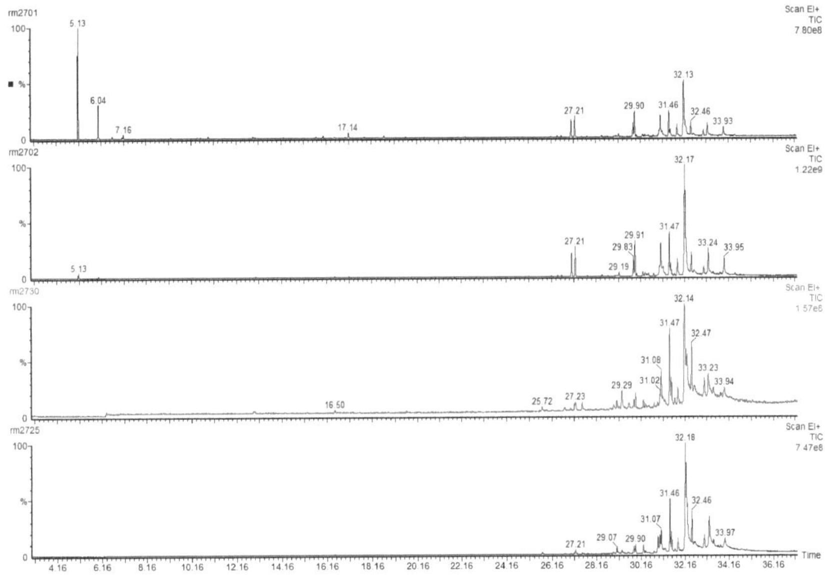

Figure 8. Gas chromatographs of fresh resin, and resin exposed to isothermal programmes at 120, 170, and 190 °C.

In the samples from an isothermal temperature of 170 °C, similar sesquiterpenes were seen at the equivalent elution times, but in lower relative abundance, indicating a greater degree of evaporation at this higher temperature. A new sesquiterpene was observed at 25.72 min, and again, the structure indicated that this could be a triene with a pimarane skeleton, for example, norpimaratriene. This could be consistent with further dehydration or degradation of the original diterpene components. For the 190 °C isotherm, the sample contained only the 25.71 min norpimaratriene, the 27.21 min epimanool, and the 27.52 min unidentified furan derivative.

It is clear that heat has two modes of operation on the diterpenes within larch resin—firstly, dehydration reactions, converting diterpene alcohols to di- or trienes; and secondly, slow evaporative loss occurring over long periods at higher temperatures. Within a 60 min period at 190 °C, the loss of diterpenes was significant, in addition to the previous loss of mono- and sesquiterpenes seen at lower temperatures.

4. Discussion

The GC-MS analysis of pure oleoresin from TGA/DSC studies demonstrated that the monoterpenes in pure larch resin samples were almost completely evaporated by isothermal treatment at 120 °C, and completely removed in higher isothermal treatments at 150, 170, and 190 °C. This is significant because the boiling point of pure α-pinene is 155 °C, and β-pinene, 163 °C. Both pinenes have relatively low enthalpies of vaporisation (37.8 and 38.6 kJ/mol, respectively), which may lead to the volatilisation of individual molecules below the boiling point. The vapour pressure of α- and β-pinenes is 3.5 and 2.4 mmHg, while water is 17.5 mmHg. Water appears to be closely linked to the volatility of terpenes; for example, Banerjee suggests three stages in the evaporation of monoterpenes from sawdust during high-temperature drying, relating to the moisture content of the material and the stage in the drying process [12].

In addition, this study has also shown that sesquiterpenes are driven off readily at temperatures from 150 °C upwards, despite the boiling point of these compounds being higher, e.g., 204 °C for longifolene and 208 °C for germacrene. This is consistent with the work reported by Granström for sesquiterpenes in high-temperature (170 to 200 °C) kiln-drying of softwoods, where up to 20% sesquiterpenes were identified in the kiln emissions alongside monoterpenes [29]. In the TGA output from the TGA/DSC runs, it is obvious that the weight change due to the loss of terpenes is not sudden, as might be expected for the evaporation of a pure compound, accompanied by a sharp endotherm in the DSC trace. Neither was it a series of such evaporation events, as would indicate a sequence of components from a separable mixture. Instead, the TGA records a gradual weight loss over the full 60-minute isotherm. This implies a strong interaction between the various components, which is also likely, given the azeotropic tendencies of the various monoterpenes as discussed above and as indicated in the studies by Banerjee et al. [11,12].

Relatively little has been written regarding the diterpene composition of larch oleoresin. Holmbom et al. [21] derivatised the oleoresin and analysed the trimethyl silyl (TMS) derivatives of the resin acids; as a result, the authors observed a profile dominated by isopimaric acid (17.4%) and abietic acid (11.3%), with various other resin acids. However, no alkene diterpenes were listed in the identified compounds for the oleoresin studied. Sato and co-workers [23] analysed diterpenes from branch bark without derivatisation, isolating the diterpenes in pure form using dichloromethane. Similarly, in this present study, no derivatisation step was performed to maximise the knowledge gained about the raw oleoresin compounds before and after the thermal treatment or the thermal simulation. The absence of TMS derivatisation is likely to have suppressed the quantity of carboxylic acid diterpenoids observed in this experiment. However, the direct study of mono- and sesquiterpene profiles was effective, and the losses of both components during heating were clearly observed.

Within this study, the most noticeable change in the diterpene region of the 150 °C isotherm sample was the formation of two new peaks at 25.25 and 25.48 min. The untreated resin did not contain these peaks. These were identified as a thunbergene and a pimaradiene, which likely formed by the dehydration of thunbergol and pimaradienol during heating. The dehydration of thunbergol to thunbergene is widely known in museum and fine art curation [18,30], where Venice turpentine (larch rosin from European larch) was widely used in oil paintings of old masters. The dehydration of pimaradienol to piamaradiene is less widely reported; however, the mass spectrum of the second new peak observed at 25.48 min contained the characteristic mass fragments for pimara 8,15-diene.

The temperatures used in the DSC experiments and the thermal modification kiln were lower than is common in conventional thermal modification systems (typically 180 to 230 °C), where chemical changes in the wood-cell-wall components are sought. But this mild and moderate thermal modification process was conducted at a sufficiently high temperature to mobilise and volatilise a significant proportion of the larch resins. The temperatures studied (120 to 190 °C) were higher than those investigated by Grüll and co-workers [31], where 70 °C with and without partial vacuum was tested for European larch to reduce resin bleed in coated larch for exterior applications. They were also higher than the work by Kačik and co-workers [32], who used 60 °C and 120 °C to reduce the monoterpene content of fir to limit attack by wood-boring insects. DSC results indicate that 70 °C would be insufficient to mobilise the liquid component of the larch resin.

5. Conclusions

This study reports chemical profiling experiments to explain the changes which occur when larch resin dries during the mild thermal modification process. This indicates potential for the mild modification to reduce the dulling of blades in paling or other secondary machining processes, increasing efficiency and product quality. After mild thermal modification, the chemical composition of resin had changed as expected, showing a reduction in monoterpenes such as α- and β-pinenes, which have a boiling point below the treatment temperature. The loss of short-chain alkanes and fatty acids is believed to explain the solidification or 'curing' of larch resin, which occurs during thermal modification; however, relatively few short-chain alkanes were detected even in the GC-MS analysis of fresh resin. The chief differences observed were the evaporation of mono- and sesquiterpenes.

Additionally, changes in the diterpene composition were observed, relating to chemical reactions occurring to the terpene alcohols under the action of heat.

The use of TGA/DSC combined with GC-MS has allowed for the evaporation processes seen in mild thermal modification to be observed independently of the chemical composition changes which occur within wood. A near-complete removal of monoterpenes occurred in the 120 °C isotherm schedule. Both mono- and sesquiterpenes were removed in the schedules at 150 °C and above. The profile of terpenes present in the dried oleoresin from 150 °C isothermal runs was found to be similar to that harvested from mildly thermally treated larch timber, while the 120 °C isothermal run resembled resin from the Day 1 pre-treatment step. This experiment revealed that the monoterpenes and sesquiterpenes are driven off at temperatures which are lower than their boiling points, in a process which is believed to be related to the evaporation of water from within the resin. This corresponds with previously reported observations of terpene evaporation from solid wood during high-temperature drying processes.

Author Contributions: Conceptualisation, M.J.S.; methodology, M.J.S., R.M., and A.D.; validation, M.J.S. and R.M.; formal analysis, M.J.S., R.M., and A.D.; investigation, M.J.S. and A.D.; writing—original draft preparation, M.J.S.; writing—review and editing, M.J.S., A.D., and R.M.; visualisation, M.J.S. All authors have read and agreed to the published version of the manuscript.

Funding: This research was commissioned by Coed Cymru Cyf. using funds awarded through the Rural Development Plan for Wales Supply Chain Efficiencies Scheme administered by the Welsh Government.

Data Availability Statement: Data is contained within the article.

Acknowledgments: The authors would like to extend thanks to staff and former staff at the BioComposites Centre who assisted in the completion of this work, including Gee-Sian Leung for instructions on using GC-MS. In addition, we thank Tabitha Binding, formerly of Coed Cymru Cyf., for input and assistance with the thermal modification activities.

Conflicts of Interest: The authors declare that the research was conducted in the absence of any commercial or financial relationships that could be construed as a potential conflict of interest.

References

1. Brasier, C.; Webber, J. Sudden larch death. *Nature* **2010**, *466*, 824–825. [CrossRef] [PubMed]
2. Spear, M.; Binding, T.; Jenkins, D.; Nicholls, J.; Ormondroyd, G. Mild thermal modification to enhance the machinability of larch. In Proceedings of the 7th European Conference on Wood Modification, Lisbon, Portugal, 10–12 March 2014.
3. Spear, M.J.; Binding, T.; Nicholls, J.; Jenkins, D.; Ormondroyd, G.A. Physical properties of UK grown larch subjected to mild and moderate thermal modification processes. In Proceedings of the 11th Meeting of the Nordic European Network for Wood Sciences and Engineering, Poznan, Poland, 14–15 September 2015; Perdoch, W., Broda, M., Eds.; pp. 98–104.
4. Esteves, B.; Pereira, H. Wood modification by heat treatment: A review. *Bioresources* **2009**, *4*, 370–404. [CrossRef]
5. Fritz, B.; Lamb, B.; Westberg, H.; Folk, R.; Knighton, B.; Grimsrud, E. Pilot- and full-scale measurements of VOC emissions from lumber drying of inland Northwest species. *For. Prod. J.* **2004**, *54*, 50–56.
6. Dahlen, J.; Prewitt, L.; Schmulsky, R.; Jones, D. Hazardous air pollutants and volatile organic compounds emitted during kiln drying of Southern pine lumber to interior and export moisture specifications. *For. Prod. J.* **2011**, *61*, 229–234. [CrossRef]
7. Pang, S. Emissions from kiln drying of *Pinus radiata* timber: Analysis, recovery and treatment. *Dry. Technol.* **2012**, *30*, 1099–1104. [CrossRef]
8. Manninen, A.M.; Pasanen, P.; Holopainen, J.K. Comparing the VOC emissions between air-dried and heat-treated Scots pine wood. *Atmos. Environ.* **2002**, *36*, 1763–1768. [CrossRef]
9. Xu, J.; Zhang, Y.; Shen, Y.; Li, C.; Wang, Y.; Ma, Y.; Sun, W. New perspective on wood thermal modification: Relevance between the evolution of chemical structure and physical-mechanical properties, and online analysis of release of VOCs. *Polymers* **2019**, *11*, 1145. [CrossRef] [PubMed]
10. Granström, K. Emissions of monoterpenes and VOCs during drying of sawdust in a spouted bed. *For. Prod. J.* **2003**, *53*, 43–55.
11. Banerjee, S.; Su, W.; Wild, M.P.; Otwell, L.P.; Hittmeier, M.E.; Nichols, K.M. Wet line extension reduces VOCs from softwood drying. *Environ. Sci. Technol.* **1998**, *32*, 1303–1307. [CrossRef]
12. Banerjee, S. Mechanisms of terpene release during sawdust and flake drying. *Holzforschung* **2001**, *55*, 413–416. [CrossRef]
13. Peters, J.; Fischer, K.; Fischer, S. Characterisation of emissions from thermally modified wood and their reduction by chemical treatment. *BioResources* **2008**, *3*, 491–502. [CrossRef]

14. Herrera, R.; da Silva, D.T.; Llano-Ponte, R.; Labidi, J. Characterization of pine wood liquid and solid residues generated during industrial hydrothermal treatment. *Biomass Bioenergy* **2016**, *95*, 174–181. [CrossRef]
15. Hyttinen, M.; Masalinweijo, M.; Kalliokoski, P.; Pasanen, P. Comparison of VOC emissions between air-dried and heat-treated Norway spruce (*Picea abies*), Scots pine (*Pinus sylvesteris*) and European aspen (*Populus tremula*) wood. *Atmos. Environ.* **2010**, *44*, 5028–5033. [CrossRef]
16. Kotilainen, R.A.; Toivanen, T.-J.; Alén, R.J. FTIR monitoring of chemical changes in softwood during heating. *J. Wood Chem. Technol.* **2000**, *20*, 307–320. [CrossRef]
17. Sjöström, E. *Wood Chemistry: Fundamentals and Applications*; Academic Press: San Diego, CA, USA, 1993; 293p.
18. Mills, J.S. Diterpenes of *Larix* oleoresins. *Phytochemistry* **1973**, *12*, 2407–2412. [CrossRef]
19. Derrick, M. Fourier Transform Infrared spectral analysis of natural resins used in furniture finishes. *J. Am. Inst. Conserv.* **1988**, *28*, 43–56. [CrossRef]
20. Banthorpe, D.V. Terpenoids. In *Natural Products: Their Chemistry and Biological Significance*; Longman Scientific and Technical: Harlow, UK, 1994; pp. 289–359.
21. Holmbom, T.; Reunanen, M.; Fardim, P. Composition of callus resin of Norway spruce, Scots pine, European larch and Douglas fir. *Holzforschung* **2008**, *62*, 417–422. [CrossRef]
22. Holmbom, B.; Avela, E.; Pekkala, S. Capillary gas-chromatography-mass spectrometry of resin acids in tall oil rosin. *J. Am. Oil Chem. Soc.* **1974**, *51*, 397–399. [CrossRef]
23. Sato, M.; Seki, K.; Kita, K.; Moriguchi, Y.; Hashimoto, M.; Yunoki, K.; Ohnishi, M. Comparative analysis of diterpene composition in the bark of the hybrid larch F_1, *Larix gmelinii* var. japonica × *L. kaempferi* and their parent trees. *J. Wood Sci.* **2009**, *55*, 32–40. [CrossRef]
24. Spear, M.; Binding, T.; Dimitriou, A.; Cowley, C.; Ormondroyd, G. Chemical changes in mild thermal treatment of larch: To what extent to they differ from full thermal modification? In Proceedings of the 8th Eighth European Conference on Wood Modification, Helsinki, Finland, 26–27 October 2015.
25. Babushok, V.I.; Linstrom, P.J.; Zenkevich, I.G. Retention indices for frequently reported compounds of plant essential oils. *J. Phys. Chem. Ref. Data* **2011**, *40*, 043101. [CrossRef]
26. Silverstein, R.M.; Bassler, G.C.; Morrill, T.C. *Spectrometric Identification of Organic Compounds*, 4th ed.; Wiley: Hoboken, NJ, USA, 1982.
27. Chernenko, G.F.; Kobzar, E.A.; Salakhutdinov, N.F.; Schmidt, E.N.; Bagrayanskaya, I.Y.; Galitov, Y.V. Transformations of terpinoids on synthetic zeolites. *Khimiya Prir. Soedin.* **1991**, *5*, 657–667; Translated Plenum Publishing Corp, 1992.
28. Kosyukova, L.V.; Khorguani, T.V. Retention indices of diterpenes isolated from resins of coniferous trees. *Zhurnal Anal. Khimii (J. Anal. Chem. USSR)* **1989**, *44*, 1309–1313.
29. Granström, K.M. Emissions of sesquiterpenes from spruce sawdust during drying. *Eur. J. Wood Wood Prod.* **2009**, *67*, 343–350. [CrossRef]
30. Russo, M.V.; Avino, P. Characterization and identification of natural terpenic resins employed in "Madonna con bambino e angeli" by Antonello da Messina using Gas Chromatography-Mass Spectrometry. *Chem. Cent. J.* **2012**, *6*, 59. [CrossRef]
31. Grüll, G.; Forsthuber, B.; Truskaller, M.; Illy, A. Methods to reduce resin exudation from Larch wood. In *Advances in Modified and Functional Bio-Based Surfaces, Proceedings of the COST Action FP1006, Thessaloniki, Greece, 8–9 April 2015*; Bangor University: Bangor, UK, 2015.
32. Kačik, F.; Velkova, V.; Šmira, P.; Nasswettrova, D.; Reinprecht, L. Release of terpenes from fir wood during its long-term use and in thermal treatment. *Molecules* **2012**, *17*, 9990–9999. [CrossRef]

forests

MDPI

Article

Densification of Delignified Wood: Influence of Chemical Composition on Wood Density, Compressive Strength, and Hardness of Eurasian Aspen and Scots Pine

Przemysław Mania [1,*], Carlo Kupfernagel [2] and Simon Curling [3]

1 Department of Wood Science and Thermal Techniques, Faculty of Forestry and Wood Technology, Poznań University of Life Sciences, Wojska Polskiego 38/42, 60-627 Poznan, Poland
2 Insitut für Holztechnologie Dresden gGmbH, Zellescher Weg 24, 01217 Dresden, Germany; carlo.kupfernagel@ihd-dresden.de
3 BioComposites Centre, Bangor University, Deiniol Road, Bangor LL57 2UW, UK; s.curling@bangor.ac.uk
* Correspondence: przemyslaw.mania@up.poznan.pl

Abstract: The densification of solid wood is a well-studied technique that aims to increase the strength and hardness of the material by permanently compressing the wood tissue. To optimise the densification process in this study, a pre-treatment with sodium sulphite was used (delignification). With delignification prior to densification, one achieves higher compression ratios and better mechanical properties compared to densification without pre-treatment. The reactivity of syringyl (dominant in hardwoods) and guaiacyl (dominant in softwoods) lignin towards delignification is different. The influences of this difference on the delignification and densification of softwoods and hardwoods need to be investigated. This study aimed to densify wood after delignification and investigate how variations in chemical composition between coniferous and deciduous species affect the densification process. Scots pine and Eurasian aspen specimens with a similar initial density were investigated to study the influence of the different lignin chemistry in softwoods and hardwoods on the densification process. Both timbers were delignified with sodium sulphite and sodium hydroxide and subsequently densified. While the delignification was twice as efficient in aspen than in pine, the compression ratios were almost identical in both species. The Brinell hardness and compressive strength showed a more significant increase in aspen than in Scots pine; however, one exception was the compressive strength in a radial direction, which increased more effectively in Scots pine. Scanning electron microscopy (SEM) revealed the microstructure of densified aspen and Scots pine, showing the crushing and collapse of the cells.

Keywords: delignification; densification; guaiacyl; syringyl; hardness; wood

Citation: Mania, P.; Kupfernagel, C.; Curling, S. Densification of Delignified Wood: Influence of Chemical Composition on Wood Density, Compressive Strength, and Hardness of Eurasian Aspen and Scots Pine. *Forests* **2024**, *15*, 892. https://doi.org/10.3390/f15060892

Academic Editor: Miklós Bak

Received: 18 April 2024
Revised: 16 May 2024
Accepted: 16 May 2024
Published: 21 May 2024

1. Introduction

Density is one of the most important characteristics of wood, determining its physical and mechanical properties. Numerous studies show that increasing wood density contributes to higher material stiffness, strength, and hardness [1–4]. The density of the wood depends on the tree species, growth conditions, age of tree, and sampling location in the trunk. Within one species, the density may vary considerably due to variable anatomical features, such as the number of vessels and their diameters, wood fibres, proportion of latewood, or the thickness of cell walls [1,5–7].

As a high density usually improves the physio-mechanical properties, many attempts have been made to manufacture artificially densified wood products [1,8–10]. The densification process reduces the void volume in wood by collapsing the vessels and fibres. The various methods available for this task include compressing wood in hydraulic presses, using heat and steam, impregnating wood with thermosetting resins before densification, or combining either of these methods [8,11]. The above techniques improve key mechanical parameters such as hardness, stiffness, and modulus of elasticity [12–15]. Moreover, the set-recovery, i.e., dimensional stability, can be improved by resin impregnation prior to densification [16] or thermo–hydro mechanical treatments [17].

Thermomechanical wood densification is one of the well-established methods that uses elevated temperatures in the wood stabilisation process, during which a press with heated plates is used. To minimise internal stresses that would occur if the moisture content of the wood was too high or too low, it is necessary to condition the timber until it reaches a moisture content of approx. 8%–15% [18]. Subsequently, the wood is compressed using heated plates with a temperature between 130 and 180 °C. The compression force is maintained as long as the wood cools below 100 °C. The temperature of the heated plates and pressing time significantly affect the densified wood's hardness, strength, and dimensional stability [19].

The first patent related to wood densification dates back to 1900 [1,11,17]. Initially, research focused on different densification methods, but over time, more attention was paid to secondary properties, such as the dimensional stability after densification [20]. In the literature, the most important parameters are moisture content before densification, pressing time and temperature, and the degree of densification, i.e., how much the wood was compacted compared to the initial dimensions [13,21]. Between 1930 and 1960, wood densification combined with chemical and heat treatments was investigated because earlier methods lacked dimensional stability and did not fully exploit the compaction capabilities [22]. Densification of wood after a chemical pre-treatment was primarily aimed at reducing set-recovery and increasing the dimensional stability of wood.

Wood densification is carried out on both low and high-density species. As a result, low-density grades can be used where they have not been considered so far due to their otherwise poor strength properties. The properties of wood species with higher density improve further, and their dimensions can be reduced [23–26].

Lignin, the cell wall kit, is a heterogeneous aromatic compound that imparts rigidity to the wood, whereas cellulose is the main load-bearing component in tensile tests [27]. Hence, lignin plays an important role during the densification process, where compressive forces are applied. Hardwood lignin consists of syringyl (S) and guaiacyl (G) units linked by a series of ether and carbon–carbon bonds. In contrast, softwood lignin is essentially composed of G units and is less reactive in the sulphite process, making it more difficult to remove from wood tissue. The share of lignin in typical softwood species ranges from 26% to 32% and in regular hardwood from 20% to 25% [28]. Delignification affects the wood structure, dimensional stability, and bond strength of cellulose [29,30]. Using delignification before densification improves the mechanical properties beyond the level achieved by densification alone. This is due to the breaking and reconstruction of new hydrogen bonds [30–33].

The influence of variable chemical composition, specifically the lignin type, on the combined delignification and densification process remains poorly understood. This work aimed to densify wood after delignification and investigate how variations in chemical composition between coniferous and deciduous species affect the densification process. The effectiveness of the densification process was assessed by analysing selected physical and mechanical parameters (wood density, Brinell hardness, compressive strength).

2. Materials and Methods

2.1. Wood Specimen

The wood of Eurasian aspen (*Populus tremula* L.) and Scots pine (*Pinus sylvestris* L.) were used in the research. The choice of species was based on their lignin's chemical composition and almost identical initial density. The lignin in pine wood is characterised by a guaiacyl system and a lower reactivity in the sulphite process, which makes it difficult to remove it from the wood tissue, in contrast to aspen wood, which has lignin with greater reactivity, mainly with a syringyl system. The samples were cut according to the anatomical directions to the dimensions of 35 mm in the tangential direction, 25 mm in the radial direction, and a length of 25 mm. A total of 53 clear wood specimens per species were subjected to the delignification and densification process. Before the delignification process, the samples were oven-dried at 103 ± 2 °C. The density of each specimen before and after the densification process was determined using the stereometric method [34].

2.2. Analysis of the Chemical Wood Composition and Attenuated Total Reflectance–Fourier Transform Infrared Spectroscopy (FTIR-ATR)

To analyse the chemical composition of the wood, the specimens were ground using a laboratory mill (Fritsch type Pulverisette 15, Fritsch GmbH, Idar-Oberstein, Germany). The resulting wood powder fraction of 0.5–1.0 mm was utilised for chemical analyses. Cellulose content was determined employing the Kürschner–Hoffer method [35], while pentosan content was assessed following the Technical Association of Pulp and Paper Industry (TAPPI) T 223 cm-01 standard [36]. Acid-insoluble lignin content was tested according to TAPPI T 222 om-06 standard [37], involving the use of 72% sulfuric acid to hydrolyse and solubilise carbohydrates. Each analysis was conducted a minimum of three times per specimen. The chemical composition was evaluated both before and after the delignification process.

FTIR-ATR spectra of both control and densified wood specimens were captured using a Nicolet 8700 FTIR instrument (Thermo Scientific Instrument Co., Waltham, MA, USA) equipped with GladiATR vision (Pike Technologies, Madison, WI, USA). Sample preparation involved milling in a laboratory mill, with subsequent drying of the fraction that passed through a mesh size with a 250 μm nominal sieve overnight at 103 ± 2 °C. The FTIR analysis was conducted on the resulting dry powder. Each spectrum was an average of 32 scans, covering wave numbers from 4000 to 400 cm^{-1}. Background spectra were taken after every two samples to reduce spectral noise and account for peaks associated with CO_2 and water vapour resulting from environmental variations.

2.3. Delignification and Densification of Wood

Delignification was performed using the laboratory method, according to Song et al. [38] and Mania et al. [39]. Pine and aspen specimens were placed in a round-bottomed flask, which was flooded with an aqueous solution of 2.5 M NaOH and 0.4 M Na_2SO_3. The flask was placed in a heating mantle and connected to a reflux condenser. The liquid solution was boiled by the heating mantle and kept in this state for nine hours. After this time, the specimens were transferred to boiling distilled water and rinsed several times from lignin decomposition products.

After delignification, specimens were placed individually in the central part of specially prepared steel clamps. The steel clamps consisted of two 7 mm thick steel flat bars connected together with four M10 metric screws. All samples were oriented to be densified in the radial direction. Then, the set prepared this way was placed on a testing machine, Zwick ZO50TH (Zwick/Roell, Ulm, Germany). Densification was carried out under the stress of 5 $N{\cdot}mm^{-2}$ and the lowest possible speed of exerting the machine force. The nuts at the steel clamps were tightened during the test to minimise post-deformation recovery. After densification, the samples (still in clamps) were left for one week, during which the tightening of the nuts was monitored continuously. The porosity (C) of wood was calculated before and after densification as in Equation (1):

$$C = \left(1 - \frac{\rho_D}{1540}\right) \cdot 100\ (\%), \quad (1)$$

where C is the porosity of wood, 1540 is the cell wall density in $kg{\cdot}m^{-3}$, and ρ_D is the oven-dry density of the specimen in $kg{\cdot}m^{-3}$.

2.4. Compression and Brinell Hardness Test

Compression tests were performed according to the standard of the International Organization for Standardization—ISO 13061-5 [40] in all three anatomical directions using a numerically controlled test machine, Zwick Z050TH (Zwick/Roell, Ulm, Germany). Ten specimens of each wood species were tested before and after densification. The stress at the proportionality limit, hence, the compressive strength perpendicular to the grain, was measured in the radial (RcR) and tangential (RcT) direction. As the determination of destructive force in compression tests within the radial and tangential directions (for further wood densification) was unfeasible, the force at the limit of proportionality was instead identified. The testing machine's software facilitated this by extrapolating it from the stress–strain curve, specifically at the point where the tangent line diverged from the actual curve (Figure 1). The point described above, corresponding to the proportionality limit, was circled in the figure. Failure stress (RcL) was also determined for the longitudinal direction. After the strength determination, wood moisture content (MC) was determined on selected control and densified samples in accordance with the standard [41]. The moisture content of the samples during the test was similar and averaged 9.5%. However, the MC of the densified samples was slightly higher and was close to 10.5%, despite the similar conditions of sample seasoning.

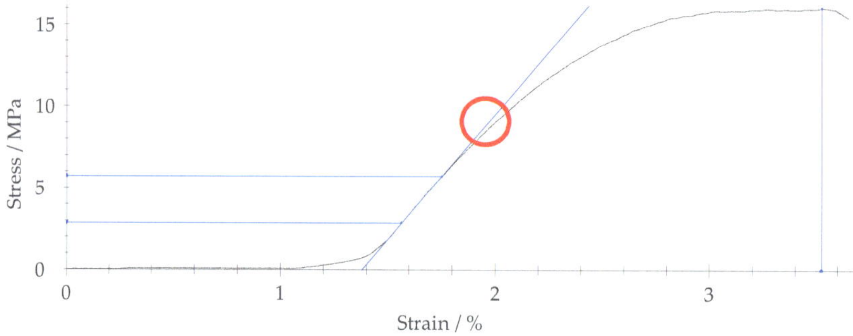

Figure 1. Stress–strain relationship during wood compression test.

Following the guidelines outlined in the European Standard—EN 1534 standard [42], hardness in the radial direction was assessed using the Brinell method. This involved employing a 10 mm diameter steel ball subjected to a 1 kN load. The maximum force of 1 kN was reached within 15 s, maintained for 30 s, and then gradually reduced to zero within the subsequent 15 s. Brinell hardness was calculated as in Equation (2):

$$HB = \frac{2F}{\pi D\left(D - \sqrt{D^2 - d^2}\right)} \quad (2)$$

where HB is the Brinell hardness [N·mm^{-2}], F is the nominal force [N], D is the diameter of the steel ball (10 mm), and d is the mean diameter of the residual indentation [mm]. Brinell hardness values were determined for both control and densified specimens in the radial direction.

2.5. Scanning Electron Microscopy (SEM)

The densified wood and the controls were prepared for scanning electron microscopy (SEM) observations as follows. The controls were soaked in water for 24 h to soften the tissue. Subsequently, the cross sections of the controls were "smoothed" using a rotary microtome with disposable blades to make serval cuts with a thickness of 10 μm. The densified wood was cut in an air-dry state to avoid set-recovery during water soaking. After smoothing, the controls were dried in an oven set to 103 ± 2 °C overnight. Before microtome cutting, the densified wood was prepared in a cone-like shape to minimise the surface that was presented to the blade (approximately 1 mm^2). Then, several cuts with a thickness of 1 μm were made. All samples were placed on carbon tape for SEM observations

(HITACHI, TM4000, Tokyo, Japan), which were performed using an accelerating voltage of 15 kV and a detector for backscattered electrons.

2.6. Statistics

The experimental data underwent analysis utilising STATISTICA 13.3 software by TIBCO Software Inc. (Palo Alto, CA, USA) employing analysis of variance (ANOVA). Subsequently, Tukey's HSD test discerned noteworthy disparities among the mean values of the properties investigated within the control group, as well as densified pine and aspen wood samples. All comparison tests were conducted at a significance level of 0.05.

3. Results and Discussion

3.1. Chemical Composition

Table 1 shows the lignin content for aspen and pine samples before and after delignification and the lignin loss during the process. Among the controls, aspen contained significantly less lignin than pine, which is consistent with the data provided in the literature [43,44].

Table 1. Lignin content of lignin in pine and aspen wood before and after the delignification process.

Specimens	Percentage of Lignin [%]	Reduction of Lignin [%]
Aspen (control)	23.4	39.7
Aspen after chemical treatment	14.1	
Scots pine (control)	29.8	21.1
Scots pine after chemical treatment	23.5	

After delignification, the lignin content was reduced by 39.7% in aspen and by 21.1% in pine (Table 1). This reduction is statistically significant, as shown by single-factor analysis of variance (ANOVA). As mentioned earlier, softwood lignin is more difficult to remove from the wood tissue, therefore the reduction in the lignin content is much lower in Scots pine compared to aspen.

3.2. Attenuated Total Reflectance–Fourier Transform Infrared Spectroscopy

The surface chemical changes caused by delignification and densification were determined by FTIR-ATR (Figure 2). The spectra in reflectance mode are normalised to wavenumber 1033 cm^{-1}, showing some qualitative differences between samples.

Comparing the densified Scots pine to the control, it is evident that the 1730 cm^{-1} peak disappears after densification. Since this peak is exclusive to carbonyl groups in hemicelluloses (esters, carboxylic acids, aldehydes), it can be assumed that hemicelluloses were partly removed or heavily deacetylated during delignification. In Scots pine, wavenumbers 1269 and 1225 cm^{-1} represent the C-O stretching in G-lignin and S-Lignin, respectively [45]. After densification, the former peak becomes more intense compared to the latter, further indicating the partial degradation of S-Lignin. Additionally, in Scots pine, the peaks at 798 and 777 cm^{-1} disappear after densification. These bands correspond to C-H deformations in the aromatic ring of lignin and their disappearance is the result of delignification.

Similar to pine, the aspen wood shows signs of partly hemicellulose degradation at wavenumber 1732 cm^{-1}. Moreover, the 1233 cm^{-1} peak shows a substantial decrease in aspen wood. This peak mainly represents the C-O stretching in S-lignin, constituting around 50% of the total lignin in aspen. The depletion of this band shows that S-lignin is removed more efficiently than G-lignin and supports the data provided in Table 1. Comparing the two species, it can be said that hemicelluloses are removed or at least deacetylated, cellulose is mainly unaffected, and S-lignin is removed more effectively than G-lignin.

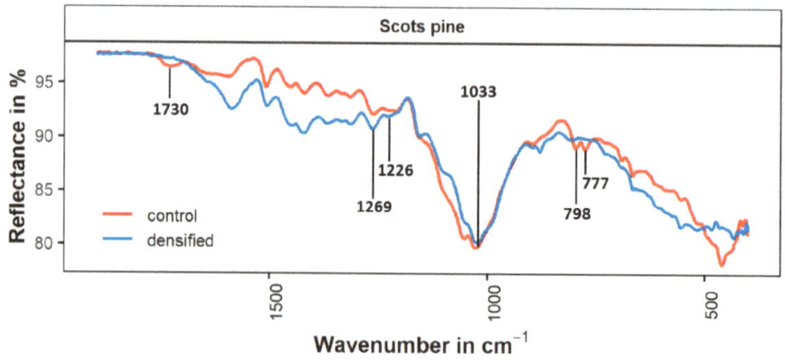

(a)

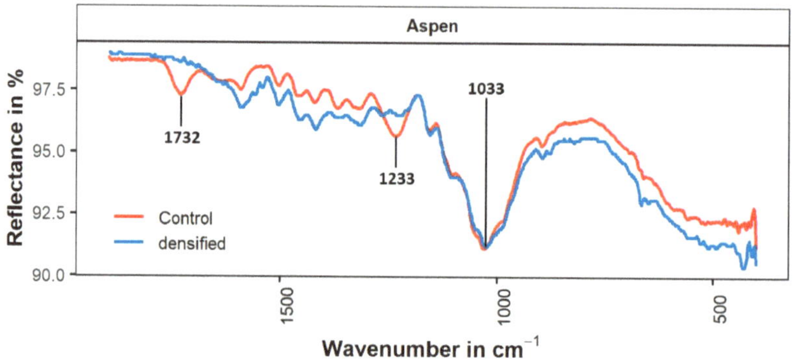

(b)

Figure 2. FTIR spectra of Scots pine (**a**) and aspen (**b**) wood before and after delignification and densification.

3.3. Densification

Table 2 shows the average density of aspen and Scots pine samples both before and after delignification and densification. Aspen wood, which had an average initial density of 458 kg·m^{-3}, was densified to 1142 kg·m^{-3}. Control Scots pine wood had an average density of 462 kg·m^{-3} and was densified to 1146 kg·m^{-3}. The compression ratio (calculated by dividing the density after densification by the density of the wood before) was 2.5 for both woods studied. Hence, both species could be densified to a similar extent. The compression ratio at a similar level for aspen, spruce, and beech wood was obtained by Schwarzkopf (2021) [16]. Tenorio et al. (2022) [46], by densifying several species of tropical wood, obtained a much lower density in the range of 600–1100 kg·m^{-3}.

The density of aspen controls and densified samples was more variable than in pine, but the average was similar. ANOVA analysis showed statistically significant differences in density within species. Almost identical compression ratio values result from a comparable density of control wood, regardless of the type of wood species. The use of wood with a very similar density also contributed to an almost identical change in the porosity of the tissue. In aspen, the porosity decreased from 69.4% in the controls to 23.9% after densification. In pine, it decreased from 69.2% to 23.6%. Therefore, the porosity reduction amounted to about 65%, regardless of the species. There were no statistically meaningful distinctions detected among wood species regarding porosity reduction. ANOVA analysis revealed statistically insignificant variations in porosity within the individual species.

Table 2. Wood density (ρ) before and after the densification process.

Material	Number of Specimens	ρmin	ρmean	ρmax	Standard Deviation (±SD)
		[kg·m^{-3}]			
Aspen (control)	53	429.3	458.4	486.8	18.86
Densified aspen	53	985.3	1141.7	1268.1	64.76
Scots pine (control)	53	450.5	461.9	478.4	9.49
Densified Scots pine	53	1082.9	1145.5	1204.8	27.80

3.4. Brinell Hardness

Brinell hardness tests are presented with basic statistical parameters in Table 3. The indentation was applied in the radial direction. Hardness measurements in other anatomical directions were unfeasible due to the sample geometry of the densified samples. The thickness of the specimens was smaller than the steel ball used for indentation (10 mm).

Table 3. Wood Brinell hardness (HB) before and after the densification process.

Material	Number of Specimens	HBmin	HBmean	HBmax	Standard Deviation (±SD)
		[N·mm^{-2}]			
Aspen (control)	10	9.1	12.4	15.3	1.20
Densified aspen	10	75.5	92.8	143.1	20.30
Scots pine (control)	10	11.2	16.7	24.9	2.12
Densified Scots pine	10	65.4	88.6	121.1	14.79

The average hardness of densified aspen and pine wood is similar. However, there is a large discrepancy between the minimum and maximum values. Within the aspen control, the hardness ranges from 9.1 to 15.3 N·mm^{-2}; after densification, these values increase to 75.5–143.1 N·mm^{-2}. The variable hardness relates to the natural variation in the density of aspen wood. The pine control ranges from 11.2 to 24.9 N·mm^{-2}, increasing to 65.4–121.1 N·mm^{-2} after densification. A single-factor analysis of variance (ANOVA) has shown that the Brinell hardness of densified aspen does not significantly differ from densified Scots pine. However, the relative increase in the average hardness is higher in aspen (748%) than in pine (531%). Other authors also achieved a similar level of hardness improvement. Pelit and Emiroglu [47] obtained an increase in HB for aspen that was more than threefold for densified and styrene-treated wood. A several-fold increase in hardness in surface-densified pine wood was observed by Scharf et al. (2022) [48]. Research on spruce wood also showed a significant increase in the hardness of wood densified and treated with acetylation [49].

3.5. Compressive Strength in Three Anatomical Directions

Table 4 presents the average values of the compressive strength of aspen and Scots pine wood in longitudinal (L), tangential (T), and radial (R) directions.

During the test, short scratches can be observed along the fibres in Scots pine, which increase in size to reveal the damaged zone. When wood is compressed in such a way, changes in cell membranes occur even before the appearance of visible deformations; these membranes are deformed, causing shearing and, finally, crushing of the cell.

The compressive strength along the grain (L) was higher than the compressive strength across the grain (R and T). The control specimens of both species had almost identical strength longitudinal direction. After densification, the strength along the grain increased by 179% in aspen but only by 143% in pine. In the tangential direction, the average strength of aspen increased by 493% and that of pine only by 179%.

Table 4. Wood compressive strength before and after the densification process.

Material	Number of Specimens	L	T	R
			[N·mm^{-2}]	
Aspen (control)	10	47.8 ± 2.8	1.6 ± 0.2	5.8 ± 0.4
Densified aspen	10	85.3 ± 20.7	7.9 ± 2.2	82.7 ± 19.5
Scots pine (control)	10	47.5 ± 1.6	2.8 ± 0.4	5.3 ± 0.5
Densified Scots pine	10	67.8 ± 13.7	5.0 ± 1.0	130.9 ± 17.7

L—longitudinal, T—tangential, and R—radial directions; ±SD.

For aspen, the strength in the radial direction increased by 14 times compared to the control. For pine, it increased by almost 25 times. However, this is also the direction of densification, so such a significant increase was expected. The compressive strength in the radial direction was even higher in pine than in the longitudinal direction. In aspen wood, the values for both directions are in a similar range.

Compared to the literature, the relative strength increase in this study is low. Song et al. (2018) [38] increased the compressive strength in the tangential direction from 2.6 N·mm^{-2} to 87.6 N·mm^{-2}. In the current study, however, samples buckled very quickly during the compressive strength test. Buckling is, therefore, most likely the cause of a low-strength increase in the longitudinal and tangential directions. Buckling did not occur in the radial compression test, where a massive increase in strength was achieved. As in the case of hardness, a single-factor analysis of variance (ANOVA) was performed for each anatomical direction. The study aims to indicate whether differences between species after densification are statistically significant or not (Table 5). When analysing the ANOVA results, it can be seen that statistically significant differences occur only in the radial compressive strength of wood (the direction of densification).

Table 5. ANOVA of compressive strength in three anatomical directions in densified wood species.

Source of Variation	SSB	df	MSB	SSE	df	MSE	F	p
Longitudinal	998.057	1	998.057	3363.650	11	305.786	3.263	0.09822 ^{ns}
Tangential	25.288	1	25.288	68.981	10	6.898	3.665	0.08455 ^{ns}
Radial	27,824.922	1	27,824.922	16,661.836	46	362.213	76.819	0.00000 ^{s}

SSB—sum squares between groups; MSB—mean squares between groups; SSE—sum squares within groups; MSE—mean squares within groups; df—degrees of freedom; F—value of test function; p—level of significance; ^{s}—significant differences, ^{ns}—not significant differences.

3.6. Scanning Electron Microscopy (SEM) Observations

Figure 3 shows micrographs of the material before and after the densification process. The Scots pine wood tissue (Figure 3a) consists of early and late wood tracheids arranged in radial rows and parenchyma cells arranged perpendicular to the grain. The diffuse porous hardwood (Figure 3c) shows vessels, evenly distributed wood fibres, and wood parenchyma cells. Figure 3b,d show both wood species after the densification process. Both show a significant reduction in wood porosity. In pine wood (Figure 3b), outlines of cells (tracheids) can still be observed without clear cell lumens. There are also visible cracks, which run mainly along the wood rays.

The ray parenchyma cells differ in structure and shape from the tracheids and have different thicknesses of the cell walls. The stress is applied in the direction of the parenchyma cell axis by radially compressing the wood. Due to their anatomy and the direction of applied force, parenchyma cells are compressed to a lesser extent than tracheids. A visible crack may also result from the presence of horizontal resin canals in this place. It is an element of wood tissue which, in the case of pine, is mainly composed of thin-walled cells and can reach a diameter of up to 200 μm. The difference in the degree of densification causes internal stress and makes parenchyma cells a likely place of crack initiation.

More cracks were seen in aspen wood than in Scots pine. However, what appears as cracks might be collapsed vessels. Residual lumens fibres can also be observed. In both cases, however, residual pore spaces are visible. The densification process does not entirely close the lumens of wood cells, bringing them closer together. It should be added that the wood tissue in both cases was cut in an air-dry state with wood moisture not exceeding 6%.

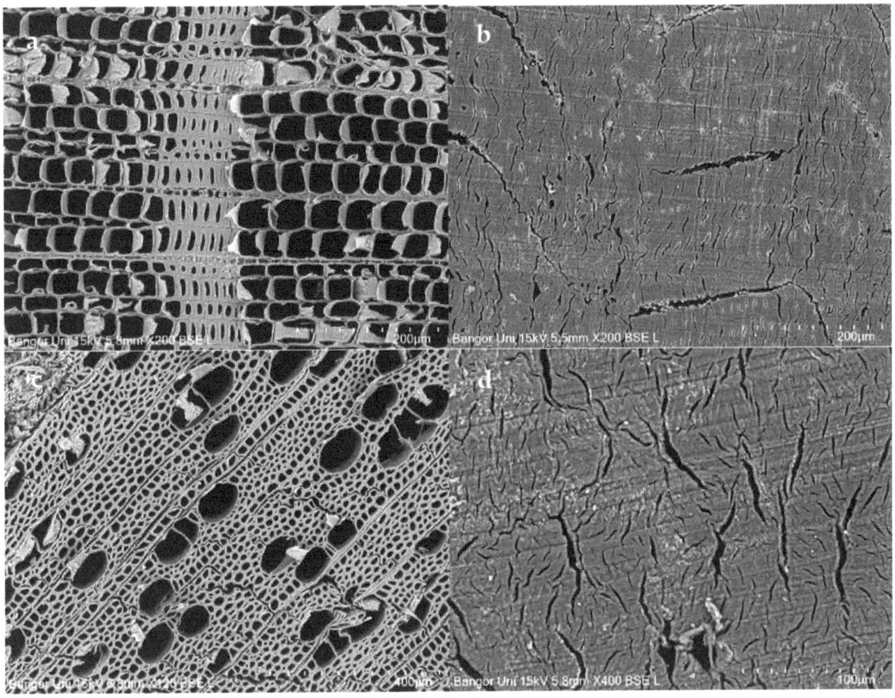

Figure 3. SEM micrographs of pine and aspen wood before and after the densification process. (**a**) Scots pine control, (**b**) densified Scots pine, (**c**) aspen control, (**d**) densified aspen.

4. Conclusions

This study has shown that Eurasian aspen and Scots pine specimens could be densified to a similar extent, showing an almost identical compaction ratio of 2.5. The delignification process reduced the lignin content in aspen samples from 23.4% to 14.1%, which resulted in a loss of lignin of 39.7%. In Scots pine specimens with naturally higher lignin content, delignification reduced the lignin content from 29.8% to 23.5%, corresponding to a lignin loss of 21.1%. However, the different chemical compositions and degree of delignification of the samples did not contribute to any significant differences in the following densification process.

The densification significantly increased the compressive strength of both aspen and Scots pine in all directions compared to the controls. Densified aspen exhibited a higher increase than Scots pine in compressive strength in the longitudinal and tangential directions. In the radial direction, Scots pine showed a higher increase than aspen. In pine, the stress increase exceeded 24 times, compared to 14 times in aspen. This significant rise in compressive stress (not observed in the hardness test) could stem from the consistent microstructure of pine wood. Densification diminishes the heterogeneity within annual growth rings, equalising the density of earlywood with that of latewood. This effect is less pronounced in aspen, a diffuse-porous species with lower heterogeneity compared to coniferous species. When the rise in wood strength lags behind the increase in density or shows only slightly more improvement, cell wall destruction occurs. Latewood pine, with its cell structure and notably lesser deformations during densification compared to earlywood, is less susceptible to mechanical cell wall destruction effects.

Author Contributions: P.M. was responsible for the research design, data collection, analysis, and manuscript writing. The conceptualisation and supervision were conducted by P.M. and S.C. The data visualisation was performed by P.M. and C.K. Each author made contributions to the data analysis, review, editing, and assisted with proofreading. All authors have read and agreed to the published version of the manuscript.

Funding: The research was funded by the Polish National Science Center grant: 2021/05/X/NZ9/00801.

Data Availability Statement: The data presented in this study are available on request from the corresponding author.

Acknowledgments: The authors would like to thank all the staff of BioComposites Centre (Bangor, Wales) for their help. Many thanks also to Mikołaj Ziętek for assistance in preparing specimens and carrying out determinations.

Conflicts of Interest: The authors declare no conflicts of interest.

References

1. Kolman, F.; Cote, J.R. *Principles of Wood Science and Technology: I Solid Wood*; Springer: Berlin/Heidelberg, Germany, 1968.
2. Niklas, K.J.; Spatz, H.-C. Worldwide Correlations of Mechanical Properties and Green Wood Density. *Am. J. Bot.* **2010**, *97*, 1587–1594. [CrossRef]
3. Jungstedt, E.; Montanari, C.; Östlund, S.; Berglund, L. Mechanical Properties of Transparent High Strength Biocomposites from Delignified Wood Veneer. *Compos. Part A Appl. Sci. Manuf.* **2020**, *133*, 105853. [CrossRef]
4. Xu, E.; Wang, D.; Lin, L. Chemical Structure and Mechanical Properties of Wood Cell Walls Treated with Acid and Alkali Solution. *Forests* **2020**, *11*, 87. [CrossRef]
5. Thomas, D.S.; Montagu, K.D.; Conroy, J.P. Changes in Wood Density of Eucalyptus Camaldulensis Due to Temperature—The Physiological Link between Water Viscosity and Wood Anatomy. *For. Ecol. Manag.* **2004**, *193*, 157–165. [CrossRef]
6. Jaakkola, T.; Mäkinen, H.; Saranpää, P. Wood Density in Norway Spruce: Changes with Thinning Intensity and Tree Age. *Can. J. For. Res.* **2005**, *35*, 1767–1778. [CrossRef]
7. Ross, R.J. *Wood Handbook: Wood as an Engineering Material*; U.S. Department of Agriculture, Forest Service, Forest Products Laboratory: Madison, WI, USA, 2021.
8. Kutnar, A.; Šernek, M. Densification of Wood. *Zb. Gozdarstva Lesar.* **2007**, *82*, 53–62.
9. Fang, C.-H.; Cloutier, A.; Jiang, Z.-H.; He, J.-Z.; Fei, B.-H. Improvement of Wood Densification Process via Enhancing Steam Diffusion, Distribution, and Evaporation. *BioResources* **2019**, *14*, 3278–3288. [CrossRef]
10. Tenorio, C.; Moya, R. Development of a Thermo-Hydro-Mechanical Device for Wood Densification Adaptable to Universal Testing Machines and Its Evaluation in a Tropical Species. *J. Test. Eval.* **2021**, *49*, 2597–2608. [CrossRef]
11. Sandberg, D.; Kutnar, A.; Karlsson, O.; Jones, D. *Wood Modification Technologies*; CRC Press: Boca Raton, FL, USA, 2021; ISBN 978-1-351-02822-6.
12. Sandberg, D.; Navi, P. *Introduction to Thermo-Hydro-Mechanical (THM) Wood Processing*; Växjö universitet Skog & Trä: Växjö, Sweden, 2007.
13. Skyba, O.; Schwarze, F.; Niemz, P. Physical and Mechanical Properties of Thermo-Hygro-Mechanically (THM)-Densified Wood. *Wood Res.* **2009**, *54*, 1–18.
14. Fu, Q.; Cloutier, A.; Laghdir, A. Optimization of the Thermo-Hygromechanical (THM) Process for Sugar Maple Wood Densification. *BioResources* **2016**, *11*, 8844–8859. [CrossRef]
15. Bao, M.; Huang, X.; Jiang, M.; Yu, W.; Yu, Y. Effect of Thermo-Hydro-Mechanical Densification on Microstructure and Properties of Poplar Wood (*Populus tomentosa*). *J. Wood Sci.* **2017**, *63*, 591–605. [CrossRef]
16. Schwarzkopf, M. Densified Wood Impregnated with Phenol Resin for Reduced Set-Recovery. *Wood Mater. Sci. Eng.* **2021**, *16*, 35–41. [CrossRef]
17. Navi, P.; Sandberg, D. *Thermo-Hydro-Mechanical Wood Processing*; CRC Press: Boca Raton, FL, USA, 2012.
18. Diawanich, P.; Matan, N.; Kyokong, B. Evolution of Internal Stress during Drying, Cooling and Conditioning of Rubberwood Lumber. *Holz Als Roh-Und Werkst.* **2010**, *68*, 1–12. [CrossRef]
19. Seborg, R.M. *Heat-Stabilized Compressed Wood (Staypak)*; Report (Forest Products Laboratory (U.S.)); Forest Products Laboratory, Forest Service, U.S. Department of Agriculture: Madison, WI, USA, 1956.

20. Wehsener, J.; Bremer, M.; Haller, P.; Fischer, S. Bending Tests of Delignified and Densified Poplar. *Wood Mater. Sci. Eng.* **2023**, *18*, 42–50. [CrossRef]
21. Fang, C.-H.; Mariotti, N.; Cloutier, A.; Koubaa, A.; Blanchet, P. Densification of Wood Veneers by Compression Combined with Heat and Steam. *Eur. J. Wood Prod.* **2012**, *70*, 155–163. [CrossRef]
22. Morsing, N. *Densification of Wood.: The Influence of Hygrothermal Treatment on Compression of Beech Perpendicular to Gain*; BYG-Rapport; Technical University of Denmark: Kongens Lyngby, Denmark, 1998.
23. Ito, Y.; Tanahashi, M.; Shigematsu, M.; Shinoda, Y. Compressive-Molding of Wood by High-Pressure Steam-Treatment: Part 2. Mechanism of Permanent Fixation. *Holzforschung* **1998**, *52*, 217–221. [CrossRef]
24. Blomberg, J.; Persson, B.; Blomberg, A. Effects of Semi-Isostatic Densification of Wood on the Variation in Strength Properties with Density. *Wood Sci. Technol.* **2005**, *39*, 339–350. [CrossRef]
25. Sandberg, D.; Haller, P.; Navi, P. Thermo-Hydro and Thermo-Hydro-Mechanical Wood Processing: An Opportunity for Future Environmentally Friendly Wood Products. *Wood Mater. Sci. Eng.* **2013**, *8*, 64–88. [CrossRef]
26. Kadivar, M.; Gauss, C.; Ghavami, K.; Savastano, H., Jr. Densification of Bamboo: State of the Art. *Materials* **2020**, *13*, 4346. [CrossRef]
27. Sun, R. Lignin Source and Structural Characterization. *ChemSusChem* **2020**, *13*, 4385–4393. [CrossRef]
28. Sjostrom, E. *Wood Chemistry: Fundamentals and Applications*; Elsevier: Amsterdam, The Netherlands, 2013; ISBN 978-0-08-092589-9.
29. Wang, Q.; Xiao, S.; Shi, S.Q.; Cai, L. Effect of Light-Delignification on Mechanical, Hydrophobic, and Thermal Properties of High-Strength Molded Fiber Materials. *Sci. Rep.* **2018**, *8*, 955. [CrossRef]
30. Wang, J.; Liu, J.; Li, J.; Zhu, J.Y. Characterization of Microstructure, Chemical, and Physical Properties of Delignified and Densified Poplar Wood. *Materials* **2021**, *14*, 5709. [CrossRef]
31. Cheng, S.; Huang, A.; Wang, S.; Zhang, Q. Effect of Different Heat Treatment Temperatures on the Chemical Composition and Structure of Chinese Fir Wood. *BioResources* **2016**, *11*, 4006–4016. [CrossRef]
32. Jakob, M.; Gaugeler, J.; Gindl-Altmutter, W. Effects of Fiber Angle on the Tensile Properties of Partially Delignified and Densified Wood. *Materials* **2020**, *13*, 5405. [CrossRef]
33. Kuai, B.; Wang, Z.; Gao, J.; Tong, J.; Zhan, T.; Zhang, Y.; Lu, J.; Cai, L. Development of Densified Wood with High Strength and Excellent Dimensional Stability by Impregnating Delignified Poplar by Sodium Silicate. *Constr. Build. Mater.* **2022**, *344*, 128282. [CrossRef]
34. *ISO 13061-2*; Physical and Mechanical Properties of Wood—Test Methods for Small Clear Wood Specimens—Part 2: Determination of Density for Physical and Mechanical Tests. International Organization for Standardization: Geneva, Switzerland, 2014.
35. Tribulová, T.; Kačík, F.; Evtuguin, D.V.; Čabalová, I.; Ďurkovič, J. The Effects of Transition Metal Sulfates on Cellulose Crystallinity during Accelerated Ageing of Silver Fir Wood. *Cellulose* **2019**, *26*, 2625–2638. [CrossRef]
36. Tappi, T. *223 Cm-01 (2001) Pentosans in Wood and Pulp*; US Technical Association of Pulp and Paper Industry: Atlanta, GA, USA, 2001.
37. Tappi, T. *222 Om-06 (2006) Acid Insoluble Lignin in Wood and Pulp*; US Technical Association of Pulp and Paper Industry: Atlanta, GA, USA, 2006.
38. Song, J.; Chen, C.; Zhu, S.; Zhu, M.; Dai, J.; Ray, U.; Li, Y.; Kuang, Y.; Li, Y.; Quispe, N. Processing Bulk Natural Wood into a High-Performance Structural Material. *Nature* **2018**, *554*, 224–228. [CrossRef]
39. Mania, P.; Wróblewski, M.; Wójciak, A.; Roszyk, E.; Moliński, W. Hardness of Densified Wood in Relation to Changed Chemical Composition. *Forests* **2020**, *11*, 506. [CrossRef]
40. *ISO 13061-5:2020*; Physical and Mechanical Properties of Wood—Test Methods for Small Clear Wood Specimens—Part 5: Determination of Strength in Compression Perpendicular to Grain. International Organization for Standardization: Geneva, Switzerland, 2020.
41. *ISO 13061-1:2014*; Physical and Mechanical Properties of Wood—Test Methods for Small Clear Wood Specimens—Part 1: Determination of Moisture Content for Physical and Mechanical Tests. International Organization for Standardization: Geneva, Switzerland, 2014. Available online: https://standards.iteh.ai/catalog/standards/iso/c2b4d64b-f874-4026-94d4-3faa1ce1729e/iso-13061-1-2014 (accessed on 9 May 2024).
42. *EN 1534:2010*; Wood Flooring—Determination of Resistance to Indentation—Test Method. Slovenian Institute for Standardization: Ljubljana, Slovenia, 2011. Available online: https://standards.iteh.ai/catalog/standards/cen/53e9be09-7fc1-4178-b222-9d78be718593/en-1534-2010 (accessed on 3 April 2024).
43. Prosiński, S. *Chemia Drewna, Wyd. 2*; PWRiL: Warszawa, Poland, 1984; 475p.
44. Novaes, E.; Kirst, M.; Chiang, V.; Winter-Sederoff, H.; Sederoff, R. Lignin and Biomass: A Negative Correlation for Wood Formation and Lignin Content in Trees. *Plant Physiol.* **2010**, *154*, 555–561. [CrossRef]
45. Liu, C.F.; Xu, F.; Sun, J.X.; Ren, J.L.; Curling, S.; Sun, R.C.; Fowler, P.; Baird, M.S. Physicochemical Characterization of Cellulose from Perennial Ryegrass Leaves (*Lolium perenne*). *Carbohydr. Res.* **2006**, *341*, 2677–2687. [CrossRef]
46. Tenorio, C.; Moya, R.; Starbird-Perez, R. Effect of Steaming and Furfuryl Alcohol Impregnation Pre-Treatments on the Spring Back, Set Recovery and Thermal Degradation of Densified Wood of Three Tropical Hardwood Species. *Eur. J. Wood Wood Prod.* **2023**, *81*, 467–480. [CrossRef]
47. Pelit, H.; Emiroglu, F. Density, Hardness and Strength Properties of Densified Fir and Aspen Woods Pretreated with Water Repellents. *Holzforschung* **2021**, *75*, 358–367. [CrossRef]

48. Scharf, A.; Neyses, B.; Sandberg, D. Hardness of Surface-Densified Wood. Part 1: Material or Product Property? *Holzforschung* **2022**, *76*, 503–514. [CrossRef]
49. Guo, J.; Wang, C.; Li, C.; Liu, Y. Effect of Acetylation on the Physical and Mechanical Performances of Mechanical Densified Spruce Wood. *Forests* **2022**, *13*, 1620. [CrossRef]

forests

Article

Silver Nanoparticles and Chitosan Oligomers Composites as Poplar Wood Protective Treatments against Wood-Decay Fungi and Termites

Eleana Spavento [1], María Teresa de Troya-Franco [2], Luis Acuña-Rello [3], Mónica Murace [1], Sara M. Santos [2], Milagros Casado-Sanz [3], Roberto D. Martínez-López [3], Jesús Martín-Gil [4], Javier Álvarez-Martínez [4] and Pablo Martín-Ramos [4,*]

1 Wood Research Laboratory (LIMAD), School of Agrarian and Forestry Sciences, National University of La Plata, Diag. 113 N° 469, La Plata B1904, Argentina; eleana.spavento@agro.unlp.edu.ar (E.S.); monica.murace@agro.unlp.edu.ar (M.M.)

2 Institute of Forest Sciences (ICIFOR) and Crop Protection Department, National Institute of Agrarian and Food Research and Technology (INIA)—Superior Council for Scientific Research (CSIC), Ctra. Coruña km. 7, 28040 Madrid, Spain; troya@inia.csic.es (M.T.d.T.-F.); santos@inia.csic.es (S.M.S.)

3 Timber Structures and Wood Technology Research Group, Department of Agricultural and Forestry Engineering, ETSIIAA, University of Valladolid, Avda. Madrid 44, 34004 Palencia, Spain; maderas@iaf.uva.es (L.A.-R.); mmcasado@uva.es (M.C.-S.); robertodiego.martinez@uva.es (R.D.M.-L.)

4 Advanced Materials Laboratory, Department of Agricultural and Forestry Engineering, ETSIIAA, University of Valladolid, Avda. Madrid 44, 34004 Palencia, Spain; jesus.martin.gil@uva.es (J.M.-G.); javier.alvarez.martinez@uva.es (J.Á.-M.)

* Correspondence: pmr@uva.es

check for updates

Citation: Spavento, E.; de Troya-Franco, M.T.; Acuña-Rello, L.; Murace, M.; Santos, S.M.; Casado-Sanz, M.; Martínez-López, R.D.; Martín-Gil, J.; Álvarez-Martínez, J.; Martín-Ramos, P. Silver Nanoparticles and Chitosan Oligomers Composites as Poplar Wood Protective Treatments against Wood-Decay Fungi and Termites. *Forests* **2023**, *14*, 2316. https://doi.org/10.3390/f14122316

Academic Editors: Miklós Bak and Morwenna Spear

Received: 30 October 2023
Revised: 16 November 2023
Accepted: 22 November 2023
Published: 25 November 2023

Abstract: This study focuses on *Populus* ×*euramericana* (Dode) Guinier, a globally distributed fast-growing tree. Despite its valuable wood, it exhibits low durability. The aim of this study was to assess the efficacy of a binary composite comprising silver nanoparticles (AgNPs) and chitosan oligomers (COS) in protecting *P.* ×*euramericana* 'I-214' wood against degradation caused by xylophagous fungi and termites through vacuum-pressure impregnation. The test material was carefully selected and conditioned following the guidelines of EN 350:2016, and impregnation was carried out in accordance with EN 113-1:2021. Five concentrations of AgNPs–COS composites were utilized. Biodeterioration resistance was evaluated based on EN 350:2016 for white (*Trametes versicolor* (L.) Lloyd) and brown (*Coniophora puteana* (Schumach.) P.Karst.) rot fungi, and EN 117:2012 for subterranean termites (*Reticulitermis grassei* Clément). The durability class and use class were assigned following EN 350:2016 and EN 335:2013, respectively. In comparison to the untreated control, the binary solution at its highest concentration (AgNPs 4 ppm + COS 20 $g \cdot L^{-1}$) demonstrated a notable reduction in weight loss, decreasing from 41.96 ± 4.49% to 30.15 ± 3.08% for white-rot fungi and from 41.93 ± 4.33% to 27.22 ± 0.66% for brown rot fungi. Furthermore, the observed termite infestation shifted from "heavy" to "attempted attack", resulting in a decrease in the survival rate from 53.98 ± 10.40% to 26.62 ± 8.63%. Consequently, the durability classification of *P.* ×*euramericana* I-214 witnessed an enhancement from "Not durable" to "Slightly" and "Moderately durable" concerning decay fungi and termites, respectively. These findings expand the potential applications of this wood and substantiate the advantages of employing this environmentally friendly treatment.

Keywords: AgNPs; *Coniophora puteana*; durability; nanocomposite; pressure-vacuum treatment; *Populus* ×*euramericana*; *Reticulitermes grassei*; *Trametes versicolor*; use class

1. Introduction

The inherent resistance of wood to decay induced by fungi, termites and other insects, and marine borers, as defined in EN 350:2016 [1], establishes its natural durability. This property enables the identification of the optimal utilization scenario for wood or the specification of the necessary protective measures corresponding to its anticipated utilization class.

Xylophagous fungi, responsible for brown and white rot, along with subterranean termites, stand out as the most aggressive agents causing damage to wood in service. These fungi induce structural alterations, impacting the inherent resistance of wood through enzymatic attacks on the polymeric fraction of the cell wall, encompassing cellulose, hemicelluloses, and lignin. In turn, termites contribute to the structural weakening of wood through a dual process of mechanical chewing and enzymatic digestion of lignocellulosic components. This enzymatic digestion is facilitated by symbiotic microorganisms residing in their digestive tract [2–4].

Populus ×euramericana (Dode) Guinier (=*Populus ×canadensis* Moench; poplar) clone 'I-214' holds significant representativeness and economic importance in Spain [5]. Nevertheless, the inherent low natural durability of its wood constrains its potential applications and usage scenarios [6]. Thus, there is a need to develop treatments aimed at enhancing its resistance to deterioration.

The application of preservative substances to wood through vacuum-pressure treatment has long been recognized as an effective strategy for enhancing wood durability. However, numerous chemical compounds commonly employed for this purpose exhibit limitations, particularly in terms of environmental sustainability and human health considerations. Consequently, there is a critical need to explore innovative technologies and alternative preservatives, diverging from those traditionally utilized, to broaden the spectrum of wood applications and extend its useful life, especially in scenarios where natural durability may not be sufficient.

In the realm of wood enhancement, nanotechnology is emerging as a transformative force, focusing on improving the properties of wood and wood products, particularly in terms of dimensional stability and resistance to microorganism attacks. This advancement involves the impregnation or coating of wood, with its standout feature being its capability to deeply infiltrate the wood substrate, facilitated by its minute size (smaller than the voids in the wood cell wall). This penetration allows nanotechnology to effectively modify the surface chemistry, yielding a more potent and enduring impact [7–9].

Recent investigations have delved into the utilization of various nano-composites, such as copper-, zinc-, and silver-based composites, among others, aiming to augment wood resistance against decay fungi and termites. Notable studies by Shiny et al. [10], Lykidis et al. [11], Terzi et al. [12], Mantanis et al. [13], Akhtari and Nicholas [14], and Clausen et al. [15], among others, have yielded promising results in this regard.

The application of silver nanoparticles (AgNPs) has become increasingly popular for enhancing diverse wood properties, including resistance to biological degradation, water absorption capacity, and dimensional stability. These improvements are achieved through both depth and surface treatments, spanning various concentrations of AgNPs [16–24].

For wood processing and protection purposes, the selection of AgNPs is underpinned by their distinct advantages over alternative options, including the aforementioned metal nanoparticles, chemical agents, and industrial-waste-derived antiseptics. These advantages encompass antimicrobial properties, durability enhancement, and a diminished environmental impact. In contrast to copper-based NPs or nanocomposites, AgNPs exhibit a more versatile antimicrobial application across diverse biological agents and disciplinary fields, showcasing efficacy on various materials and in different application formats [25]. Additionally, they demonstrate superior protective efficacy over time and at varying concentrations, although outcomes at lower concentrations may vary due to factors such as size, shape, application methods, and resistance to leaching [18,26–30]. Furthermore, the use of silver nanoparticles does not interfere with the deposition process or introduce defects in wood paint, offering a distinct advantage over copper nanoparticles that may potentially introduce harmful residues and defects [26]. Concerning zinc-based NPs or nanocomposites, although some studies underscore their effectiveness as a biocide, additional research is essential to fully comprehend their potential advantages and limitations, as well as the underlying mechanisms governing their protective capabilities [31]. In the broader context of wood preservatives, AgNPs are deemed more environmentally friendly than traditional alternatives containing toxic chemicals that pose risks to human health and the environment [32]. Despite the notable drawback of AgNPs related to their higher synthesis and production cost, this limitation can be mitigated by exploring the minimum effective concentration and employing alternative synthesis strategies [33,34].

Chitosan, a natural and biodegradable polymer derived from the deacetylation of chitin, and its oligomers (COS) have also been investigated as protective solutions for wood due to their antimicrobial properties and harmlessness to human health and the environment. Their solutions, whether employed alone or in conjunction with nanoparticles (e.g., AgNPs), are primarily aimed at enhancing wood's durability against rotting fungi, yielding promising outcomes [18,19,35,36]. Chitosan's advantageous characteristics, including its low molecular weight, high water solubility, and elevated viscosity, contribute to its stabilizing properties. These attributes enable its efficient utilization in deep impregnation processes through vacuum-pressure application, with acceptable binding values [37]. Simultaneously, the chemical structure of chitosan, featuring amino ($-NH_2$) and hydroxyl ($-OH$) functional groups, equips it with chelating and reducing properties in the synthesis of AgNPs. This structural composition plays a crucial role in preventing the agglomeration of AgNPs [38].

In light of the aforementioned considerations, and with the current market increasingly emphasizing the use of wood products that align with the structural requirements outlined in Eurocode 5 (EN 1995-1-1:2016 [39]) concerning physical–mechanical performance and durability, nanotechnology emerges as a viable solution to enhance the inherent durability of wood, particularly in the case of *P.* ×*euramericana*. Therefore, building upon a preceding study [40], the primary objective of this research was to assess the efficacy of a binary composite consisting of AgNPs and COS in combating biodeterioration caused by white and brown rot fungi, as well as subterranean termites, in wood from *P.* ×*euramericana* 'I-214' treated through vacuum-pressure impregnation. Special emphasis has been placed on determining the minimum effective concentration of the preservative solution, aiming for economic efficiency.

The findings of this research contribute valuable insights to the field of wood preservation, aligning with the contemporary trend of seeking protective products that demonstrate greater consideration for environmental sustainability and human health [3,31,41].

2. Materials and Methods

2.1. Wood Material

Twenty-year-old *P.* ×*euramericana* 'I-214' wood originating from a commercial plantation situated in Quintanilla de Sollamas (42°36′00″ N, 05°49′00″ W), Castile-Leon, Spain, was utilized in this study. The material was received in the form of beams measuring 50 mm × 150 mm × 3000 mm, which were conditioned in the laboratory at approximately 65% relative humidity and 20 °C temperature, to achieve an equilibrium moisture content of 12 ± 2% (determined using a 606-1 digital hygrometer (Testo SE & Co. KGaA; Titisee-Neustadt, Germany)). The sampling and selection of test specimens (TSs) adhered to EN 350:2016 [1] standards for commercial sawn timber. In line with the findings by Spavento et al. [6], who found no significant differences between the inner and outer sections of wood, and considering that the selected beams of this commercial timber exhibited no evidence of heartwood, the TSs, in this case, were not differentiated based on the inner and outer sections of each beam. The beams were subsequently cut into defect-free specimens measuring 15 × 25 × 50 mm (thickness, width, and length, respectively), adhering to the standards EN 113-1-2:2021 [42,43] and EN 117:2012 [44] for decay fungi and termites, respectively.

In accordance with the validation requirements of the aforementioned standards, *Fagus sylvatica* L. (beech) and *Pinus sylvestris* L. (pine, sapwood) defect-free timber, adhering to the same dimensions and sampling and selection criteria as outlined for the TSs, were employed as reference timber (RTs) for decay fungi and termites, respectively.

The treatments and number of repetitions per treatment are detailed in Table 1.

Table 1. Concentrations of each solution to obtain the binary composites and repetitions.

Treatments	Species	AgNPs (ppm)	COS ($g{\cdot}L^{-1}$)	Decay Fungi (Repetitions/ Fungal Species)	Termite (Repetitions)
Untreated	Poplar [a]	–	–	30	9
	Poplar [b]	–	–	30	9
	Beech [c]	–	–	30	–
	Pine [d]	–	–	–	9
AgNPs–$COS_{(4–20)}$	Poplar	4	20	30	9
AgNPs–$COS_{(2–10)}$		2	10	30	9
AgNPs–$COS_{(1–5)}$		1	5	30	9
AgNPs–$COS_{(0.5–2.5)}$		0.5	2.5	30	9
AgNPs–$COS_{(0.25–1.25)}$		0.25	1.25	30	9
	Total repetitions			480	72

[a] Control test specimen; [b] solvent (water) control specimen; [c,d] reference timber for rot fungi and termites, respectively.

2.2. Silver Nanoparticles and Chitosan Oligomers Composite Preparation

Chitosan oligomers were produced through enzymatic degradation following the procedure outlined by Buzón-Durán et al. [45], with adjustments as per the methods described by Ho et al. [46] and Santos-Moriano et al. [47]. In this process, 100 g of high-molecular-weight chitosan powder (CAS No. 9012-76-4; 310,000–375,000 Da), obtained from Caldic Ibérica S.L. (Barcelona, Spain), was dissolved in 5000 mL of Milli-Q water by incorporating 100 g of citric acid (CAS No. 77-92-9) with continuous stirring at 60 °C. Following complete dissolution, 1.67 $g{\cdot}L^{-1}$ of Neutrase® 0.8 L (supplied by Novozymes, Bagsvaerd, Denmark) was added to initiate the degradation of polymer chains. The solution underwent stirring at 40–60 °C for 12 h. Subsequently, it underwent a 5 min ultrasonication process in 1 min periods (employing a probe-type UIP1000hdT ultrasonicator; Hielscher, Teltow, Germany; 1000 W, 20 kHz), maintaining the temperature between 40 and 60 °C. At the conclusion of the procedure, a solution with a pH approximately equal to 4.5–5 and a molecular weight less than 2000 Da (determined through viscosity measurements) was obtained.

Silver nanoparticles were synthesized using an ultrasonication method. The process involved mixing 0.08 g of silver nitrate with 200 mL of *Chamaemelum nobile* (L.) All. solution (5% *v/v*) as a reducing agent. The mixture was stirred and heated at 40–60 °C under UV light until the solution transitioned from colorless to pale yellow, eventually intensifying (pH = 4.5–5). The yellowish solution underwent sonication for 3–5 min and was left to stand for at least 24 h in a refrigerator at 5 °C. The resulting AgNPs were characterized by transmission electron microscopy using a JEOL (Akishima, Tokyo, Japan) JEM-FS2200 HRP microscope.

Following EN 113-1:2021 [42] guidelines, a sufficient quantity of the highest concentration solution was prepared for both COS and AgNPs, aiming to generate, through dilution, a series of five concentrations spread around the anticipated toxic value (see Table 1). The concentrations for AgNPs, adhering to the reference method from EUCAST [48], were chosen based on previous in vitro findings as reported by Silva-Castro et al. [19,35].

For the binary solution, the COS and AgNP solutions were mixed, as specified in Table 1. Subsequently, ultrasonication for 5 min was applied to facilitate the formation of smaller and monodisperse COS particles [46,49].

2.3. Impregnation Method

The impregnation process followed the guidelines of EN 113-2:2021 [43], with certain modifications, such as adjustments to pressure conditions aimed at enhancing the absorption of treatment solutions (as indicated in the following paragraph). Prior to impregnation, each set of TSs for the AgNPs–COS treatment and the water control underwent oven drying until they reached an initial dry mass (m_0). Subsequently, they were stored in a desiccator to maintain dryness until the impregnation process.

The impregnation procedure involved placing the TSs in the treatment vessel, which was then positioned in the vacuum vessel equipped with a vacuum pump and stopcock. The pressure was reduced to 0.7 kPa and maintained for 15 min. Following this, the stopcock connected to the vacuum pump was closed, and another stopcock was opened to allow the preservative solution to be drawn into the treatment vessel within the vacuum environment. Air was then introduced, restoring the vacuum vessel to atmospheric pressure. Subsequently, another stopcock was opened to increase the pressure to 8 bar, deviating from the atmospheric pressure stipulated in EN 113-2:2021 [43]. This adjustment was made to replicate the Bethell industrial process and was sustained for a period of 2 h. Afterward, the TSs were extracted from the treatment vessel, excess liquid was removed by gently blotting with absorbent paper, and the TSs were immediately weighed to the nearest 0.01 g to determine the mass after impregnation (m_1).

Net absorption and retention were calculated using Equations (1) and (2), respectively.

$$NA = \frac{m_1 - m_0}{v}, \quad (1)$$

$$R = NA \times \frac{C}{100}, \quad (2)$$

where NA is the net absorption in $kg{\cdot}m^{-3}$; m_1 is the final mass (post-impregnation) in kg; m_0 is the initial dry mass (pre-impregnation) in kg; v is the volume in m^3; R is the retention of the active principle, AgNPs–COS, in $kg{\cdot}m^{-3}$; and C is the concentration in %.

2.4. Wood-Decay Fungi Resistance Tests

The assessment of poplar resistance to decay fungi was conducted following EN 113-1:2021 [42]. *Trametes versicolor* (L.) Lloyd (strain CTB 863 A) and *Coniophora puteana* (Schumach.) P.Karst. (strain BAM Ebw.15) served as the white and brown rot fungi, respectively. Each strain was cultivated in agar–malt medium (1.5% agar, 3% malt) in 350 mL glass dishes and incubated at 22 ± 2 °C with 70 ± 5% relative humidity (RH). Once covered by fungal tissue, the TSs and reference timber RT were introduced into the rot dishes, with two specimens per flask, after determining the initial dry mass (m_i) and autoclaving. Following 16 weeks under controlled temperature and humidity conditions (22 ± 2 °C and 70 ± 5% RH), the samples were removed from the dishes, conditioned (superficial mycelium extraction), and oven-dried at 103 ± 2 °C to ascertain the final dry mass (m_f). Subsequently, the mass loss percentage (ML) was calculated using Equation (3).

$$ML = \frac{m_i - m_f}{m_i} \times 100 \quad (3)$$

where ML is the mass loss in %; m_i is the initial dry mass in g; and m_f is the final dry mass in g.

2.5. Termite Resistance Tests

The evaluation of poplar wood resistance to subterranean termites was conducted in accordance with standard EN 117:2012 [44], utilizing *Reticulitermes grassei* Clément as the biological material. The termite colonies consisted of 250 workers, accompanied by a small number of nymphs and soldiers, maintaining a ratio of 1 to 5%, mirroring the proportion found in the original colony from which the workers were sourced.

Each colony was situated in glass containers with a substrate of moist sand, comprising 1 part water and 8–10 parts sand. Additionally, 0.5 g of *Pinus* sp. wood shavings were evenly scattered on the substrate as the initial food source for the colonies. A glass ring was introduced against one of the vertical walls of the container, partially extending above the substrate surface. The containers were then incubated in a culturing chamber at 26 ± 2 °C with 70 ± 5% RH for a period of 2 to 4 days until the colony achieved uniformity and vitality. Following this, a wood specimen (TS or RT) was placed on the glass ring in each container, and the containers were further incubated in the chamber for 8 weeks, with periodic checks on the substrate humidity.

After the incubation period, the wood specimens were removed from each container, cleaned, and examined. The count of living termites (workers, nymphs, and soldiers) was recorded, and the attack ratings were visually assessed based on EN 117:2012 [44] criteria: (0) no attack; (1) attempted attack; (2) light attack; (3) medium attack; and (4) heavy attack. The survival rate (SR, %) was estimated using the ratio of the initial number of workers to the number of survivors. For the mean visual rating (VR), if more than 50% of samples fell into a specific class (e.g., heavy attack), the entire sample batch was assigned that class. The same methodology was applied to other attack criteria. Although not specified by the standard, the percentage of mass loss for the specimens was also evaluated using Equation (3).

2.6. *Durability and Use Classes Assignment*

The criteria for assessing resistance to deterioration induced by decay fungi and termites, along with their corresponding durability class (DC), are outlined in Table 2 (EN 350:2016 [1]). The use class was estimated in accordance with standard EN 335:2013 [50].

Table 2. Durability class (DC) against the attack of decay fungi and termites.

Decay Fungi			Termites		
DC	**Description**	**Median ML (%) EN 350:2016**	**DC**	**Description**	**Attack Level * EN 350:2016**
1	Very durable	≤5	D	Durable	>90% '0 o 1' and maximum 10% '2' *
2	Durable	>5 to ≤10	M	Moderately durable	<50% '3, 4'
3	Moderately durable	>10 to ≤15			
4	Slightly durable	>15 to ≤30	S	Not durable	>50% '3, 4'
5	Not durable	>30			

* 90% of test specimens classified as '0' or '1', with a maximum of 10% of test specimens classified as '2', and no test specimens classified as '3' and '4'.

2.7. *Microstructural Alterations and Microanalysis with SEM–EDS*

To investigate fungal degradation and termite attack, and to study the multi-elemental composition of the wood, small sections (approximately 2 × 3 mm) from specimens subjected to *T. versicolor* and *C. puteana* for 16 weeks and to *R. grassei* for 8 weeks were analyzed using scanning electron microscopy (SEM) and an energy-dispersive X-ray spectrometer (EDS). The specimens included those treated with AgNPs at 4 ppm + COS at 20 $g \cdot L^{-1}$ (provided that it was the concentration that led to the best results in protection tests) and untreated control test specimens.

Specimen sections, obtained from both the transverse and longitudinal planes, were prepared with disposable blades, and mounted on slides using bifacial tape. Observations were conducted using an ESEM Quanta 250 FEG (FEI Company, Thermo Fisher Scientific; Waltham, MA, USA) microscope equipped with an electron probe microanalysis unit and an Oxford Instruments (Abingdon, UK) 5DD X-Act model EDS. The analyses were performed at the Scanning Electron Microscopy and Microanalysis Service of the LIMF (Facultad de Ingeniería, Universidad Nacional de La Plata, Buenos Aires, Argentina).

2.8. Statistical Analysis

Statistical analyses were performed using R software (version 4.2.2) [51]. Of the total number of trials carried out by wood species and treatments (Table 1), 236, 234, and 72 data values were subjected to analysis for white, brown, and termite rot, respectively. The statistical assumptions of independence, normality, and homoscedasticity were thoroughly assessed. Normality and homoscedasticity were evaluated using the Shapiro–Wilks test and either the Bartlett or Levene test, respectively. Based on these assessments, a comparative analysis of linear statistics (ANOVA) and robust statistics (Welch) with (or without, as appropriate) the bootstrapping trimmed means method was applied to determine the equality of medians and means, respectively, and to identify homogeneous groups.

3. Results

3.1. Impregnation and Efficacy of Treatments against Biological Agents

The solutions of the binary composite formulated at various concentrations of AgNPs and COS are illustrated in Figure 1a. The characterization of the AgNPs revealed a predominantly spherical morphology with an average diameter of 20 ± 6 nm (Figure 1b).

Figure 1. (**a**) AgNPs–COS solutions; (**b**) TEM micrograph of the synthesized AgNPs.

Table 3 presents the results of the impregnation process along with the density values of samples for each treatment. Figure 2 illustrates, for each biological agent, the retention behavior of the active principle (AgNPs–COS) concerning the concentrations applied, revealing a significant and exponentially increasing relationship between the two.

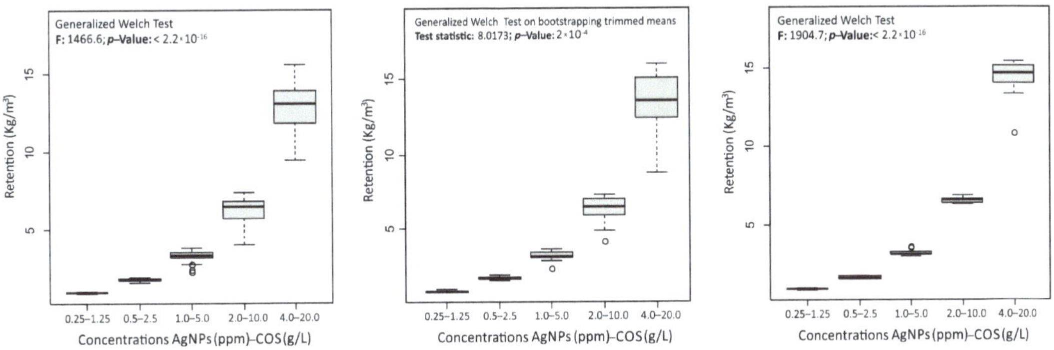

Figure 2. Retention vs. treatment in poplar wood by biological agent. From left to right: white rot, brown rot, and termite.

Table 3. Descriptive analysis of density and impregnation process results by biological agent.

Treatment *	Species	Density † (kg·m^{-3}) ± IC *			Net Absorption (kg·m^{-3}) ± IC *			Retention (kg·m^{-3}) ± IC *		
		Wr **	Br **	T **	Wr **	Br **	T **	Wr **	Br **	T **
Untreated	Poplar [a]	418.50 ± 18.24	436.57 ± 18.29	411.11 ± 17.72	–	–	–	–	–	–
	Poplar [b]	419.01 ± 15.50	426.64 ± 16.15	383.16 ± 22.32	717.17 ± 15.66	733.40 ± 10.89	750.11 ± 13.77		–	–
	Beech [c]	659.08 ± 6.41	669.67 ± 13.41	–	–	–	–	–	–	–
	Pine [d]	–	–	467.10 ± 13.34	–	–	–	–	–	–
AgNPs–COS$_{(4–20)}$	Poplar	376.75 ± 12.08	372.93 ± 12.78	403.17 ± 16.92	643.59 ± 27.06	680.64 ± 32.82	713.22 ± 46.82	12.87 ± 0.54	13.62 ± 0.66	14.27 ± 14.27
AgNPs–COS$_{(2–10)}$		391.59 ± 13.33	401.43 ± 16.04	409.27 ± 37.16	613.41 ± 32.46	632.37 ± 29.11	654.60 ± 13.22	6.14 ± 0.33	6.33 ± 0.29	6.55 ± 0.13
AgNPs–COS$_{(1–5)}$		404.02 ± 13.01	405.91 ± 13.74	417.04 ± 31.86	651.02 ± 26.44	647.63 ± 20.08	646.00 ± 29.17	3.26 ± 0.13	3.24 ± 0.10	3.23 ± 0.15
AgNPs–COS$_{(0.5–2.5)}$		374.11 ± 11.73	385.01 ± 12.72	367.69 ± 16.39	691.55 ± 12.82	694.69 ± 15.04	649.14 ± 29.17	1.73 ± 0.03	1.74 ± 0.04	1.62 ± 0.06
AgNPs–COS$_{(0.25–1.25)}$		374.39 ± 8.93	377.42 ± 10.13	388.66 ± 24.84	728.79 ± 9.85	691.87 ± 15.20	726.17 ± 20.34	0.91 ± 0.03	0.87 ± 0.02	0.91 ± 0.03

[a] Control test specimen; [b] solvent (water) control specimen; [c,d] reference timber for decay fungi and termites (beech and pine, respectively); * IC = robust confidence intervals; ** Wr = white rot, Br = brown rot, T = termite. † Density measured with 12% moisture content.

The assessment of treatment efficacy (untreated vs. treated wood) against *T. versicolor* (Wr), *C. puteana* (Br), and *R. grassei* (T) is summarized in Table 4, Table 5, and Table 6, respectively. The validity of the tests was confirmed in the reference treatments (RTs). For *T. versicolor* and *C. puteana*, this validation was corroborated in beech samples through weight loss exceeding 15% and 20%, respectively (according to EN 113-1:2021 [42]). In the case of *R. grassei*, the validation was conducted in pine samples using a visual attack criterion (VR) of '4' and a survivor rate (SR) exceeding 50% (as per EN 117:2012 [44]).

Table 4. Comparative analysis of mass loss due to white rot for each treatment.

Treatments	Species	ML * (%) ± CI	S–W Test *	B Test *	ANOVA	Homogenous Groups **
			p-Value	*p*-Value	*p*-Value	
Untreated	Poplar [a]	41.96 ± 4.49	0.194	0.089	7.44×10^{-5}	a
	Poplar [b]	40.46 ± 3.06	0.502			a
	Beech [c]	42.01 ± 4.97	0.274			a
AgNPs–COS$_{(4–20)}$	Poplar	30.15 ± 3.08	0.650			b
AgNPs–COS$_{(2–10)}$		36.76 ± 3.87	0.047			ab
AgNPs–COS$_{(1–5)}$		41.49 ± 4.80	0.207			a
AgNPs–COS$_{(0.5–2.5)}$		40.69 ± 4.66	0.753			a
AgNPs–COS$_{(0.25–1.25)}$		45.44 ± 4.48	0.682			a

[a] Control test specimen; [b] solvent (water) control specimen; [c] reference timber (beech); * ML = mass loss, CI = confidence interval, S–W = Shapiro–Wilk test, B = Bartlett test; ** Significant differences between treatments are indicated by different letters (Tukey test $p < 0.05$).

Table 5. Comparative analysis of mass loss due to brown rot for each treatment.

Treatments	Species	ML * (%) ± CI	S–W Test * p-Value	Lv Test * p-Value	W Test * p-Value	Homogenous Groups **
Untreated	Poplar [a]	41.93 ± 4.33	0.085	1.604×10^{-7}	2×10^{-4}	cd
	Poplar [b]	47.86 ± 3.26	0.048			b
	Beech [c]	40.57 ± 3.46	0.001			d
AgNPs–$COS_{(4–20)}$	Poplar	27.22 ± 0.66	0.000			a
AgNPs–$COS_{(2–10)}$		49.90 ± 3.26	0.000			b
AgNPs–$COS_{(1–5)}$		47.69 ± 4.25	0.048			bc
AgNPs–$COS_{(0.5–2.5)}$		50.34 ± 3.44	0.000			b
AgNPs–$COS_{(0.25–1.25)}$		48.99 ± 3.78	0.043			b

[a] Control test specimen; [b] solvent (water) control specimen; [c] reference timber (beech); * ML = mass loss, CI = robust confidence interval, S–W = Shapiro–Wilk Test, Lv = Levene Test, W = Welch Test; ** Significant differences between treatments are indicated by different letters (post hoc test on bootstrapping trimmed means, $p < 0.05$).

Table 6. Comparative analysis of visual rating, survival rate, and mass loss caused by termites for each treatment.

Treatments	Species	VR *	SR * (%)	ML * (%) ± CI	S–W Test * p-Value	B Test * p-Value	ANOVA p-Value	Homogenous Groups ** SR	Homogenous Groups ** ML
Untreated	Poplar [a]	4	53.98 ± 10.40	15.04 ± 1.20	0.898	0.0495	5.65×10^{-12}	a	b
	Poplar [b]	4	55.57 ± 3.67	22.29 ± 3.40	0.868			a	a
	Pine [c]	4	56.34 ± 9.03	13.85 ± 2.48	0.150			a	b
AgNPs–$COS_{(4–20)}$	Poplar	1	26.62 ± 8.63	7.95 ± 1.03	0.897			b	d
AgNPs–$COS_{(2–10)}$		4	34.03 ± 7.65	8.55 ± 2.08	0.836			b	d
AgNPs–$COS_{(1–5)}$		4	36.43 ± 8.29	9.82 ± 2.36	0.147			b	cd
AgNPs–$COS_{(0.5–2.5)}$		4	42.85 ± 6.67	14.94 ± 2.56	0.093			ab	bc
AgNPs–$COS_{(0.25–1.25)}$		4	50.19 ± 4.74	11.94 ± 2.19	0.365			ab	b

[a] Control test specimen; [b] solvent (water) control specimen; [c] reference timber (pine); * VR = visual rating, SR = survival rate, ML = mass loss, CI = confidence interval, S–W: Shapiro–Wilk test, B = Bartlett test; ** Significant differences between treatments are indicated by different letters ($p < 0.05$).

Consistently, across all cases, the formulation featuring 4 ppm AgNPs and 20 $g \cdot L^{-1}$ of COS was the only treatment that significantly enhanced resistance to decay caused by Wr, Br, and T. This particular formulation led to noteworthy reductions in weight loss compared to untreated poplar wood (control test specimen), with percentage values of 28, 35, and 47%, respectively. In the case of T, the degree of attack was also diminished, reaching level '1', corresponding to a survival rate of approximately 27%, showcasing a declining trend with increasing concentration (refer to Figure 3).

Based on the obtained results and considering the criteria for assigning durability class (DC) outlined in EN 350:2016 [1] (Table 2), poplar wood impregnated with the solution containing 4 ppm AgNPs and 20 $g \cdot L^{-1}$ of COS demonstrated enhanced resistance against *C. puteana*, progressing from DC '5' (not durable, as per Spavento et al. [6]) to DC '4' (not very durable). Although there was no observed improvement in DC against *T. versicolor* attack, it nearly reached the threshold between DC '5' and DC '4' (30%). Furthermore, resistance to termite attack was enhanced, moving from DC 'S' (not durable) to DC 'M' (moderately durable).

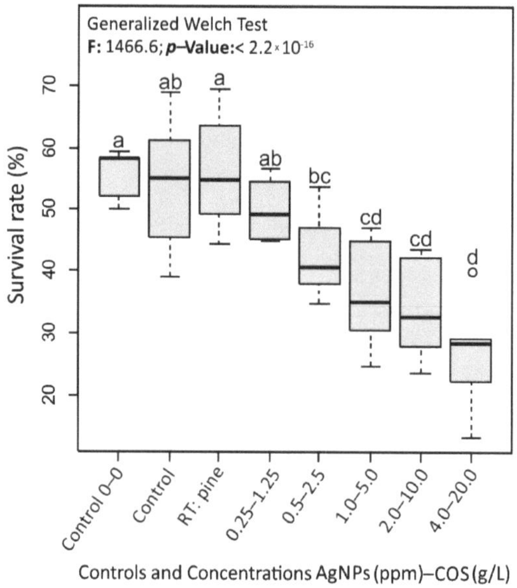

Figure 3. Survival rate behavior. Significant differences between treatments are indicated by different letters ($p < 0.05$).

3.2. Microstructural Alterations and Material Microanalysis

Wood samples subjected to both xylophagous fungal strains, with and without treatment (AgNPs 4 ppm + COS 20 g·L^{-1} and control test specimen, respectively), exhibited microstructural alterations, confirming the diagnosed degradation based on mass loss.

In sections of specimens attacked by *T. versicolor*, predominant alterations included the thinning and deformation of the vessel wall, holes in radial parenchyma, the collapse of the ray sectors, the presence of erosion trails/channels in all cell types, and the erosion of the ray-vessel pits and fibers. Notably, a high concentration of mycelium was observed in the vessels and sectors of tissue that had completely collapsed, displaying a fibrillar aspect (see Figure 4). The material exhibited a corky consistency, more pronounced in the control specimens.

In specimens exposed to *C. puteana*, identified alterations comprised tissue deformation, fractures in all cell types, and the presence of mycelium mainly in the vessels of untreated wood (refer to Figure 5). These samples were characterized by their brown coloration and brittleness.

In all instances, the impregnated material exhibited lower severity and extent of microstructural alterations in the tissue, a finding consistent with the recorded percentages of mass loss.

Furthermore, white deposits were identified in wood sections from both treatments (AgNPs 4 ppm + COS 20 g·L^{-1} and control test specimen) and both biological agents (Wr and Br). These deposits, observed as isolated particles or forming aggregates, were present in various cell types and associated with evidence of colonization and fungal degradation. They were notably abundant in the vessels of the control wood attacked by *C. puteana* (see Figure 5B,C).

Concerning the material subjected to termite attack, SEM-identified alterations, such as mechanical damage and the tearing of the woody tissue, aligned with descriptions by Spavento et al. [6]. These alterations were less severe in the impregnated wood, consistent with its lower shrinkage and greater resistance observed when pressed between the fingers.

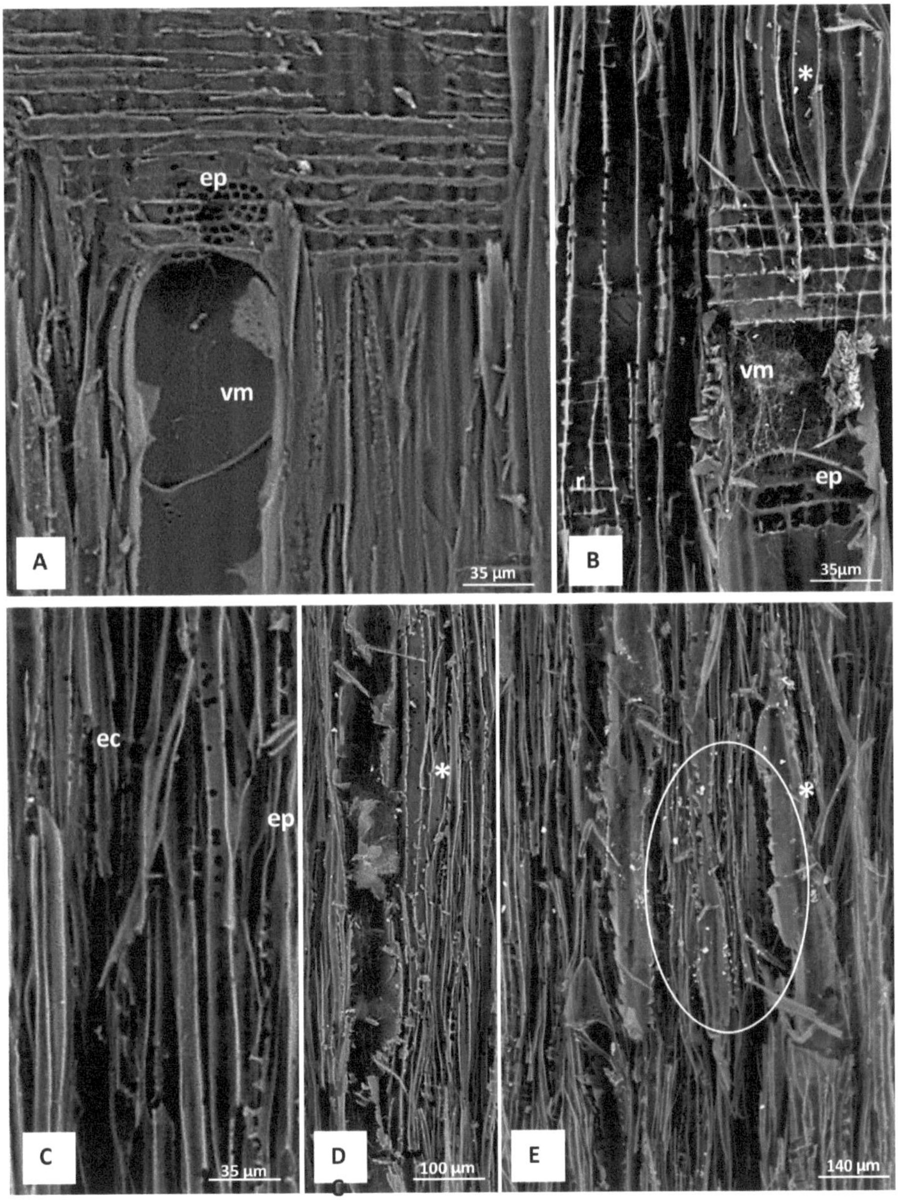

Figure 4. Microstructural alterations in wood exposed to *Trametes versicolor* (L.) Lloyd. (**A**) Impregnated wood. Scarce mycelium in the vessel and incipient erosion in the vessel-radius pits (radial longitudinal view). (**B–E**) Untreated wood. (**B**) Abundant mycelium in vessel element, erosion in ray-vessel pits, collapse of radial parenchyma cells, and deposits in fiber wall (radial longitudinal view). (**B**) Erosion path/channel and eroded pits in fibers (tangential longitudinal view). (**D,E**) Fibrillar appearance of tissue and presence of deposits (radial longitudinal view and tangential longitudinal view, respectively). vm = Vessel mycelium, ep = eroded pit; r = parenchymal radius; ec = erosion path/channel, * = deposits.

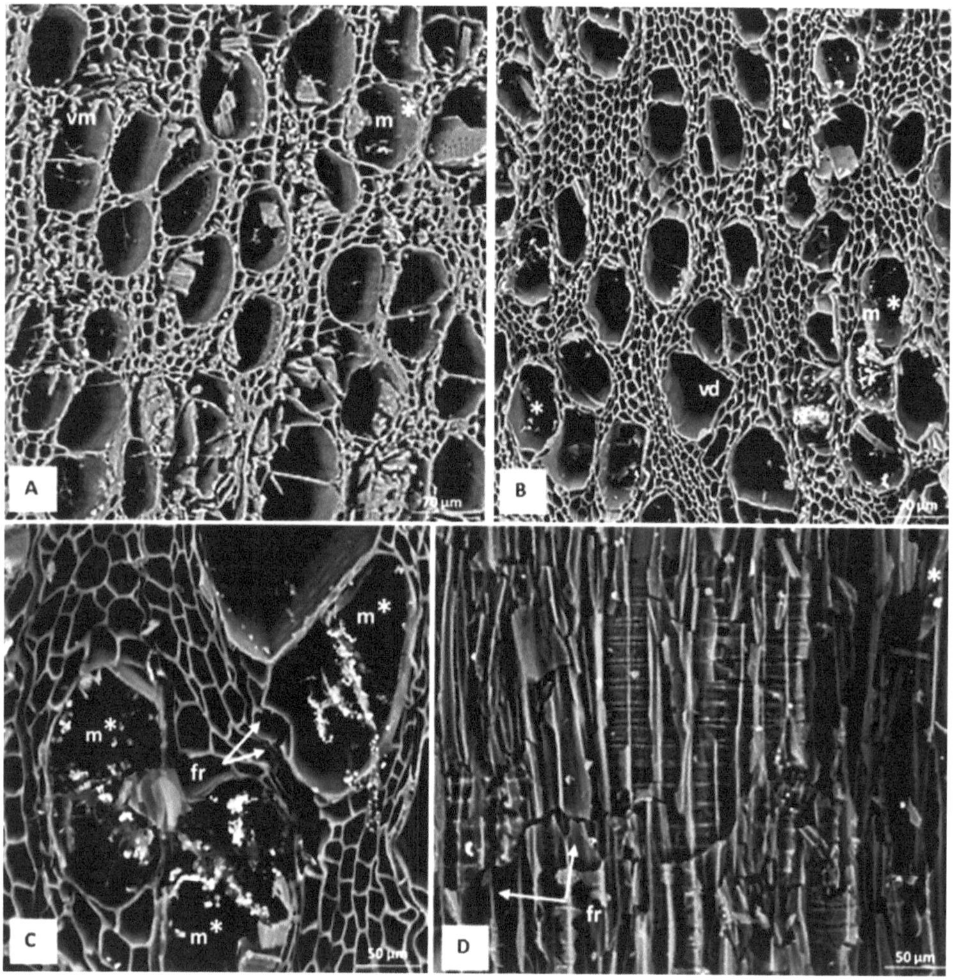

Figure 5. Microstructural alterations in wood exposed to *Coniophora puteana* (Schumach.) P.Karst. (**A**) Impregnated wood. Vessels with mycelium and sparse deposits (transverse view). (**B–D**) Untreated wood. (**B**,**C**) Mycelium and deposits in vessels, general aspect, and detail, respectively, and fractures in fiber (transverse view). Note the deformation of the vessels with respect to the treated material and the higher concentration of mycelium and deposits. (**D**) Fracture perpendicular and transverse to the grain, and deposits (longitudinal radial view). v = Vessel element, m = mycelium, d = deformation, fr = fracture, * = deposits.

With EDS, AgNPs were not identified in the bulk material treated with the binary composite. In the control and impregnated samples exposed to both types of rot, the elements present in different sectors of the tissue (vessels, fibers, and rays with and without evidence of attack) were C and O with the highest concentrations, followed by Cu, Ca, K, Zn and, to a lesser extent, Cl, P, Na, Si, S, and Mg. In all the treatments, a trend towards an increase in Ca concentration was identified in the areas with evidence of colonization and fungal degradation and the presence of deposits. Particularly accentuated was the increase in Ca in the vessel lumens of the control specimens exposed to brown rot (Figure 6).

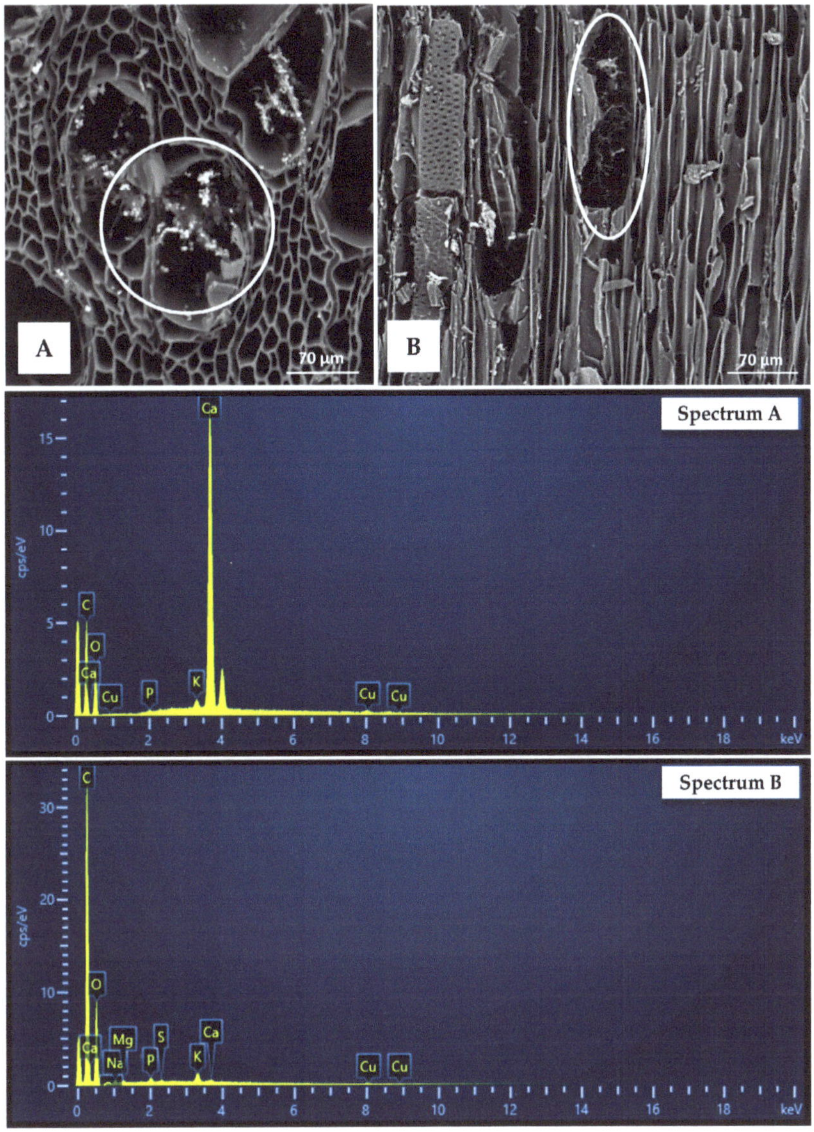

Figure 6. Sections of specimens exposed to brown rot analyzed with SEM–EDS along with the EDS spectra of the circled regions. (**A**) Deformed vessels with mycelium and deposits (transverse view). (**B**) Sector of vessel element with abundant mycelium lacking deposits (tangential longitudinal view). Note the difference in Ca percentage.

4. Discussion

4.1. Efficacy of Treatments against Biological Agents

The effectiveness of the binary solution containing AgNPs–COS on *P.* ×*euramericana* 'I-214' wood from Spain was discernible only at the highest concentration (4 ppm + 20 g·L^{-1}), particularly against *C. puteana* and *R. grassei*.

In terms of efficacy against *T. versicolor*, the results obtained in this study were less favorable than those reported by Casado et al. [18] in poplar wood treated through vacuum-pressure impregnation of AgNPs solutions at 5 ppm after 16 weeks (ML = 8.94% vs. 30.15% in this study). Moya et al. [20,21] also observed enhanced resistance to degradation by *T. versicolor* in 12 tropical commercial wood species treated with a solution of AgNPs. Weight loss percentages ranged from 7 to 11% after four months, which were lower than those reported herein, but it is worth noting that a significantly higher concentration (50 ppm AgNPs + ethylene glycol) was employed. However, their results were notably less promising against the brown rot causal agent *Lenzites acutus* Berk. (=*Cellulariella acuta* (Berk.) Zmitr. & Malysheva), with weight loss values in the 8 to 35% range (vs. 27.22% for *C. puteana* at 4 ppm in this study). These researchers also noted improved dimensional stability and reduced water absorption capacity in the treated wood. Pařil et al. [30] reported similar outcomes in beech and pine wood treated with AgNP solutions, with weight losses below 3%, although at much higher AgNP concentrations (1000 ppm and 3000 ppm), when tested against *T. versicolor* and *Poria placenta* (Fr.) Cooke (=*Rhodonia placenta* (Fr.) Niemelä, K.H.Larss. & Schigel; brown rot), respectively. Considering the literature, the relatively low efficacy against *T. versicolor* observed in this study could be attributed to the chosen AgNP concentrations, which were substantially lower than those employed by the cited authors, particularly through deep impregnation.

Regarding the mechanism of action, the reduction in weight loss in both white rot and brown rot cases might be attributed to different factors. AgNPs exposed to high humidity exhibit remarkable antifungal activity. In this process, metallic silver undergoes oxidation in the presence of water, leading to the release of ions. These silver ions, when in solution, significantly impact the exoenzymatic activity of both white and brown rot fungi. As elucidated by Dorau et al. [24], their specific influence is notable in the activity of cellulase enzymes produced by decay fungi. Given the inherently slow nature of this chemical reaction transformation, the particle size assumes critical importance in inhibiting fungal growth. Consequently, smaller particle sizes result in a higher specific surface area and enhanced oxidation effectiveness, providing more effective protection against these spoilage agents through the sustained slow release of silver ions [27]. Additionally, AgNPs, particularly those of smaller sizes, can induce the generation of reactive oxygen species (ROS) and free radicals within cells, leading to oxidative stress and, ultimately, cell death. These AgNPs enter cells through proton pumps, disrupting their function and that of the electron transport chain, resulting in an increased production of ROS and subsequent damage to proteins, lipids, and nucleic acids [34,52,53]. Furthermore, AgNPs have been identified to hinder the DNA replication process. Another mechanism involves the interaction between AgNPs and thiol groups of proteins, leading to the deactivation of the latter [54]. AgNPs may also adhere to cell surfaces, compromising membrane integrity and function [55]. Importantly, these mechanisms are considered to operate simultaneously, representing multiple modes of action rather than functioning in isolation [54].

Concerning termites, the existing literature on nanotechnology as a strategy to enhance resistance against this form of deterioration primarily focuses on coniferous wood. Additionally, studies explore the application of chemical elements such as zinc oxides, borates, and copper oxides, either independently or in combination with AgNPs, usually at concentrations not surpassing 2% [13,14,23]. However, the outcomes from these investigations align closely with the findings obtained in the current study.

Regarding the mortality rate, the one observed with the AgNPs-COS (4 ppm–20 $g \cdot L^{-1}$) solution (73.38%, SR = 26.62%) closely resembled those documented by Green and Arango [23] for 0.5% nano zinc oxide + Ag (76%). However, it was lower than the rate reported by Mantanis et al. [13] for formulations containing 2% nano zinc oxide (>90%).

As per Green and Arango [23], instances where a high survival rate was observed could be attributed to repellency or avoidance rather than direct toxicity. It is worth considering that, in general, the effectiveness of AgNPs against biological degradation agents might hinge on nanoparticle characteristics, such as size and shape, influencing their performance in wood protection applications concerning adhesion and internalization [15,56]. Can et al. [57] propose that achieving effective protection through the slow release of silver ions requires particles to be less than 100 nm in size. Additionally, Bugnicourt et al. [58] suggest that spherical-shaped nanoparticles, as obtained in this study (Figure 1b), enhance the internalization into cells. Conversely, Jafernik et al. [56] point out that non-spherical NPs might have a greater impact on transmission, circulation, transfer, and distribution.

The stability of the solution is another crucial aspect, alongside the shape and size of the nanoparticles. As previously mentioned, the characteristics of chitosan (COS), including high water-solubility and viscosity, coupled with its chemical structure featuring chelating and reducing groups, play a crucial role in preventing agglomeration. This combination ensures a more effective stabilization of the solution [35,38,59–62].

Chitosan oligomer nanoparticles (COS NPs) have been explored as wood protective solutions, either alone or in combination with other nanoparticles (e.g., AgNPs), to enhance the durability of wood against mold fungi, brown-rot, and white-rot fungi, consistently yielding promising results in various studies [18,19,35,36,63–65].

The inference that the decrease in weight loss and the subsequent improvement in resistance to degradation, especially against the brown-rot strain and termites in *P.* ×*euramericana* 'I-214' wood from Spain, is linked to the combined biocidal action of AgNPs–COS and appears valid based on the information provided. The prolonged efficacy observed after 16 or 8 weeks of trials, respectively, is attributed to the size and shape of the AgNPs and the stabilizing property of COS. This combination leads to a slow and/or controlled release of AgNPs, the primary bioactive compound of the binary solution, within the wood. This suggests a comparative advantage of the AgNPs–COS solution, aligning with the observations of Casado et al. [18].

Moreover, the promising results in improving durability class against deterioration caused by *C. puteana* are noteworthy. This improvement is significant, given the detrimental impact of brown rot on wood's resistance capacity, a critical property influencing its useful life in structural applications.

4.2. Microstructural Alterations and Material Microanalysis

The microstructural alterations observed in the wood subjected to different treatments (control and AgNPs–COS (4 ppm + 20 $g \cdot L^{-1}$)) and fungi (white rot and brown rot) align with the degradation identified by mass loss and the type of rot to which they were exposed. The modifications observed in test specimens degraded by *T. versicolor* and *C. puteana* are consistent with the diagnostic characteristics of white and brown rot, respectively.

White rots typically involve the degradation of the main polymers of the cell wall (holocellulose and lignin), resulting in structural alterations, a fibrillar aspect, and corky to spongy consistency in advanced stages of rot. These traits were detected in the material exposed to *T. versicolor*. On the other hand, brown rots involve holocellulose degradation and partial oxidation of lignin, leading to a brittle consistency, cubic fracture pattern, and brown color. These characteristics were identified in the test specimens exposed to *C. puteana*.

Concerning the EDS results, the absence of detectable AgNPs in the treated wood could be attributed to their low concentration in the binary solution (4 ppm). Elemental microanalysis findings, particularly the identification of peaks or increases in calcium concentration in sectors with evidence of colonization and degradation and the presence of deposits, may result from the precipitation of this element due to the biochelation phenomena caused by fungal acids. Xylophagous fungi, particularly those responsible for brown rot, produce high amounts of oxalic acid and chelating polycarboxylic acids, leading to the immobilization of metals in the form of insoluble oxalate crystals. The production of calcium oxalate is associated with wood degradation [66,67]. Rudakiya and Gupte [67] corroborate the presence of oxalate crystals, particularly with a high percentage of calcium.

The notable reduction in calcium percentage observed in wood sectors displaying evidence of decay and the absence of deposits may be attributed to a decrease in oxalic acid concentration. Several authors have noted that the disappearance of calcium oxalate crystals correlates with a decline in oxalic acid levels. Mechanisms such as nutrient depletion, metabolic changes, the assimilation of oxalic acid as a carbon source, and the enzymatic degradation of oxalic acid are proposed explanations for the absence of calcium oxalate crystals and the subsequent decline in calcium percentage [66,68,69].

4.3. Applications and Limitations of the Study, and Further Research

In concordance with established research in wood preservation, the outcomes of this study underscore the potential of emerging technologies as eco-friendly alternatives to mitigate wood deterioration, extend its service life, and broaden the utility of less naturally durable wood, such as *P.* ×*euramericana* clone I-214.

While the results show promise, improving the durability class against the aggressive termites and brown rot strain, and notably reducing mass loss with the white-rot strain, a limitation surfaced in the inability to widen the classification gap compared to untreated wood. This highlights the necessity for the further refinement of treatment concentrations.

Aligned with the structural requirements set by Eurocode 5 (EN 1995-1-1:2016 [39]), ongoing research is dedicated to a comprehensive evaluation of the treated wood's resistance, considering the solutions and concentrations presented in this work. This evaluation encompasses the wood's resilience against deterioration caused by micro-fungi soft rot, as well as its elastic-resistant behavior pre- and post-treatment and fungal exposure. Moreover, to confirm treatment effectiveness and assess its broad applicability and reproducibility, there is a consideration to extend the study to other wood species with low natural durability and wide distribution (*Salix* spp., *Pinus pinaster* Aiton, and *Quercus* spp., among others), along with exploring fungal strains of significance in the wood industry (*Gloephylum* spp. and *Picnoporus* spp., among others).

5. Conclusions

The binary solution of AgNPs–COS, at concentrations of 4 ppm and 20 $g \cdot L^{-1}$, respectively, significantly enhanced the resistance to decay induced by *T. versicolor*, *C. puteana*, and *R. grassei* in *P.* ×*euramericana* 'I-214' wood of Spanish origin when impregnated by vacuum pressure. Particularly noteworthy was the increase in the durability class achieved against *C. puteana* and *R. grassei*. Scanning electron microscopy studies provided visual confirmation of the treatment's effectiveness, revealing discernible differences between the control and treated wood in the severity and extent of tissue degradation. This work contributes significant evidence supporting the effectiveness against the biological deterioration of the AgNPs–COS composite at low concentrations when introduced via vacuum pressure. This treatment aligns with the characteristics of an 'ideal preservative', showcasing the capacity to respond effectively to preservation methods, exerting an impact on various organisms at low concentrations of the active principle, featuring slow release and prolonged action, and demonstrating environmental friendliness, among other desirable attributes. Furthermore, valuable insights are provided regarding its impact on degradation caused by a strain of *C. puteana*, an aspect that has been less explored to date but holds importance considering the influence of such deterioration on the wood's structural resilience.

Author Contributions: Conceptualization, E.S., M.T.d.T.-F., L.A.-R. and P.M.-R.; methodology, E.S., M.T.d.T.-F., M.M., S.M.S., M.C.-S., J.M.-G. and J.Á.-M.; formal analysis, E.S. and L.A.-R.; investigation, E.S., M.T.d.T.-F., S.M.S., R.D.M.-L., J.M.-G., J.Á.-M. and P.M.-R.; resources, E.S., M.T.d.T.-F., M.C.-S., R.D.M.-L., J.M.-G. and J.Á.-M.; writing—original draft preparation, E.S., M.T.d.T.-F., L.A.-R., M.M. and P.M.-R.; writing—review and editing, E.S. and P.M.-R.; visualization, M.C.-S., R.D.M.-L. and J.Á.-M.; supervision, M.T.d.T.-F., L.A.-R. and P.M.-R.; funding acquisition, E.S., M.T.d.T.-F. and M.C.-S. All authors have read and agreed to the published version of the manuscript.

Funding: This research was conducted as part of a postdoctoral stay funded by the BEC.Ar Program of the Ministry of Education, Argentina.

Data Availability Statement: Data are contained within the article.

Acknowledgments: The authors express their gratitude to the technical staff of the Wood Technology Laboratory at the University of Valladolid (UVa) and the National Research Institute and Agrarian and Food Technology—Institute of Forestry Sciences (INIA-ICIFOR) in Spain for their valuable assistance during the experiments. Special thanks are extended to Alberto Santiago Aliste for conducting the transmission electron microscopy (TEM) characterization of the silver nanoparticles (AgNPs) at the Microscopy Unit of Parque Científico UVa.

Conflicts of Interest: The authors declare no conflict of interest. The funders had no role in the design of the study; in the collection, analyses, or interpretation of data; in the writing of the manuscript; or in the decision to publish the results.

References

1. *EN 350:2016*; Durability of Wood and Wood-Based Products—Testing and Classification of the Durability to Biological Agents of Wood and Wood-Based Materials. European Committee for Standardization: Brussels, Belgium, 2016; p. 67.
2. Martín, J.A.; López, R. Biological deterioration and natural durability of wood in Europe. *Forests* **2023**, *14*, 283. [CrossRef]
3. Zabel, R.A.; Morrell, J.J. Wood deterioration agents. In *Wood Microbiology*, 2nd ed.; Zabel, R.A., Morrell, J.J., Eds.; Academic Press: San Diego, CA, USA, 2020; pp. 19–54. [CrossRef]
4. Tucker, C.L.; Koehler, P.G.; Pereira, R.M. Development of a method to evaluate the effects of eastern subterranean termite damage to the thermal properties of building construction materials (Isoptera: Rhinotermitidae). *Sociobiology* **2008**, *51*, 589–600.
5. Garnica, J. La importancia del chopo en la industria. In Proceedings of the Jornadas de Salicáceas-V Congreso Internacional de Salicáceas, Talca, Chile, 13–17 November 2017; p. 22.
6. Spavento, E.; Murace, M.; Acuña Rello, L.; Monteoliva, S.-E.; Troya Franco, M.T.D. Susceptibility of *Populus × euramericana* 'I-214'of Spanish origin to xylophagous attacks: Durability tests for its possible inclusion in European standard. *For. Syst.* **2019**, *28*, e008. [CrossRef]
7. Papadopoulos, A.N. Nanotechnology and wood science. *Nanomaterials* **2023**, *13*, 691. [CrossRef] [PubMed]
8. Bi, W.; Li, H.; Hui, D.; Gaff, M.; Lorenzo, R.; Corbi, I.; Corbi, O.; Ashraf, M. Effects of chemical modification and nanotechnology on wood properties. *Nanotechnol. Rev.* **2021**, *10*, 978–1008. [CrossRef]
9. Papadopoulos, A.N.; Kyzas, G.Z. Nanotechnology and wood science. In *Interface Science and Technology*; Kyzas, G.Z., Mitropoulos, A.C., Eds.; Elsevier: Amsterdam, The Netherlands, 2019; Volume 30, pp. 199–216.
10. Shiny, K.S.; Sundararaj, R.; Mamatha, N.; Lingappa, B. A new approach to wood protection: Preliminary study of biologically synthesized copper oxide nanoparticle formulation as an environmental friendly wood protectant against decay fungi and termites. *Maderas. Ciencia y Tecnología* **2019**, *21*, 347–356. [CrossRef]
11. Lykidis, C.; De Troya, T.; Conde, M.; Galván, J.; Mantanis, G. Termite resistance of beech wood treated with zinc oxide and zinc borate nanocompounds. *Wood Mater. Sci. Eng.* **2018**, *13*, 45–49. [CrossRef]
12. Terzi, E.; Kartal, S.N.; Yılgör, N.; Rautkari, L.; Yoshimura, T. Role of various nano-particles in prevention of fungal decay, mold growth and termite attack in wood, and their effect on weathering properties and water repellency. *Int. Biodeterior. Biodegrad.* **2016**, *107*, 77–87. [CrossRef]
13. Mantanis, G.; Terzi, E.; Kartal, S.N.; Papadopoulos, A.N. Evaluation of mold, decay and termite resistance of pine wood treated with zinc- and copper-based nanocompounds. *Int. Biodeterior. Biodegrad.* **2014**, *90*, 140–144. [CrossRef]
14. Akhtari, M.; Nicholas, D. Evaluation of particulate zinc and copper as wood preservatives for termite control. *Eur. J. Wood Wood Prod.* **2013**, *71*, 395–396. [CrossRef]
15. Clausen, C.A.; Kartal, S.N.; Arango, R.A.; Green, F. The role of particle size of particulate nano-zinc oxide wood preservatives on termite mortality and leach resistance. *Nanoscale Res. Lett.* **2011**, *6*, 427. [CrossRef] [PubMed]
16. Dai, X.; Qi, Y.; Luo, H.; He, Z.; Wei, L.; Dong, X.; Ma, X.; Yang, D.-Q.; Li, Y. Leachability and anti-mold efficiency of nanosilver on poplar wood surface. *Polymers* **2022**, *14*, 884. [CrossRef] [PubMed]
17. Iqtedar, M.; Mirza, N.; Aihetasham, A.; Iftikhar, S.; Kaleem, A.; Abdullah, R. Termiticidal activity of mycosynthesized silver nanoparticles from *Aspergillus fumigatus* BTCB15. *Revista Mexicana de Ingeniería Química* **2020**, *19*, 1201–1211. [CrossRef]
18. Casado, S.; Silva, C.; Ponce, H.; Martín, R.; Martín, G.; Acuña, R. White-rot fungi control on populus spp. Wood by pressure treatments with silver nanoparticles, chitosan oligomers and propolis. *Forests* **2019**, *10*, 885. [CrossRef]
19. Silva-Castro, I.; Casados-Sanz, M.; Alonso-Cortés, A.; Martín-Ramos, P.; Martín-Gil, J.; Acuña-Rello, L. Chitosan-based coatings to prevent the decay of *Populus* spp. wood caused by *Trametes versicolor*. *Coatings* **2018**, *8*, 415. [CrossRef]
20. Moya, R.; Rodriguez-Zuñiga, A.; Berrocal, A.; Vega-Baudrit, J. Effect of silver nanoparticles synthesized with NPsAg-ethylene glycol ($C_2H_6O_2$) on brown decay and white decay fungi of nine tropical woods. *J. Nanosci. Nanotechnol.* **2017**, *17*, 5233–5240. [CrossRef]
21. Moya, R.; Berrocal Jiménez, A.; Rodríguez Zúñiga, A.; Vega Baudrit, J.; Chaves Noguera, S. Effect of silver nanoparticles on white-rot wood decay and some physical properties of three tropical wood species. *Wood Fiber Sci.* **2014**, *46*, 527–538.
22. Kartal, S.; Green, F.; Clausen, C. Do the unique properties of nanometals affect leachability or efficacy against fungi and termites? *Int. Biodeterior. Biodegrad.* **2009**, *63*, 490–495. [CrossRef]

23. Green, F.; Arango, R.A. Wood protection by commercial silver formulations against eastern subterranean termites. In Proceedings of the International Research Group on Wood Protection 38th Annual Meeting, Jackson Hole, WY, USA, 20–24 May 2007; p. 07-30422.
24. Dorau, B.; Arango, R.; Green, F. An investigation into the potential of ionic silver as a wood preservative. In Proceedings of the 2nd Woodframe Housing Durability and Disaster Issues Conference, Madison, WI, USA, 6–8 November 2004; pp. 133–145.
25. Yudaev, P.; Mezhuev, Y.; Chistyakov, E. Nanoparticle-containing wound dressing: Antimicrobial and healing effects. *Gels* **2022**, *8*, 329. [CrossRef]
26. Calovi, M.; Coroneo, V.; Rossi, S. Antibacterial efficiency over time and barrier properties of wood coatings with colloidal silver. *Appl. Microbiol. Biotechnol.* **2023**, *107*, 5975–5986. [CrossRef]
27. Piętka, J.; Adamczuk, A.; Zarzycka, E.; Tulik, M.; Studnicki, M.; Oszako, T.; Aleksandrowicz-Trzcińska, M. The application of copper and silver nanoparticles in the protection of *Fagus sylvatica* wood against decomposition by *Fomes fomentarius*. *Forests* **2022**, *13*, 1724. [CrossRef]
28. Aleksandrowicz-Trzcińska, M.; Szaniawski, A.; Olchowik, J.; Drozdowski, S. Effects of copper and silver nanoparticles on growth of selected species of pathogenic and wood-decay fungi in vitro. *For. Chron.* **2018**, *94*, 109–116. [CrossRef]
29. Bak, M.; Németh, R. Effect of different nanoparticle treatments on the decay resistance of wood. *BioResources* **2018**, *13*, 7886–7899. [CrossRef]
30. Pařil, P.; Baar, J.; Čermák, P.; Rademacher, P.; Prucek, R.; Sivera, M.; Panáček, A. Antifungal effects of copper and silver nanoparticles against white and brown-rot fungi. *J. Mater. Sci.* **2016**, *52*, 2720–2729. [CrossRef]
31. Arpanaei, A.; Fu, Q.; Singh, T. Nanotechnology approaches towards biodeterioration-resistant wood: A review. *J. Bioresour. Bioprod.* 2023; *in press, corrected proof.* [CrossRef]
32. EPA. Overview of Wood Preservative Chemicals. Available online: https://www.epa.gov/ingredients-used-pesticide-products/overview-wood-preservative-chemicals (accessed on 16 November 2023).
33. Ali, S.; Chen, X.; Ahmad, S.; Shah, W.; Shafique, M.; Chaubey, P.; Mustafa, G.; Alrashidi, A.; Alharthi, S. Advancements and challenges in phytochemical-mediated silver nanoparticles for food packaging: Recent review (2021–2023). *Trends Food Sci. Technol.* **2023**, *141*, 104197. [CrossRef]
34. Mansoor, S.; Zahoor, I.; Baba, T.R.; Padder, S.A.; Bhat, Z.A.; Koul, A.M.; Jiang, L. Fabrication of silver nanoparticles against fungal pathogens. *Front. Nanotechnol.* **2021**, *3*, 679358. [CrossRef]
35. Silva-Castro, I.; Martín-García, J.; Diez, J.J.; Flores-Pacheco, J.A.; Martín-Gil, J.; Martín-Ramos, P. Potential control of forest diseases by solutions of chitosan oligomers, propolis and nanosilver. *Eur. J. Plant Pathol.* **2017**, *150*, 401–411. [CrossRef]
36. Matei, P.M.; Martín-Ramos, P.; Sánchez-Báscones, M.; Hernández-Navarro, S.; Correa-Guimaraes, A.; Navas-Gracia, L.M.; Rufino, C.A.; Ramos-Sánchez, M.C.; Martín-Gil, J. Synthesis of chitosan oligomers/propolis/silver nanoparticles composite systems and study of their activity against *Diplodia seriata*. *Int. J. Polym. Sci.* **2015**, *2015*, 864729. [CrossRef]
37. Woźniak, M.; Gromadzka, K.; Kwaśniewska-Sip, P.; Cofta, G.; Ratajczak, I. Chitosan–caffeine formulation as an ecological preservative in wood protection. *Wood Sci. Technol.* **2022**, *56*, 1851–1867. [CrossRef]
38. Mirda, E.; Idroes, R.; Khairan, K.; Tallei, T.E.; Ramli, M.; Earlia, N.; Maulana, A.; Idroes, G.M.; Muslem, M.; Jalil, Z. Synthesis of chitosan-silver nanoparticle composite spheres and their antimicrobial activities. *Polymers* **2021**, *13*, 3990. [CrossRef] [PubMed]
39. *UNE-EN 1995-1-1:2016*; Eurocode 5: Design of Timber Structures—Part 1-1: General—Common Rules and Rules for Buildings. Asociación Española de Normalización: Madrid, Spain, 2016.
40. Spavento, E.; Troya, M.T.; Casado-Sanz, M.; Santos, S.M.; Martín-Gil, J.; Martín-Ramos, P.; Robertson, L.; Acuña-Rello, L. Evaluation of the efficacy of silver nanoparticles and chitosan oligomer composites as poplar wood protective treatments against termites. In Proceedings of the IRG54 Annual Meeting, Cairns, Australia, 28 May–1 June 2023; p. IRG/WP 23-40963.
41. Bossert, D.; Geers, C.; Placencia Peña, M.I.; Volkmer, T.; Rothen-Rutishauser, B.; Petri-Fink, A. Size and surface charge dependent impregnation of nanoparticles in soft- and hardwood. *Chemistry* **2020**, *2*, 361–373. [CrossRef]
42. *EN 113-1:2021*; Durability of Wood and Wood-Based Products—Test Method against Wood Destroying Basidiomycetes—Part 1: Assessment of Biocidal Efficacy of Wood Preservatives. European Committee for Standardization: Brussels, Belgium, 2021; p. 31.
43. *EN 113-2:2021*; Durability of Wood and Wood-Based Products—Test Method against Wood Destroying Basidiomycetes—Part 2: Assessment of Inherent or Enhanced Durability. European Committee for Standardization: Brussels, Belgium, 2021; p. 29.
44. *EN 117:2012*; Wood Preservatives—Determination of Toxic Values against Reticulitermes Species (European Termites) (Laboratory Method). European Committee for Standardization: Brussels, Belgium, 2012; p. 22.
45. Buzón-Durán, L.; Martín-Gil, J.; Pérez-Lebeña, E.; Ruano-Rosa, D.; Revuelta, J.L.; Casanova-Gascón, J.; Ramos-Sánchez, M.C.; Martín-Ramos, P. Antifungal agents based on chitosan oligomers, ε-polylysine and *Streptomyces* spp. secondary metabolites against three *Botryosphaeriaceae* species. *Antibiotics* **2019**, *8*, 99. [CrossRef] [PubMed]
46. Ho, K.W.; Ooi, C.W.; Mwangi, W.W.; Leong, W.F.; Tey, B.T.; Chan, E.-S. Comparison of self-aggregated chitosan particles prepared with and without ultrasonication pretreatment as Pickering emulsifier. *Food Hydrocoll.* **2016**, *52*, 827–837. [CrossRef]
47. Santos-Moriano, P.; Fernandez-Arrojo, L.; Mengibar, M.; Belmonte-Reche, E.; Peñalver, P.; Acosta, F.N.; Ballesteros, A.O.; Morales, J.C.; Kidibule, P.; Fernandez-Lobato, M.; et al. Enzymatic production of fully deacetylated chitooligosaccharides and their neuroprotective and anti-inflammatory properties. *Biocatal. Biotransform.* **2018**, *36*, 57–67. [CrossRef]

48. Arendrup, M.C.; Cuenca-Estrella, M.; Lass-Flörl, C.; Hope, W. EUCAST technical note on the EUCAST definitive document EDef 7.2: Method for the determination of broth dilution minimum inhibitory concentrations of antifungal agents for yeasts EDef 7.2 (EUCAST-AFST). *Clin. Microbiol. Infect.* **2012**, *18*, E246–E247. [CrossRef]
49. Gokce, Y.; Cengiz, B.; Yildiz, N.; Calimli, A.; Aktas, Z. Ultrasonication of chitosan nanoparticle suspension: Influence on particle size. *Colloids Surf. A Physicochem. Eng. Asp.* **2014**, *462*, 75–81. [CrossRef]
50. *EN 335:2013*; Durability of Wood and Wood-Based Products—Use Classes: Definitions, Application to Solid Wood and Wood-Based Products. European Committee for Standardization: Brussels, Belgium, 2013; p. 14.
51. R Core Team. *R: A Language and Environment for Statistical Computing*; R Foundation for Statistical Computing: Vienna, Austria, 2022.
52. Mikhailova, E.O. Silver nanoparticles: Mechanism of action and probable bio-application. *J. Funct. Biomater.* **2020**, *11*, 84. [CrossRef]
53. Nel, A.; Xia, T.; Madler, L.; Li, N. Toxic potential of materials at the nanolevel. *Science* **2006**, *311*, 622–627. [CrossRef]
54. Alghuthaymi, M.A.; Almoammar, H.; Rai, M.; Said-Galiev, E.; Abd-Elsalam, K.A. Myconanoparticles: Synthesis and their role in phytopathogens management. *Biotechnol. Biotechnol. Equip.* **2015**, *29*, 221–236. [CrossRef]
55. Klaine, S.J.; Alvarez, P.J.J.; Batley, G.E.; Fernandes, T.F.; Handy, R.D.; Lyon, D.Y.; Mahendra, S.; McLaughlin, M.J.; Lead, J.R. Nanomaterials in the environment: Behavior, fate, bioavailability, and effects. *Environ. Toxicol. Chem.* **2009**, *27*, 1825–1851. [CrossRef] [PubMed]
56. Jafernik, K.; Ładniak, A.; Blicharska, E.; Czarnek, K.; Ekiert, H.; Wiącek, A.E.; Szopa, A. Chitosan-based nanoparticles as effective drug delivery systems—A review. *Molecules* **2023**, *28*, 1963. [CrossRef] [PubMed]
57. Can, A.; Sivrikaya, H.; Hazer, B.; Palanti, S. Beech (*Fagus orientalis*) wood modification through the incorporation of polystyrene-ricinoleic acid copolymer with Ag nanoparticles. *Cellulose* **2022**, *29*, 1149–1161. [CrossRef]
58. Bugnicourt, L.; Alcouffe, P.; Ladavière, C. Elaboration of chitosan nanoparticles: Favorable impact of a mild thermal treatment to obtain finely divided, spherical, and colloidally stable objects. *Colloids Surf. A Physicochem. Eng. Asp.* **2014**, *457*, 476–486. [CrossRef]
59. Babaee, M.; Garavand, F.; Rehman, A.; Jafarazadeh, S.; Amini, E.; Cacciotti, I. Biodegradability, physical, mechanical and antimicrobial attributes of starch nanocomposites containing chitosan nanoparticles. *Int. J. Biol. Macromol.* **2022**, *195*, 49–58. [CrossRef] [PubMed]
60. Nascimento, T.; Rego, C.; Oliveira, H. Potential use of chitosan in the control of grapevine trunk diseases. *Phytopathologia Mediterranea* **2007**, *46*, 218–224.
61. Venkatesham, M.; Ayodhya, D.; Madhusudhan, A.; Veera Babu, N.; Veerabhadram, G. A novel green one-step synthesis of silver nanoparticles using chitosan: Catalytic activity and antimicrobial studies. *Appl. Nanosci.* **2012**, *4*, 113–119. [CrossRef]
62. Wang, L.-S.; Wang, C.-Y.; Yang, C.-H.; Hsieh, C.-L.; Chen, S.-Y.; Shen, C.-Y.; Wang, J.-J.; Huang, K.-S. Synthesis and anti-fungal effect of silver nanoparticles-chitosan composite particles. *Int. J. Nanomed.* **2015**, *10*, 2685. [CrossRef]
63. Alfredsen, G.; Eikenes, M.; Militz, H.; Solheim, H. Screening of chitosan against wood-deteriorating fungi. *Scand. J. For. Res.* **2011**, *19*, 4–13. [CrossRef]
64. Singh, T.; Vesentini, D.; Singh, A.P.; Daniel, G. Effect of chitosan on physiological, morphological, and ultrastructural characteristics of wood-degrading fungi. *Int. Biodeterior. Biodegrad.* **2008**, *62*, 116–124. [CrossRef]
65. Torr, K.M.; Chittenden, C.; Franich, R.A.; Kreber, B. Advances in understanding bioactivity of chitosan and chitosan oligomers against selected wood-inhabiting fungi. *Holzforschung* **2005**, *59*, 559–567. [CrossRef]
66. Guggiari, M.; Bloque, R.; Aragno, M.; Verrecchia, E.; Job, D.; Junier, P. Experimental calcium-oxalate crystal production and dissolution by selected wood-rot fungi. *Int. Biodeterior. Biodegrad.* **2011**, *65*, 803–809. [CrossRef]
67. Rudakiya, D.M.; Gupte, A. Degradation of hardwoods by treatment of white rot fungi and its pyrolysis kinetics studies. *Int. Biodeterior. Biodegrad.* **2017**, *120*, 21–35. [CrossRef]
68. Dutton, M.V.; Evans, C.S. Oxalate production by fungi: Its role in pathogenicity and ecology in the soil environment. *Can. J. Microbiol.* **1996**, *42*, 881–895. [CrossRef]
69. Schwarze, F.W.M.R. Wood decay under the microscope. *Fungal Biol. Rev.* **2007**, *21*, 133–170. [CrossRef]

forests

Article

Wood Modification Using Imidazole and Succinimide: Effects on Dimensional Stability and Bending Properties

Alexander Scharf [1,*], Henric Dernegård [2], Johan Oja [3], Dick Sandberg [1] and Dennis Jones [1]

[1] Wood Science and Engineering, Luleå University of Technology, Forskargatan 1, SE-93187 Skellefteå, Sweden; dick.sandberg@ltu.se (D.S.); dennis.jones@ltu.se (D.J.)
[2] Holmen AB, Strandvägen 1, SE-11451 Stockholm, Sweden; henric.dernegard@holmen.com
[3] Norra Timber, Skeppargatan 1, SE-90403 Umeå, Sweden; johan.oja@norraskog.se
* Correspondence: alexander.scharf@ltu.se; Tel.: +46-730825153

Abstract: The modification of Scots pine sapwood (*Pinus sylvestris* L.) with the heterocyclic compounds imidazole and succinimide was investigated. Pressure-impregnation with aqueous solutions containing imidazole, imidazole + citric acid, succinimide, succinimide + citric acid, and citric acid + sorbitol (CIOL®) with solid contents of 5%, 10%, and 15% was followed by oven-curing at 220 °C for 1 h. During the treatment steps, the changes in mass, bending properties, and anti-swelling efficiency (ASE) were examined. The results indicate that solid concentrations within the range of 5% to 10% were optimal. The results seem to show that there are two differing mechanisms in the modification of imidazole and succinimide, respectively. Mass loss due to heat treatment was highest in the imidazole-treated specimens, whereas it remained low and concentration-independent in the succinimide-treated specimens. After three cycles, the ASE reached 31% for the imidazole-treated specimens and improved to 38% with the addition of citric acid. For succinimide, the ASE increased from 17% to 41%. The bending properties generally showed improvement, except for succinimide + citric acid and CIOL®, which displayed a reduced modulus of rupture. Chemical analyses are warranted to fully understand the reaction mechanisms of these treatments. The positive effects of imidazole treatment are suggested to stem from a thermal reaction between the chemical and the wood, indicated by substantial mass loss during leaching and specimen darkening. Succinimide and citric acid might exhibit polymerization with each other and with wood components, which is akin to the CIOL® process. Further research should delve into the reaction mechanisms and the impact of imidazole and succinimide on biological durability.

Keywords: anti-swelling efficiency; citric acid; mechanical properties; thermal treatment; wood protection

Citation: Scharf, A.; Dernegård, H.; Oja, J.; Sandberg, D.; Jones, D. Wood Modification Using Imidazole and Succinimide: Effects on Dimensional Stability and Bending Properties. *Forests* **2023**, *14*, 1976. https://doi.org/10.3390/f14101976

Academic Editor: Miklós Bak

Received: 31 August 2023
Revised: 20 September 2023
Accepted: 27 September 2023
Published: 29 September 2023

1. Introduction

Wood modification involves altering the structural and chemical composition of the wood cell wall using heat or chemicals, and this is primarily aimed at enhancing water-related characteristics like sorption, dimensional stability, and resistance to biological degradation [1,2].

Thermal wood modification, achieved through heat treatment in low-oxygen environments within the temperature range of 160 °C to 240 °C, induces favorable changes in the chemical composition of the cell wall [3]. The central mechanism driving alterations in the cell wall is the hydrolysis-induced degradation of hemicelluloses, yielding commonly observed byproducts such as furfural and hydroxymethylfurfural [4–6]. Lignin degradation also occurs at temperatures exceeding 220 °C [7]. Condensation reactions of oxidatively cleaved ether linkages can result in carbonyl and phenolic groups, which lead to the formation of new crosslinks [8]. The formation of various degradation products accounts for a mass loss of up to 20%. However, not all of the degradation products are volatiles, and thus, they partly remain in the wood until extracted by a solvent such as water [7]. The positive impact of heat treatment on sorption and dimensional stability arises from the reduced presence of moisture-sensitive hydroxyl groups within hemicelluloses [9], coupled with the generation of novel crosslinks facilitated by degradation products [10]. Furthermore, the lowered availability of accessible sugars and decreased equilibrium moisture content contribute to an enhanced level of biological durability [11]. The strength properties of thermally modified wood are negatively affected due to the reduced degree of polymerization of the holocelluloses [12], as is commonly exhibited during bending tests with a reduced modulus of rupture (MOR), often accompanied by a brittle failure mode. Both biological durability and a decrease in bending properties correlate to mass loss during thermal modification [13].

Chemical wood modification is based on the introduction of various chemicals into the wood. It exists in different forms with the main difference being the location of chemical deposition and the type of bonding with the cell wall [1]. Chemicals can react with groups of cell-wall polymers, blocking, e.g., hydroxyl groups, or leading to crosslinking by reacting with two hydroxyl groups. The former results in cell-wall bulking, whereas the latter additionally limits the maximum distance between the cell-wall polymers, effectively reducing the maximum swelling, i.e., providing anti-swelling efficiency (ASE) [14]. Additionally, lumen-filling treatments exists which may or may not react with the cell-wall components. Similar to thermal modification, chemical modification improves water-related properties and biological durability and can affect mechanical properties [15].

Abundant hydroxyl groups in hemicelluloses drive common esterification reactions for wood modification. Acids like acetic anhydride, succinic or maleic anhydride, polycarboxylic acids, and isocyanates are used, with acetic anhydride-based acetylation being commercialized [2]. Recently, the polycarboxylic acid citric acid gained attention due to its cost-effective availability from microbial fermentation using *Aspergillus niger* [16]. Initially applied in the cotton industry [17], polycarboxylic acid was later adapted for wood modification [18,19]. The reaction mechanism between wood and citric acid involves a two-step esterification process, where a cyclic anhydride is initially formed, followed by its reaction with hydroxyl groups in the wood to create ester linkages [20]. Wood modification with citric acid offers several advantages, such as reduced water absorption, enhanced resistance against termites and fungi, an improved modulus of elasticity (MOE), improved compression strength, and better dimensional stability [21,22]. However, it is important to note that there are some disadvantages, including a strong reduction in MOR, increased brittleness, and a yellowing of the treated wood [23]. Feng et al. [24] reported an initial ASE of 48% at a chemical load of 36%, whereas L'Hostis et al. [25] reported an initial ASE of 67% and a low leaching rate. However, the MOR decreased significantly. In order to improve the performance of citric acid-treated wood, compounds containing alcoholic hydroxyl groups leading to polymerization and increased chemical fixation can be incorporated [25–31].

The polymerization of citric acid and sorbitol in an aqueous solution was initially demonstrated by Centolella and Razor [32]. Doll et al. [33] proposed the formation of intermediates and, subsequently, a citrate sorbitol ester, as shown in Figure 1. Larnøy et al. [30] reported the polyesterification of citric acid and sorbitol in a 3:1 molar ratio for solid wood at 140 °C for 18 h at a weight-percentage gain (WPG) of 80%. Leaching tests indicated excellent leaching and fungal resistance, as higher curing temperatures led to higher conversion rates of the functional groups, resulting in a denser crosslinked network in the cell wall. Mubarok et al. [34] reported an ASE of 55% at a WPG of 30%. However, similar to citric acid treatments, the MOR and work-to-maximum load in bending decreased significantly. Beck [35] showed an ASE of 40% at WPGs of 14%–31%, with leaching rates below 2%. For more detailed information on citric acid in wood modification, readers are referred to the review by Lee et al. [23].

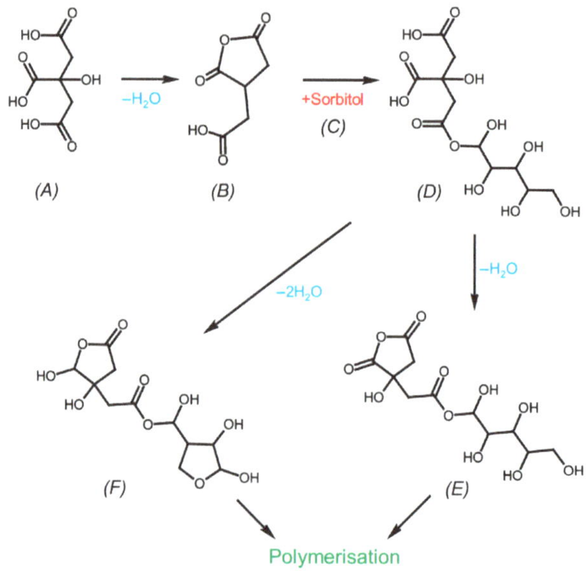

Figure 1. The reaction mechanism between citric acid (**A**) and sorbitol (**C**). Intermediate cyclic anhydride (**B**,**E**), ester (**D**), and anhydrosorbitol ring (**F**). Adapted from [30].

Imidazole, a heterocyclic compound comprising two nitrogen and three carbon atoms, possesses acidic and basic properties. It is environmentally benign and nonhazardous. Morais et al. [36] pioneered the use of imidazole for the extraction of cellulose and hemicelluloses from native wheat straw at a temperature of 170 °C. The depolymerization process yielded a range of valuable lignin-derived compounds, including vanillin, vanillic acid, and rosmarinic acid. Since then, extensive research efforts have been directed toward exploring the potential of imidazole as a solvent in the field of biomass valorization. Grylewicz et al. [37] investigated the use of imidazole in combination with glycerol for the fabrication of thermoplastic starch and wood-fiber composites, reporting enhanced moisture sorption and surface hydrophobicity properties. Del Menezzi et al. [38] reported that citric acid not only reacted with the hydroxyl groups of hemicelluloses but also that the reaction with the aliphatic chains of lignin was possible. It is, hence, possible that a combined treatment of lignin-altering imidazole and citric acid might lead to enhanced crosslinking in the wood.

Succinimide, another heterocyclic compound containing nitrogen, exhibits a carbonyl and an amide group. Succinimides have high chemical reactivity due to the presence of both the carbonyl and methylene groups [39]. It has applications in the pharmaceutical, polymer, and material industries [40]. Under certain conditions, the carbonyl group of succinimide can undergo nucleophilic additions with hydroxyl groups leading to the formation of ester linkages. The reaction is typically catalyzed by acidic conditions.

To the best of our knowledge, no previous studies have explored the application of imidazole or succinimide in wood modification. Thus, our research aimed to explore wood modification systems involving imidazole and succinimide alone and in combination with citric acid. The treatment was based on pressure impregnation and subsequent heat treatment to improve the mechanical and hygroscopic properties.

2. Materials and Methods

2.1. Materials

Scots pine (*Pinus sylvestris* L.) sawn timber harvested in Västerbotten County was obtained from a sawmill in northern Sweden (Norra Timber, Kåge, Sweden). Scots pine is an abundant and economically important species in northern Europe and relatively easy to impregnate. The timber was industrially dried to a moisture content (MC) of approx. 18%.

The chemicals used in this study were synthesis-grade >98% imidazole ($C_3N_2H_4$) powder (IoLiTec-Ionic Liquids Technologies GmbH, Heilbronn, Germany), synthesis-grade >98% succinimide ($C_4H_5NO_2$) (Sigma-Aldrich, Merck KGaA, Darmstadt, Germany), 99.9% analytical-grade citric acid ($C_6H_8O_7$), and ≥96% technical-grade D (-)-sorbitol ($C_6H_{14}O_6$) powder (VWR International AB, Stockholm, Sweden).

2.2. Specimen and Solution Preparation

Visually selected, defect-free, straight-grained sapwood specimens were prepared from the sawn timber for further pressure impregnation and heat treatment and the subsequent determination of anti-swelling efficiency (ASE) and bending properties. The specimens used for ASE test had dimensions of 21 × 20 × 10 mm (radial × tangential × longitudinal), and the specimens for the bending test had dimensions of 10 × 10 × 200 mm (radial × tangential × longitudinal). The specimens were prepared in surplus and were conditioned for 4 weeks at 20 °C and 65% relative humidity until they had a constant mass and, thus, the equilibrium moisture content (EMC) was reached. For both tests, the specimens were grouped by density, and the lowest and highest 10% were sorted out. The remaining specimens exhibited a mean density at an EMC of 480 ± 23 kg/m^3. They were randomly distributed based on density into 17 groups. Additionally, the cross-sections of the bending specimens were end-sealed with silicon, preventing uneven liquid flow during treatment, and thus securing even modification throughout the specimens.

A total of five different chemical combinations were dissolved in deionized water to produce solutions with 5, 10, and 15 wt% solid contents. The chemical combinations are shown in Table 1. The treatment CIOL® was used as a reference representing a bio-based wood modification process, which is close to commercialization [30,41,42]. Additionally, a solely heat-treated (H) and untreated control (C) group were added.

Table 1. Treatments and chemical concentration of aqueous solutions used for pressure impregnation. The number in the treatment ID states the solid concentration of the used solution. Number of specimens per group: anti-swelling efficiency test = 6, and bending test = 12.

Treatment ID	Total Concentration of Solution (wt%)	Imidazole (wt%)	Succinimide (wt%)	Citric Acid (wt%)	Sorbitol (wt%)	Heat Treatment Temperature (°C)
I5	5	5.0	-	-	-	220
I10	10	10.0	-	-	-	220
I15	15	15.0	-	-	-	220
ICA5	5	2.8	-	2.2	-	220
ICA10	10	5.6	-	4.4	-	220
ICA15	15	8.4	-	6.6	-	220
Su5	5	-	5.0	-	-	220
Su10	10	-	10.0	-	-	220
Su15	15	-	15.0	-	-	220
SuCA5	5	-	2.8	2.2	-	220
SuCA10	10	-	5.6	4.4	-	220
SuCA15	15	-	8.4	6.6	-	220
CIOL®5 [1]	5	-	-	3.8	1.2	220
CIOL®10 [1]	10	-	-	7.6	2.4	220
CIOL®15 [1]	15	-	-	11.4	3.6	220
H	-	-	-	-	-	220
C	-	-	-	-	-	-

[1] The combination of citric acid and sorbitol in a 3:1 molar ratio fixed by a curing step is carried out according to the CIOL® process [42].

2.3. Pressure Impregnation and Heat Treatment

For each group to be chemically treated, the conditioned specimens for ASE and the bending tests were placed together in a vessel and fully submerged with the respective chemical solution. The specimens were pressure-impregnated in an autoclave using a full-cell method, whereby a vacuum at 20 mbar was initially applied for 30 min, followed by 1 h pressure at 15 bar. The excess solution was wiped off with tissue paper, and the mass and dimensions were recorded. This was followed by oven-drying (open system) to 0% MC for 24 h at 70 °C and 16 h at 103 °C. The mass and dimensions of each specimen were recorded to determine solution uptake and WPG, which were calculated as follows (Equations (1) and (2)):

$$Solution\ uptake = \frac{m_1 - m_0}{m_0} - 0.11 \quad (1)$$

$$WPG = (m_2 - m_0)/m_0 \quad (2)$$

where m_0 is the initial dry mass, m_1 is the mass directly after impregnation, and m_2 is the oven-dry mass after impregnation, where m_0 was estimated for all specimens based on the EMC of the control group C. The conditioning of the specimens to EMC resulted in an MC of 11.0 ± 0.27%, determined by the gravimetric method. Additionally, the MC of 11% was subtracted from the solution uptake to account for the vapor-bound water as the specimens were impregnated in a conditioned state.

The dried specimens were tightly wrapped in aluminum foil and placed in an oven, undergoing heat treatment in an open system at 220 °C for 1 h. The mass and dimensions were recorded, and the mass loss (Δm_{HT}) and bulking coefficient (BC) after heat treatment were calculated (Equations (3) and (4)):

$$\Delta m_{HT} = (m_3 - m_0)/m_0 \quad (3)$$

$$BC = (V_3 - V_0)/V_0 \quad (4)$$

where m_3 and V_3 are the mass and wood volume after heat treatment, respectively, and V_0 is the initial dry wood volume. Referencing the mass loss during the heat treatment to the initial dry mass m_0 instead of m_2 was a deliberate choice to account for differences in WPG, providing results that are easier to compare.

The heat-treated specimens were kept for 24 h at room climate before being placed in a conditioning chamber at 20 °C and 65% relative humidity until EMC was reached.

2.4. Specimen Characterization

ASE was evaluated using a previously established procedure [43]. All specimens were subjected to repeated wet-dry cycles. The specimens underwent vacuum impregnation in deionized water for 1 h, followed by immersion in water for 72 h. After water treatment, the specimens were air-dried under ambient conditions and subsequently oven-dried for 48 h at 70 °C and 16 h at 103 °C. The mass and dimensions were measured in a wet and dry state. The ASE was calculated following the method described by Stamm [44] (Equation (5)):

$$ASE = (S_2 - S_1)/S_1 \quad (5)$$

where S_1 is the untreated volumetric swelling coefficient, and S_2 is the treated volumetric swelling coefficient between the water-saturated and dry states. This wet-dry cycle process was repeated three times to wash out any water-soluble compounds affecting the ASE. Crosslinking was evaluated by the volumetric change over the wet-dry cycles and was calculated according to Equation (6) [43]. In order to evaluate the leaching resistance of the treated wood, the mass loss due to leaching over three cycles $\Delta m_{leaching}$ was calculated (Equation (7)). It describes the share of water-soluble compounds in the wood after heat treatment, which can be either in the introduced chemicals, (thermal) degradation products, or naturally occurring water-soluble extractives. A high $\Delta m_{leaching}$ indicates the low stability of the introduced chemicals and a potential negative environmental impact during the service life of the product. Additionally, the mass loss Δm_{total} over the complete treatment and leaching process was calculated (Equation (8)). This is similar to the corrected mass loss proposed by Altgen et al. [45] and, in contrast to Δm_{HT}, includes the water-soluble compounds created during the heat treatment. If the Δm_{total} of treatment is lower than the Δm_{total} of the solely heat-treated group, chemicals are likely to be present in the wood after leaching, while a higher value would suggest that (thermal) degradation was promoted by the treatment.

$$Volumetric\ change = (V_n - V_0)/V_0 \quad (6)$$

$$\Delta m_{leaching} = (m_4 - m_3)/m_3 \quad (7)$$

$$\Delta m_{total} = (m_4 - m_0)/m_0 \quad (8)$$

where V_n is the volume in either the oven-dried or wet state in the nth wet-dry cycle, V_0 is the initial oven-dried dimension, and m_4 is the mass after three wet-dry cycles and oven-drying for 24 h at 70 °C and 16 h at 103 °C.

The MOR and the local modulus of elasticity (MOE) in bending were tested in conjunction with a four-point bending test. Specimens 10 × 10 × 200 mm in size were loaded in a universal testing machine (MTS System Corporation, Eden Prairie, MN, USA) equipped with a 10 kN load cell according to the EN 408 standard [46]. The span was 180 mm, the distance between the load points was 60 mm, and the loading rate was 0.03 mm/s.

In order to assess the statistical significance of differences between the mean values for each group, a one-way analysis of variance (ANOVA) was conducted for each measured property. Following this, a Tukey post-hoc test was performed to identify specific groups that exhibited significant differences. The significance level was set at $\alpha = 0.05$. The results of the post-hoc test are presented using a compact letter display, wherein those groups that were not significantly different from each other were assigned the same letter.

3. Results and Discussion

3.1. Impregnation and Heat Treatment

The solution uptake was 138%–150% among the groups with no significant difference between the different solid concentrations of the solutions. Figure 2 shows the WPG after impregnation and drying and the mass loss Δm_{HT} after the heat treatment determined for the ASE specimens.

The WPG behaved similarly among the different treatment groups when solutions with a 5 and 10% solid concentration were used, and an increase from 5 to 10% solid concentration in the chemical solution led to an approx. doubling in WPG. The increase from 10% to 15%, however, did not translate into a 1.5-fold increase in WPG. It may be suggested that the saturation of available sites, e.g., for hydrogen bonding, is reached at a solid concentration between 10 and 15%, likely being dependent on the molecular structure of the used chemicals. At 10% and 15%, the groups SuCA and CIOL® showed higher WPG compared to the other treatments, whereas, for imidazole, I15 showed no significant increase in uptake compared to I10.

The mass loss, Δm_{HT}, in Figure 2 comprised multiple degradative processes. Without chemical treatment, Δm_{HT} was 2.6 ± 0.4% due to thermal degradation, primarily of the hemicelluloses [7]. In the chemically treated groups (except Su), Δm_{HT} was related to the WPG with a clear difference between the treatments, with and without imidazole. On the one hand, the thermal degradation of the used chemicals likely happened since it had been noted that both imidazole and citric acid exhibit one-stage degradation, peaking at 220 °C in thermogravimetric analyses [47,48], whereas succinimide starts to thermally degrade around 250 °C [49]. On the other hand, the chemical-promoted degradation of cell wall polymers is possible. The Δm_{HT} in the imidazole treatments was partly higher than the respective WPG and was accompanied by a strong darkening of the specimens. This has been assumed to be a result of imidazole promoting the degradation of lignin during the heat treatment, a fact that has been observed in the refinery of biomass when imidazole has been present [36]. The groups treated with only succinimide showed the least mass loss independently of WPG. Additionally, the groups Su10 and Su15 exhibited a thin layer of salt deposits on the surface when the specimens were handled. It is likely that only a few reaction sites between the cell wall polymers and the succinimide existed or that no bonds were formed. The mass loss in the CIOL® can be attributed to the ester linkage of the carboxyl groups of the citric acid to the hydroxyl groups of the wood and the sorbitol, which already occurs at temperatures of 140 °C. In this process, citric acid transforms into the reactive anhydride with the release of water molecules [30,33]. In the groups ICA and SuCA, 56% of the imidazole and succinimide mass was substituted by citric acid. While the Δm_{HT} was little effected in the imidazole groups (from I to ICA), and the Δm_{HT} in succinimide groups (from Su to SuCA) increased. This can be due to the catalyzed thermal degradation of hemicelluloses and lignin in the presence of acids [50] and by the esterification of citric acid with the cell wall polymers, which involves dehydration during heat treatment. Interestingly, the mass loss was similar between CIOL® and SuCA and between I and ICA. The formation of crosslinks between imidazole, succinimide, citric acid, and the cell-wall polymers or their degradation products is possible, but chemical analyses are necessary.

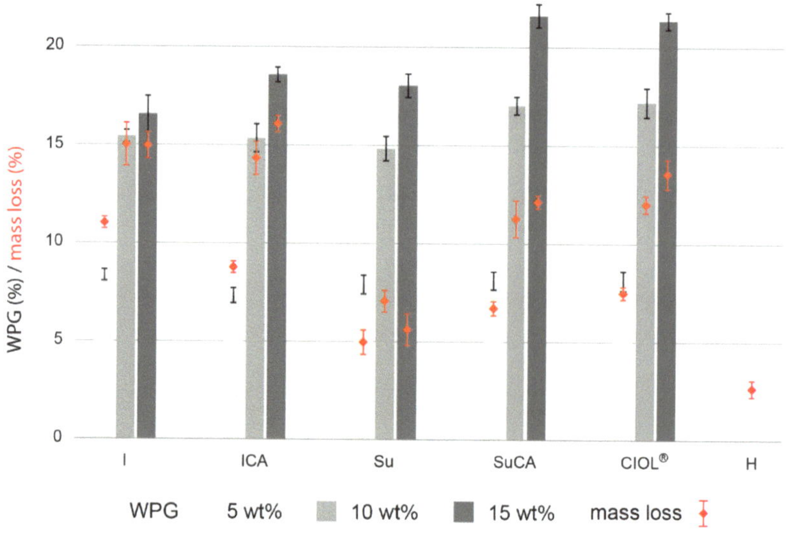

Figure 2. The mean ± standard deviation of weight percentage gain (WPG) after pressure impregnation and drying and mass loss due to heat treatment at 220 °C (Δm_{HT}) of the specimens treated with different concentrations (wt%) of the chemical solution. Δm_{HT} was calculated in relation to the initial dry mass. Treatment shorting can be seen in Table 1. The number of specimens per group = 6.

Figure 3 shows the cross-sections of one arbitrarily chosen specimen per group after the heat treatment. A color change was evident in all groups but was the strongest in groups containing imidazole, where the color was similar to carbonized wood. This indicates that imidazole might increase the amount of thermal degradation in wood and potentially lower the charring temperature of wood. Imidazole has been shown to depolymerize lignin in wheat straw, resulting in compounds such as vanillin, vanillic acid, and rosmarinic acid [36]. The groups Su, SuCA, and CIOL® showed slightly different browning compared to the heat-treated group (H), which could be due to the presence of polymerized chemicals in the wood.

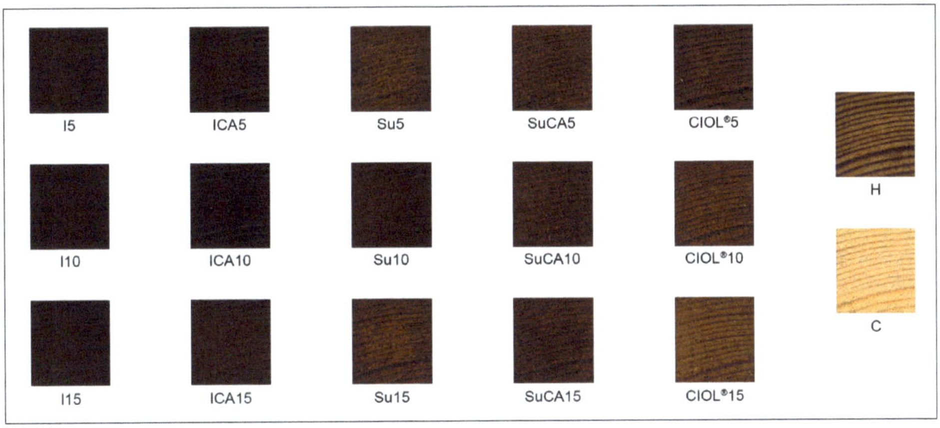

Figure 3. Cross-sectional appearance of specimens after heat treatment and the untreated control (C).

3.2. *Anti-Swelling Efficiency and Water Stability*

Table 2 shows the EMC after the treatment and the ASE after each wet-dry cycle. The EMC was 4.5%–6.3% in the chemically treated groups, which was slightly lower than the heat-treated group H. The ASE after the first cycle was high among all groups and related to the WPG. Alongside the volatile extractives, nonvolatile degradation products were formed during thermal degradation, which remain in the wood and bulk the cell wall [7]. However, after three cycles, the ASE decreased significantly, indicating that the chemicals and or water-soluble degradation products were leached out. This effect was strongest in the succinimide groups and lowest in the CIOL® groups. The low ASE for succinimide supports the theory that no modification of the wood took place. After three cycles, the ASE of the I and Su groups were not related to the initial WPG, whereas all groups containing citric acid were related to WPG. The ASE of I10 was 33.4%, and the addition of citric acid (ICA10) resulted in an ASE of 34.7%, indicating no synergistic effect of imidazole and citric acid. However, in the case of succinimide, the ASE increased from 21.0% to 37.9% via the addition of citric acid. For CIOL®, the high ASE is attributed to the formation of the insoluble polymers formed during the esterification process between citric acid and wood and citric and sorbitol [30]. Furthermore, the results are in line with work from Beck [35].

Figure 4 presents the mass loss caused by wet-dry cycling, i.e., the leaching of chemicals, degraded wood components, and water-soluble extractives, and the mass loss over the whole treatment and leaching procedure. $\Delta m_{leaching}$ correlated with the WPG in each group except for CIOL®. Combinations, including citric acid, exhibited less mass loss than the respective single compound treatments. As shown in Figure 2, different degrees of mass loss occurred during the heat treatment, which influenced $\Delta m_{leaching}$. In order to understand the mode of action of the treatments, it was, thus, more conclusive to study the total mass loss Δm_{total}. It was noted that $\Delta m_{leaching}$ in the imidazole and succinimide groups was rather similar, but the values for Δm_{total} were significantly different. Treatments involving imidazole groups exhibited a mass loss of 7%–8%, indicating that the wood components were degraded to an extended degree during the heat treatment and were washed out during the wet-dry cycling; this was supported by the results for heat treatment only, where the results showed a mass loss of 3%. The yellow/brown color of the leachate supported this.

Table 2. The mean equilibrium moisture content (EMC) and anti-swelling efficiency (ASE) of specimens treated with different concentrations (wt%) of the chemical solution followed by heat treatment. Number of specimens per group = 6.

	I			ICA			Su			SuCA			CIOL®			H	C
(wt%)	5	10	15	5	10	15	5	10	15	5	10	15	5	10	15	-	-
EMC	4.5	5.2	4.6	5.2	5.3	5.0	6.3	6.1	5.6	5.3	6.1	4.8	5.5	6.3	5.0	7.0	11.0
ASE 1st cycle	43.7	53.8	59.8	38.0	44.1	54.4	36.3	47.2	57.4	45.6	50.6	59.2	40.8	47.9	52.7	22.4	0
ASE 2nd cycle	34.5	35.2	37.1	32.7	37.0	43.4	22.7	24.2	27.2	34.2	40.5	46.5	36.4	43.4	49.9	14.7	−8.1
ASE 3rd cycle	29.6	33.4	31.3	29.2	34.7	38.2	19.0	21.0	17.4	31.6	37.9	41.2	35.0	41.5	47.5	11.1	−7.0

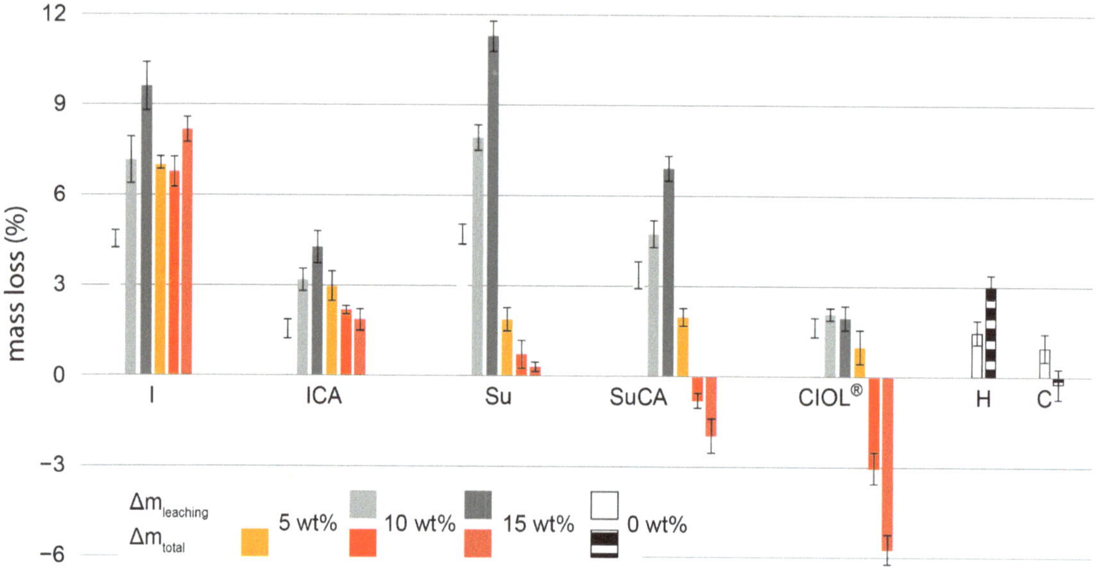

Figure 4. Mass loss after three wet-dry cycles in reference to the dry weight after the treatment ($\Delta m_{leaching}$) and to the initial dry weight (Δm_{total}). wt% corresponds to the solid concentration of the chemical solution used for pressure impregnation. The number of specimens per group = 6.

The WPG did not have a strong influence on the Δm_{total}, which was in line with the ASE values. The reductions in ASE from 50–60% in the first cycle to around 30% in the third cycle, coupled with a Δm_{total} of 7%–8% were similar to the results of thermal modification by Wentzel et al. [51]. However, if imidazole would solely promote thermal degradation, the mechanical properties would be expected to decrease, which was not the case. The addition of citric acid presented in the ICA groups drastically lowered Δm_{total} due to a lower concentration of imidazole and the presence of citric acid. It is also possible that citric acid reacted with the degradation products of the imidazole-promoted reactions, resulting in insoluble polymers and a low $\Delta m_{leaching}$ and Δm_{total}. Succinimide showed the highest $\Delta m_{leaching}$ but a Δm_{total} close to zero. When taken together with the low and WPG-independent Δm_{HT} and ASE, this indicated that succinimide either did not react with the wood components or only a few reaction sides were available. The addition of citric acid (SuCA) improved leaching resistance as well as Δm_{total}. The negative mass loss indicated that insoluble polymers were formed. It is also possible that a reaction of succinimide with the wood or the citric acid took place due to the more acidic conditions. CIOL® treatment led to little leaching and higher chemical retention due to the formation of polymers in the esterification process, and this has been shown by previous research by Larnøy et al. [30].

Figure 5 shows the volumetric change over the wet-dry cycles of the specimens treated with 15 wt% solutions, with D0 being equivalent to the BC. The chemical treatments led to a relatively low BC, with the exception of the succinimide treatments. However, except for CIOL®, the cell wall bulking reduced during the cycles as the chemicals and degradation products were leached out. The reduction in volume was smaller in the groups containing citric acid. Treatment with imidazole alone led to the smallest dimensions after three cycles, assumed to be due to the degradation of lignin. Groups containing citric acid exhibited higher dimensional stability, indicating some degree of crosslinking in ICA and SuCA. Whether these crosslinks involve imidazole or succinimide is currently uncertain. Succinimide led to a reduction in shrinking but not swelling in comparison to the unmodified wood, indicating that although cell wall bulking occurred, there was no crosslinking.

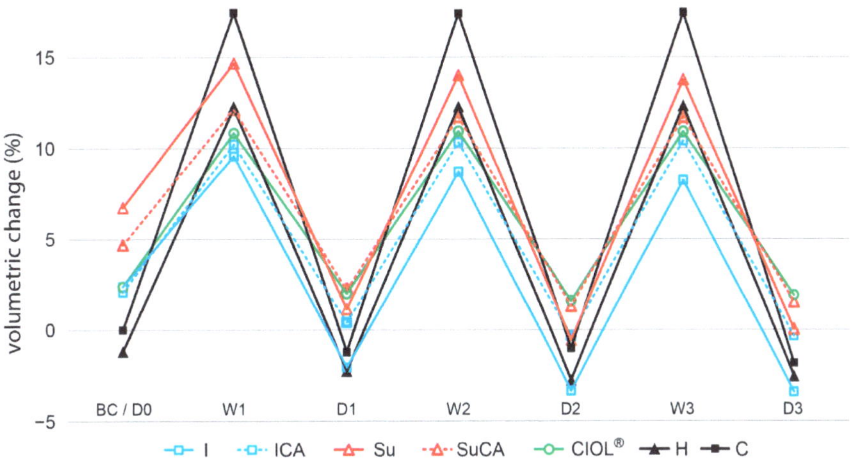

Figure 5. Mean volumetric changes in specimens treated with solutions with 15 wt% solid concentration during wet-dry cycling. Dn: oven-dried in the nth cycle; Wn: wet in the nth cycle.

3.3. Bending Performance

MOR and MOE are presented in Figure 6. The statistical analysis suggested that little significant differences existed between the groups. However, the large span of initial density in each group as a result of the even group distribution and natural variability in wood might be a reason for this, as the mean values exhibited clear differences. The mean MOR of some groups was slightly higher than the untreated control, which might be caused by a reduction in moisture content due to thermal treatment. The I and ICA groups showed high MOR at 5 and 10 wt%, with a reduction in MOR at 15 wt%. The increased degradation of cell wall polymers was likely concentrated in the lignin, as holocellulose degradation is expected to result in greater reductions in mechanical properties due to increased shear slipping between cellulose microfibrils [52]. The SuCA and CIOL® groups showed a reduction in MOR. For CIOL®, this is attributed to the acidic modification conditions that led to thermally induced mechanical degradation [34]. A similar drop in the MOR was found in SuCA but not in ICA. Interestingly, Su5 showed the highest MOR, and Su15 had one of the lowest MORs. It is possible that the combined succinimide and citric acid treatment resulted in a similar formation of crosslinks as in the CIOL® process. The MOE was less affected by the treatments. Except for the succinimide groups, the MOE seemed to be unaffected by WPG, and the slight increase in MOE is likely due to the heat treatment.

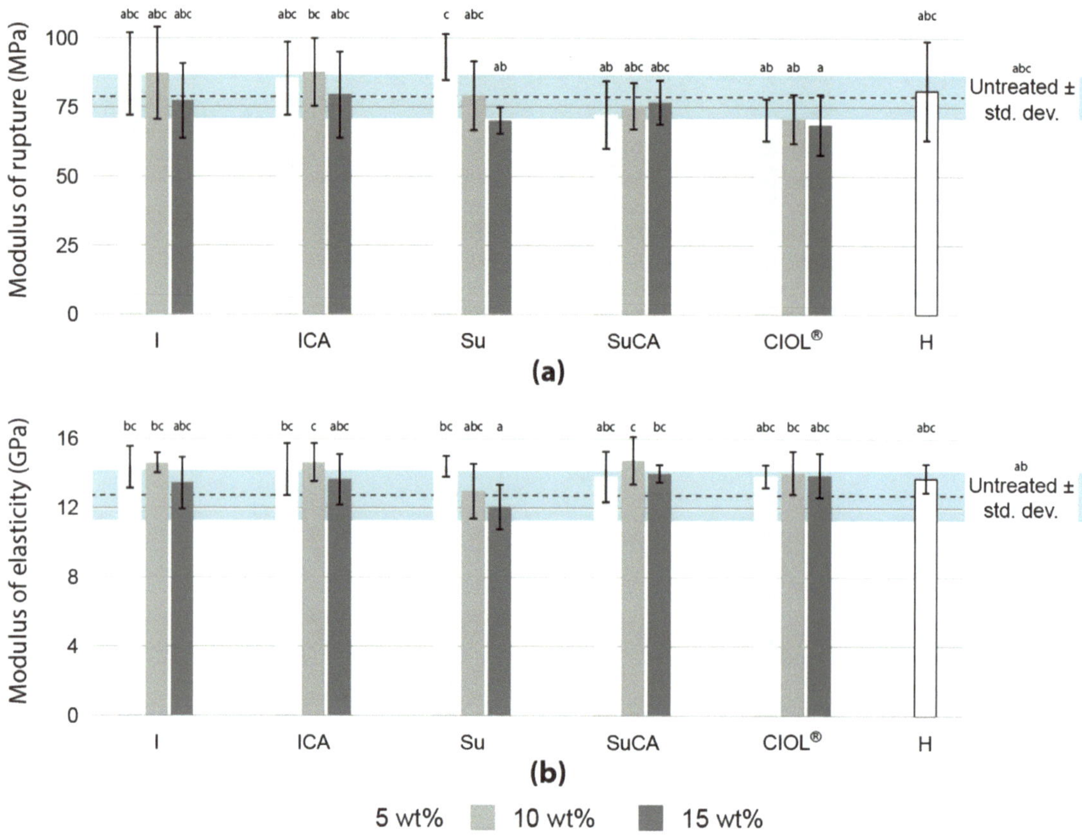

Figure 6. The mean ± standard deviation of the modulus of rupture (**a**) and modulus of elasticity (**b**) were measured for the untreated and treated specimens. wt% corresponds to the solid concentration of the chemical solution used for pressure impregnation. The result of the statistical post-hoc test is presented by the compact letter display above each bar. The number of specimens per group = 12.

The results show that the wood modification with imidazole does not negatively affect bending performance but does provide increased dimensional stability. Both the positive effect on ASE and the negative effect on MOR, are stronger in the succinimide and CIOL® treatments. In-house experience that was not included in this study indicated that some degree of cross-polymerization occurs in I, ICA, and SuCA, with ongoing work to provide definitive answers.

4. Conclusions

In order to facilitate the use of fossil-free modification reagents, this study aimed to explore wood modification systems involving imidazole and succinimide in combination with citric acid. The treatments were based on pressure impregnation and subsequent heat treatment to improve the hygroscopic properties without aggravating the mechanical performance.

The treatments with imidazole exhibited increased mass loss during heat treatment, which led to the formation of water-soluble degradation products that were leached out over the wet-dry cycles. The mass loss during the heat treatment of succinimide-containing treatments seemed to be unaffected by the chemical, and a large amount of succinimide was leachable. The total mass loss from the untreated state to the leached state of the imidazole-treated specimens was 7%–8%, exceeding the sole heat treatment mass loss of 3%. The succinimide treatment exhibited values of 0%–2%, indicating a degree of chemical retention. Substituting a part of the chemicals with citric acid improved the leaching resistance and decreased the total mass loss. It is not clear if this was solely due to the reaction of citric acid with the cell wall polymers, or if crosslinks were formed between citric acid and imidazole or succinimide. The reactions between citric acid and the imidazole-promoted lignin-derived degradation products are another possibility. Leaching resistance was highest for a combination of citric acid and sorbitol (CIOL®-process). However, in this treatment, the share of citric acid was almost twice as high as in the other treatments containing citric acid. After three cycles, the anti-swelling efficiency (ASE) reached 31% for the imidazole-treated specimens and improved to 38% with the addition of citric acid. For succinimide, the ASE increased from 17% to 41%. CIOL® exhibited an ASE of 48%.

Previous research showed that wood modification with citric acid led to a high ASE and high leaching resistance at the cost of strongly reduced mechanical properties. In the current study, the bending properties generally showed an improvement, except for succinimide at a high concentration, succinimide + citric acid, and CIOL®, which displayed a reduced modulus of rupture. This indicates that wood modification with imidazole and succinimide are based on different reaction mechanisms, where the improved hygroscopic properties of imidazole treatment does not solely originate from an enhanced thermal degradation, as the modulus of rupture should be strongly reduced. In contrast, succinimide and citric acid might polymerize with each other and any residual wood components, which is akin to the CIOL® process, as the treatments exhibited a similar reduction in modulus of rupture.

The results of this study show that wood modification with imidazole and succinimide can be utilized. However, further research should delve into analyses of the involved reaction mechanisms to optimize the chemical ratios and leaching resistance. Studies on the impact of imidazole and succinimide on biological durability are currently ongoing.

Author Contributions: Conceptualization, D.J., D.S. and H.D.; methodology, D.J.; validation, D.S., H.D. and J.O.; formal analysis, A.S.; investigation, A.S. and D.J.; data curation, A.S.; writing—original draft preparation, A.S.; writing—review and editing, A.S., D.J. and D.S.; visualization, A.S.; supervision, D.J.; project administration, D.J. and D.S.; funding acquisition, D.J. and D.S. All authors have read and agreed to the published version of the manuscript.

Funding: Support through the project CT WOOD, a center of excellence at Luleå University of Technology, and the VINNOVA project "Multifunktionella byggskivor av sågspån" (Grant no. 2022-00998) is gratefully acknowledged.

Data Availability Statement: The data presented in this study are available on request from the corresponding author. All data is supported in the paper (with statistical variance).

Conflicts of Interest: The authors declare no conflict of interest. The funders had no role in the design of the study; in the collection, analyses, or interpretation of data; in the writing of the manuscript; or in the decision to publish the results.

References

1. Hill, C.A.S. *Wood Modification: Chemical, Thermal and Other Processes*, 1st ed.; Wiley: Hoboken, NJ, USA, 2006; ISBN 978-0-470-02172-9.
2. Sandberg, D.; Kutnar, A.; Karlsson, O.; Jones, D. *Wood Modification Technologies: Principles, Sustainability, and the Need for Innovation*, 1st ed.; CRC Press: Boca Raton, FL, USA; Taylor & Francis Group: Boca Raton, FL, USA; London, UK; New York, NY, USA, 2021; ISBN 978-0-367-76782-2.
3. Esteves, B.M.; Pereira, H.M. Wood Modification by Heat Treatment: A Review. *BioRes* **2008**, *4*, 370–404. [CrossRef]
4. Klauditz, W.; Stegmann, G. Beiträge zur Kenntnis des Ablaufes, und der Wirkung thermischer Reaktionen bei der Bildung von Holzwerkstoffen. *Holz Als Roh- Und Werkstoff* **1955**, *13*, 434–440. [CrossRef]
5. Kollmann, F.; Fengel, D. Änderungen der chemischen Zusammensetzung von Holz durch thermische Behandlung. *Holz Als Roh-Und Werkstoff* **1965**, *23*, 461. [CrossRef]
6. Gérardin, P. New Alternatives for Wood Preservation Based on Thermal and Chemical Modification of Wood—A Review. *Ann. For. Sci.* **2016**, *73*, 559–570. [CrossRef]

7. Mai, C.; Militz, H. Wood Modification. In *Springer Handbook of Wood Science and Technology*; Niemz, P., Teischinger, A., Sandberg, D., Eds.; Springer Handbooks Series; Springer International Publishing: Cham, Switzerland, 2023; pp. 873–910, ISBN 978-3-030-81314-7.
8. Tjeerdsma, B.F.; Militz, H. Chemical Changes in Hydrothermal Treated Wood: FTIR Analysis of Combined Hydrothermal and Dry Heat-Treated Wood. *Holz Roh Werkst* **2005**, *63*, 102–111. [CrossRef]
9. Dirol, D.; Guyonnet, R. Durability by Rectification Process. In Proceedings of the International Research Group Wood Pre, Section 4-Processes, No IRG/WP 93-40015, Orlando, FL, USA, 16–21 May 1993.
10. Tjeerdsma, B.F.; Boonstra, M.; Pizzi, A.; Tekely, P.; Militz, H. Characterisation of Thermally Modified Wood: Molecular Reasons for Wood Performance Improvement. *Holz Als Roh- Und Werkstoff* **1998**, *56*, 149–153. [CrossRef]
11. Weiland, J.J.; Guyonnet, R. Study of Chemical Modifications and Fungi Degradation of Thermally Modified Wood Using DRIFT Spectroscopy. *Holz Als Roh- Und Werkstoff* **2003**, *61*, 216–220. [CrossRef]
12. Boonstra, M.J.; Van Acker, J.; Tjeerdsma, B.F.; Kegel, E.V. Strength Properties of Thermally Modified Softwoods and Its Relation to Polymeric Structural Wood Constituents. *Ann. For. Sci.* **2007**, *64*, 679–690. [CrossRef]
13. Kim, G.; Yun, K.; Kim, J. Effect of Heat Treatment on the Decay Resistance and the Bending Properties of Radiata Pine Sapwood. *Mater. Org.* **1998**, *32*, 101–108.
14. Ohmae, K.; Minato, K.; Norimoto, M. The Analysis of Dimensional Changes Due to Chemical Treatments and Water Soaking for Hinoki (*Chamaecyparis Obtusa*) Wood. *Holzforschung* **2002**, *56*, 98–102. [CrossRef]
15. Verma, P.; Junga, U.; Militz, H.; Mai, C. Protection Mechanisms of DMDHEU Treated Wood against White and Brown Rot Fungi. *Holzforschung* **2009**, *63*, 371–378. [CrossRef]
16. Show, P.L.; Oladele, K.O.; Siew, Q.Y.; Aziz Zakry, F.A.; Lan, J.C.-W.; Ling, T.C. Overview of Citric Acid Production from *Aspergillus niger*. *Front. Life Sci.* **2015**, *8*, 271–283. [CrossRef]
17. Yang, C.Q. FT-IR Spectroscopy Study of the Ester Crosslinking Mechanism of Cotton Cellulose. *Text. Res. J.* **1991**, *61*, 433–440. [CrossRef]
18. Bischof Vukusic, S.; Katovic, D.; Schramm, C.; Trajkovic, J.; Sefc, B. Polycarboxylic Acids as Non-Formaldehyde Anti-Swelling Agents for Wood. *Holzforschung* **2006**, *60*, 439–444. [CrossRef]
19. Zoldners, J.; Kiseleva, T. Modification of Hemicelluloses with Polycarboxylic Acids. *Holzforschung* **2013**, *67*, 567–571. [CrossRef]
20. Fang, G.; Li, J.; Xu, X. The Intermediate of Crosslinking Reaction between Wood and Polycarboxylic Acid. *Sci. Silvae Sin.* **2000**, *36*, 51–54. [CrossRef]
21. Katović, D.; Trajković, J.; Bischof Vukusic, S.; Bogoslav, S. Alternative Agents and Methods for Chemical Modification of Wood. *Drv. Ind.* **2004**, *55*, 175–180.
22. Xie, Y.; Krause, A.; Militz, H.; Turkulin, H.; Richter, K.; Mai, C. Effect of Treatments with 1,3-Dimethylol-4,5-Dihydroxy-Ethyleneurea (DMDHEU) on the Tensile Properties of Wood. *Holzforschung* **2007**, *61*, 43–50. [CrossRef]
23. Lee, S.H.; Md Tahir, P.; Lum, W.C.; Tan, L.P.; Bawon, P.; Park, B.-D.; Osman Al Edrus, S.S.; Abdullah, U.H. A Review on Citric Acid as Green Modifying Agent and Binder for Wood. *Polymers* **2020**, *12*, 1692. [CrossRef]
24. Feng, X.; Xiao, Z.; Sui, S.; Wang, Q.; Xie, Y. Esterification of Wood with Citric Acid: The Catalytic Effects of Sodium Hypophosphite (SHP). *Holzforschung* **2014**, *68*, 427–433. [CrossRef]
25. L'Hostis, C.; Thévenon, M.-F.; Fredon, E.; Gérardin, P. Improvement of Beech Wood Properties by in situ Formation of Polyesters of Citric and Tartaric Acid in Combination with Glycerol. *Holzforschung* **2018**, *72*, 291–299. [CrossRef]
26. Umemura, K.; Sugihara, O.; Kawai, S. Investigation of a New Natural Adhesive Composed of Citric Acid and Sucrose for Particleboard. *J. Wood Sci.* **2013**, *59*, 203–208. [CrossRef]
27. Essoua, G.G.; Blanchet, P.; Landry, V.; Beauregard, R. Pine Wood Treated with a Citric Acid and Glycerol Mixture: Biomaterial Performance Improved by a Bio-Byproduct. *BioResources* **2016**, *11*, 3049–3072. [CrossRef]
28. He, X.; Xiao, Z.; Feng, X.; Sui, S.; Wang, Q.; Xie, Y. Modification of Poplar Wood with Glucose Crosslinked with Citric Acid and 1,3-Dimethylol-4,5-dihydroxy ethyleneurea. *Holzforschung* **2016**, *70*, 47–53. [CrossRef]
29. Berube, M.-A.; Schorr, D.; Ball, R.J.; Landry, V.; Blanchet, P. Determination of in situ Esterification Parameters of Citric Acid-Glycerol Based Polymers for Wood Impregnation. *J. Polym. Environ.* **2018**, *26*, 970–979. [CrossRef]
30. Larnøy, E.; Karaca, A.; Gobakken, L.R.; Hill, C.A.S. Polyesterification of Wood Using Sorbitol and Citric Acid under Aqueous Conditions. *Int. Wood Prod. J.* **2018**, *9*, 66–73. [CrossRef]
31. Guo, W.; Xiao, Z.; Wentzel, M.; Emmerich, L.; Xie, Y.; Militz, H. Modification of Scots Pine with Activated Glucose and Citric Acid: Physical and Mechanical Properties. *BioRes* **2019**, *14*, 3445–3458. [CrossRef]
32. Centolella, A.P.; Razor, B.G. Polyesters of Citric Acid and Sorbitol 1972. U.S. Patent No. 3,661,955, 9 May 1972.
33. Doll, K.M.; Shogren, R.L.; Willett, J.L.; Swift, G. Solvent-Free Polymerization of Citric Acid And D-Sorbitol. *J. Polym. Sci. A Polym. Chem.* **2006**, *44*, 4259–4267. [CrossRef]
34. Mubarok, M.; Militz, H.; Dumarçay, S.; Gérardin, P. Beech Wood Modification Based on in situ Esterification with Sorbitol and Citric Acid. *Wood Sci. Technol.* **2020**, *54*, 479–502. [CrossRef]
35. Beck, G. Leachability and Decay Resistance of Wood Polyesterified with Sorbitol and Citric Acid. *Forests* **2020**, *11*, 650. [CrossRef]
36. Morais, A.R.C.; Pinto, J.V.; Nunes, D.; Roseiro, L.B.; Oliveira, M.C.; Fortunato, E.; Bogel-Łukasik, R. Imidazole: Prospect Solvent for Lignocellulosic Biomass Fractionation and Delignification. *ACS Sustain. Chem. Eng.* **2016**, *4*, 1643–1652. [CrossRef]

37. Grylewicz, A.; Spychaj, T.; Zdanowicz, M. Thermoplastic Starch/Wood Biocomposites Processed with Deep Eutectic Solvents. *Compos. Part A Appl. Sci. Manuf.* **2019**, *121*, 517–524. [CrossRef]
38. Del Menezzi, C.; Amirou, S.; Pizzi, A.; Xi, X.; Delmotte, L. Reactions with Wood Carbohydrates and Lignin of Citric Acid as a Bond Promoter of Wood Veneer Panels. *Polymers* **2018**, *10*, 833. [CrossRef]
39. Patil, M.; Rajput, S. Succinimides: Synthesis, Reaction and Biological Activity. *Int. J. Pharm. Pharm. Sci.* **2014**, *6*, 8–14.
40. Li, X.; Zhang, Z.; Suo, S.; Qiao, Y.; Du, B.; Deng, H.; Hao, Z.; Luo, H. Microstructure of Silver Coating of Cyanide-Free Brush Plating Based on Multicomponent Coordination System. *IOP Conf. Ser. Mater. Sci. Eng.* **2019**, *688*, 033047. [CrossRef]
41. Larnøy, E.; Biørnstad, J.; Treu, A. Advances in CIOL-Protected Wood—The Road Towards Commercialization. In Proceedings of the 17th annual meeting of the Northern European Network for Wood Science and Engineering, Kaunas, Lithuania, 14–15 October 2021; pp. 89–90.
42. Treu, A.; Larnøy, E.; Bjørnstad, J. CIOL®-Protection of Wood—An Update. In Proceedings of the IRG Annual Meeting, IRG/WP 22-40932, Bled, Slovenia, 29 May–2 June 2022.
43. Kim, I.; Karlsson, O.; Jones, D.; Mantanis, G.; Sandberg, D. Dimensional Stabilisation of Scots Pine (*Pinus sylvestris* L.) Sapwood by Reaction with Maleic Anhydride and Sodium Hypophosphite. *Eur. J. Wood Prod.* **2021**, *79*, 589–596. [CrossRef]
44. Stamm, A.J. *Wood and Cellulose Science*; Ronald Press: New York, NY, USA, 1964.
45. Altgen, M.; Willems, W.; Militz, H. Wood Degradation Affected by Process Conditions during Thermal Modification of European Beech in a High-Pressure Reactor System. *Eur. J. Wood Prod.* **2016**, *74*, 653–662. [CrossRef]
46. *CEN. EN 408: 2010 + A1: 2012*; Timber Structures. Structural Timber and Glued Laminated Timber. Determination of Some Physical and Mechanical Properties. European Committee for Standardization: Brussels, Belgium, 2010.
47. Umemura, K.; Ueda, T.; Kawai, S. Characterization of Wood-Based Molding Bonded with Citric Acid. *J. Wood Sci.* **2012**, *58*, 38–45. [CrossRef]
48. Trivedi, M.K.; Dahryn Trivedi, A.B.; Gunin Saikia, G.N. Physical and Structural Characterization of Biofield Treated Imidazole Derivatives. *Nat. Prod. Chem. Res.* **2015**, *3*, 1000187. [CrossRef]
49. Tudorachi, N.; Chiriac, A.P. TGA/FTIR/MS Study on Thermal Decomposition of Poly(succinimide) and Sodium Poly(aspartate). *Polym. Test.* **2011**, *30*, 397–407. [CrossRef]
50. Alén, R.; Kotilainen, R.; Zaman, A. Thermochemical Behavior of Norway Spruce (*Picea abies*) at 180–225 °C. *Wood Sci. Technol.* **2002**, *36*, 163–171. [CrossRef]
51. Wentzel, M.; Altgen, M.; Militz, H. Analyzing Reversible Changes in Hygroscopicity of Thermally Modified Eucalypt Wood from Open and Closed Reactor Systems. *Wood Sci. Technol.* **2018**, *52*, 889–907. [CrossRef]
52. Wang, D.; Fu, F.; Lin, L. Molecular-Level Characterization of Changes in the Mechanical Properties of Wood in Response to Thermal Treatment. *Cellulose* **2022**, *29*, 3131–3142. [CrossRef]

forests

MDPI

Article

Effect of Citric Acid on the Properties of Sapwood of *Pinus sylvestris* Submitted to Thermomechanical Treatment

Matheus Crisostomo [1], Cláudio Del Menezzi [1,*], Holger Militz [2], Katarzyna Kurkowiak [2], Aaron Mayer [2], Luisa Carvalho [3] and Jorge Martins [3]

1 Department of Forest Engineering, Faculty of Technology, University of Brasília, Brasília 70919-970, Brazil; matheuscc50@gmail.com

2 Wood Biology and Wood Products, University of Goettingen, Buesgenweg 4, 37077 Goettingen, Germany; hmilitz@gwdg.de (H.M.); kkurkow@uni-goettingen.de (K.K.); aaronkilian.mayer@uni-goettingen.de (A.M.)

3 Department of Wood Engineering, Polytechnic Institute of Viseu, 3504-510 Viseu, Portugal; lhcarvalho@estgv.ipv.pt (L.C.); jmmartins@estgv.ipv.pt (J.M.)

* Correspondence: cmenezzi@unb.br

Abstract: The present study aimed to evaluate the effect of citric acid on the properties of solid pine wood, which were submitted to thermomechanical treatment. A preliminary test was performed in a previous study to evaluate the influence of different temperatures of pressing and concentrations of citric acid on the physical properties of wood. After choosing the best treatments (170 °C and 5% and 10% of citric acid), the effect of these on the chemical properties (SEM, FTIR and pH); on the density profile using an X-ray microtomography and on the resistance to aging through an abrasion test were evaluated. The result of the chemical analysis showed an increase of the presence of ester functional groups, as well as better coverage and higher acidity of the surface. The density profile technique allowed us to observe the formation of peaks of density on the surface where the acid was applied. Lastly, it was also possible to verify an increase in the resistance to abrasions because of the application of citric acid.

Keywords: citric acid; thermomechanical modification; dimensional stability; solid timber

Citation: Crisostomo, M.; Del Menezzi, C.; Militz, H.; Kurkowiak, K.; Mayer, A.; Carvalho, L.; Martins, J. Effect of Citric Acid on the Properties of Sapwood of *Pinus sylvestris* Submitted to Thermomechanical Treatment. *Forests* **2023**, *14*, 1839. https://doi.org/10.3390/f14091839

Academic Editors: Morwenna Spear and Miklós Bak

Received: 15 August 2023
Revised: 1 September 2023
Accepted: 6 September 2023
Published: 9 September 2023

1. Introduction

Citric acid ($C_6H_8O_7$) is an organic acid present in citric fruits (oranges and lemons), which can be found in an anhydrous form as a white, odorless crystalline powder [1].

According to [2], it is a substance that is already widely used in the food and pharmaceutical sectors. In wood, it has been applied as a crosslinking agent [3], as an adhesive in the manufacture of panels [4] and to improve the dimensional stability of wood composites [5].

In general, the main changes caused in wood by citric acid are attributed to the esterification reactions caused between the carboxyl groups (-COOH) of the acid and the hydroxyl groups (-OH) present in the wood structure [6]. However, studies focusing on the effect of this substance on the properties of wood are still very incipient.

On the other hand, wood modification has already been evaluated by many studies using the technique of Fourier-transform infrared spectroscopy (FTIR), with a focus on the identification of esterification reactions through the evaluation of the peaks of bands 1720–1750 cm^{-1} and 1200–1300 cm^{-1} [7,8], through the technique of scanning electron microscopy, which allows to evaluate the changes provoked on the surface of the material through the analysis of micrographs [9,10], and through the technique of pH, which allows to evaluate alterations to the hydrogen ionic activity in wood [11].

Another technique that has been widely applied in many studies to evaluate changes to wood in a nondestructive manner is X-ray densitometry [12,13], which allows to obtain a density profile in a rapid and automatized way to verify how the distribution of the substance applied occurs inside of the material.

Lastly, ref. [14] verified mixed results for the effect of thermal treatments for the resistance to abrasion, with an increase of the resistance in the radial plane and a reduction in the tangential plane.

In this context, the present study aimed to evaluate the effect of the application of citric acid on densified boards of *Pinus sylvestris*, which were measured through the evaluation of the chemical properties (SEM, FTIR and pH); the density profile and the resistance to abrasion.

2. Materials and Methods

2.1. Wood Samples

Timbers of sapwood of *Pinus sylvestris* with dimensions of 380 by 15 by 2.8 cm (length × width × thickness) and 0.57 g/cm^3 of dry density were obtained at the carpentry of the Faculty of Forest Sciences and Forest Ecology of the Georg August Universität Göttingen.

The timbers were sawn into sixteen boards of 15 by 45 by 2.5 cm, conditioned in a climatization room (20 °C; 65%) to achieve a constant weight. Figure 1 presents the cutting scheme of the samples and the test blocks for the technological characterization.

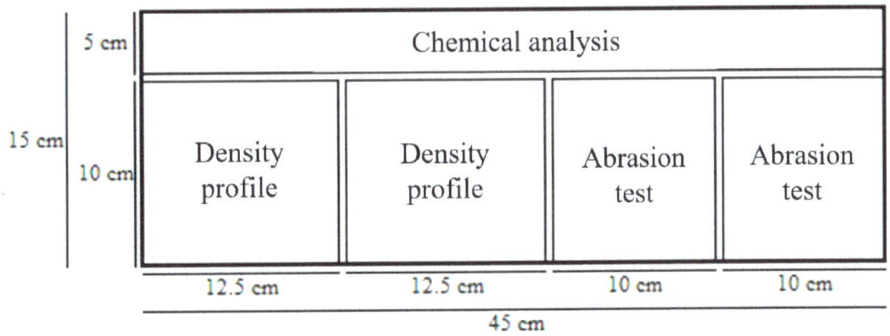

Figure 1. Cutting scheme of the samples of timber of *Pinus sylvestris* for the technological characterization.

2.2. Chemical Modification

The preparation of the solution (CA) was carried out by diluting CA in water at room temperature (+28 °C) in a 1:1 ratio, according to the methodology proposed by [4].

The application of CA was carried out using a paint roller to distribute the solution onto one of the surfaces and subsequent heating in an oven at 60 °C for 6 h, so that CA can react when adding functional chemical groups to the wood. The amount of CA applied was calculated by taking into account the mass surface volume of 450 by 150 mm (length × width) and 2.5 mm in depth. This resulted in average amounts of 0.0096 g/cm^2 and 0.0191 g/cm^2 for the concentrations of 5% and 10% of CA, respectively.

2.3. Thermomechanical Treatment

The application of the thermomechanical treatment on the samples of *Pinus sylvestris* was carried out in the carpentry of the Faculty of Forest Sciences and Forest Ecology, located in the Georg August Universität Göttingen, where it was possible to use a press with the following characteristics: pressing area of 60 by 60 cm; load capacity of 400 kN; electric resistance heating and a control panel for temperature, time and pressure adjustments. The maximum pressure applied was 50% of the estimated perpendicular compression strength of the wood: 2.6 MPa.

A steel platewith dimensions of 60 × 60 × 0.5 cm (length × width × thickness) was placed between the samples to isolate one of the surfaces of the material, allowing to apply the thermomechanical treatment on only one of the sides (Figure 2).

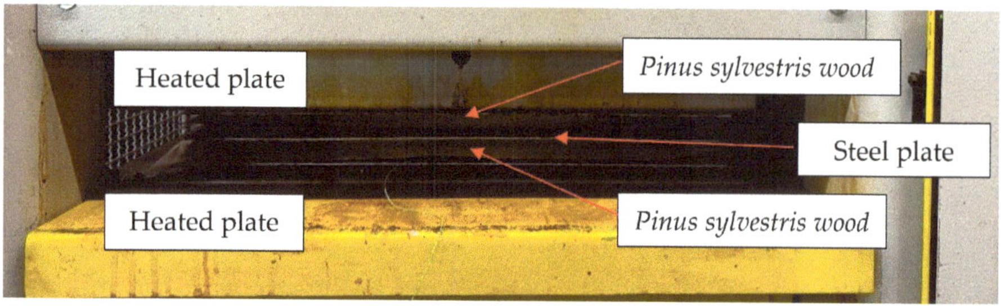

Figure 2. Application of the thermomechanical treatment on timber of *Pinus sylvestris*.

This way, the treatment (Figure 3) was divided into three steps:

I. Heating—initial temperature to treatment temperature (170 °C) and half of the full pressure (1.3 MPa);
II. Treatment—treatment temperature (170 °C) and full pressure (2.6 MPa) over a period of 10 min;
III. Posttreatment—treatment temperature (170 °C) and half-pressure release (1.30 MPa) over a period of 5 min and with all the pressure released over a period of 5 min.

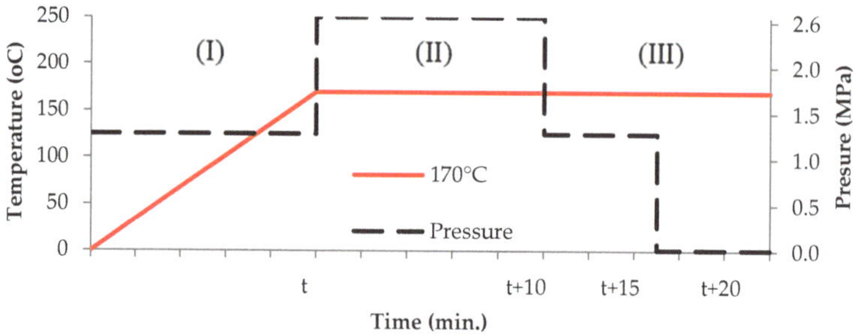

Figure 3. Schedule of the densification process.

The total time of the treatment was 29 min and 20 s.

2.4. Chemical Characterization

2.4.1. Fourier-Transform Infrared Spectroscopy

The spectroscopy test was performed using a BRUKER Alpha spectrophotometer and the mid-infrared range (MIR), which range covers waves from 4000 cm^{-1} to 300 cm^{-1} with a resolution of 1 cm^{-1}, with the aid of a diffuse reflectance accessory. To obtain the samples, it was necessary to obtain small veneers from the material, which were clamped in the spectrometer.

The measurements were carried out on 3 different points per sample for both the control and treated samples, generating spectra in the absorbance mode. This way, using OPUS 7.5 software, the spectra was generated with the average of ten measurements for each treatment. For comparison, the methodology proposed by [15] was used, which consists of the normalization of the average spectra in relation to a band height of 906 cm^{-1}, assigned by [16], to the wagging motion of the hydrogen atom on the C-1 position of the glucose ring in cellulose, which does not suffer alterations after heat application.

2.4.2. Scanning Electron Microscopy

The scanning electron microscopy test was realized using a PHENOMWORLD Phenom XL spectrometer, which allowed to obtain high-resolution images of all samples prior to and after applying the evaluated treatments. The images were obtained using a voltage of 15 kV of tension, a scale of 200 μm and a magnification of 300×. For this, it was necessary to prepare 4 samples ($1 \times 1 \times 1$ cm^3) for each treatment, which were covered by a layer of gold and silver with vacuum impregnation, according to the methodology proposed by [17].

2.4.3. pH

The pH test was performed using an INOLAB 7110 pH meter, which allowed to verify changes to the acidity of the wood prior to and after applying the evaluated treatments. For this, it was necessary to prepare 3 samples ($5 \times 5 \times$ width of the treatment in cm^3).

The pH was measured through the evaluation of the activity of the hydrogen ion, which was measured by placing an electrode on the surface of the sample, according to the following steps:

(I) Deposition of a drop of water over the wood surface;
(II) Placement of the electrode in the wet area;
(III) Measurement of the pH 2 min after the placement of the electrode to allow the stabilization of the parameter.

2.5. Density Profile

The density profile analysis was carried out through X-ray microcomputed tomography (XμCT) using a Grecon density profiler and the mid-infrared range (MIR), which range covers waves from 4000 cm^{-1} to 800 cm^{-1} with a resolution of 1 cm^{-1}, with the aid of a diffuse reflectance accessory.

For this, it was necessary to prepare samples ($50 \times 50 \times$ width of the treatment in mm^3), which were inserted into the machine in a way that the X-ray ran longitudinally through the sample, and the density profile was obtained in the radial direction. The density profile was obtained using a resolution of 0.05 mm and a voltage of 33 kV.

2.6. Abrasion Test

The resistance to abrasion test was performed in accordance with [18] using a Taber Abraser. For this, 5 test blocks of 10 by 10 cm (length × width) for each treatment were exposed to a load of 1000 g, velocity of 60 rotations per minute (rWL) and 1000 cycles.

The wear rate (WR) was calculated in accordance to Equation (1).

$$WR = [(Mi - Mf)/N] \times 100, \tag{1}$$

where WR = wear rate (%), Mi = initial mass (g), Mf = final mass (g) and N = number of cycles.

The abrasion rate (Δt) was calculated in accordance with Equation (2).

$$\Delta t = [(ti - tf)/ti] \times 100, \tag{2}$$

where Δt = abrasion rate (%), ti = initial thickness (mm) and tf = final thickness (mm).

2.7. Data Analysis

The data were evaluated using IBM SPSS 21 (Statistical Package for Social Sciences) software.

The effect of the amount of citric acid on the specific mass and other physical properties in the preliminary test were evaluated using analysis of variance (ANOVA), with a subsequent comparison of the means performed by Tukey's test at the $\alpha = 0.05$ level of significance.

The same criteria were adopted to evaluate the influence of citric acid on wood aging (abrasion test), on the chemical analysis through Fourier-transform infrared reflectance spectroscopy (FTIR) and in the density profile through X-ray microcomputed tomography (XμCT).

3. Results

3.1. Thermomechanical Treatment

Table 1 presents the specific mass of *Pinus sylvestris* before and after the application of the treatments and the average values of the compression rate (CR), densification rate (DR) and mass loss (ML).

Table 1. Densification variables of the treatments tested.

Treatment	Properties				
	DR (%)	CR (%)	ML (%)	Initial Specific Mass (g/cm^3)	Final Specific Mass (g/cm^3)
170 °C $CA_{0\%}$	20.08 a	23.35 a	7.97 b	0.59 a	0.71 a
170 °C $CA_{5\%}$	23.57 ab	25.03 a	7.36 ab	0.56 a	0.70 a
170 °C $CA_{10\%}$	29.17 b	26.98 a	7.18 a	0.58 a	0.74 b

DR: densification rate, CR: compression rate and ML: mass loss. Values followed by the same letter in the same column are not statistically different, according to Tukey's test.

The application of citric acid led to a significant improvement of the DR, with an increase between 17.43% (CA5%) and 45.29% (CA10%) when compared to the samples densified without citric acid.

On the other hand, the CR presented a tendency to increase due to the application of citric acid, with an improvement of 7.21% and 15.70% for the amounts of 5% and 10% of the substance, respectively.

The ML presented a significant difference for all treatments evaluated, with a tendency to reduce between 7.65% and 9.91% as a consequence of the application of citric acid.

Lastly, the specific mass presented a significant difference for the samples treated with $CA_{10\%}$ when compared to the samples densified without citric acid and with $CA_{5\%}$. This improvement was a consequence of a higher reduction in the width of the sample represented by a higher CR (29.17%) and a lesser ML (7.18%) than the other treatments.

3.2. Chemical Analysis

3.2.1. Fourier-Transform Infrared Spectroscopy

Figure 4 shows the spectra for all the evaluated treatments, which were obtained using the Fourier-transform infrared spectroscopy technique. The analysis of these spectra aimed to identify evidence of esterification reactions, considering as a hypothesis a possible link between the hydroxyl groups of wood and carboxyl groups of citric acid.

By analyzing the spectra, it was possible to observe two different patterns by having the spectrum of the untreated sample as a reference. For the densified and 5% of citric acid treatments, a decrease in the height of the peaks was observed, maybe because the amount was not enough to promote a significant change.

On the other hand, it was possible to verify that, for the treatment of 10% of citric acid, the behavior was the opposite, with the spectra presenting higher heights in both spectra (1374 cm^{-1} and 1735 cm^{-1}) when compared to the untreated wood.

Focusing on the effect of citric acid, it was possible to observe that the higher concentration of citric acid (10%) resulted in a higher esterification rate. According to the heights of the peaks in both spectra (1374 cm^{-1} and 1735 cm^{-1}), it is possible to infer not only that there was a greater amount of esterification reactions but also that this higher amount of reactions was able to partially compensate the thermal degradation promoted by the thermomechanical treatment.

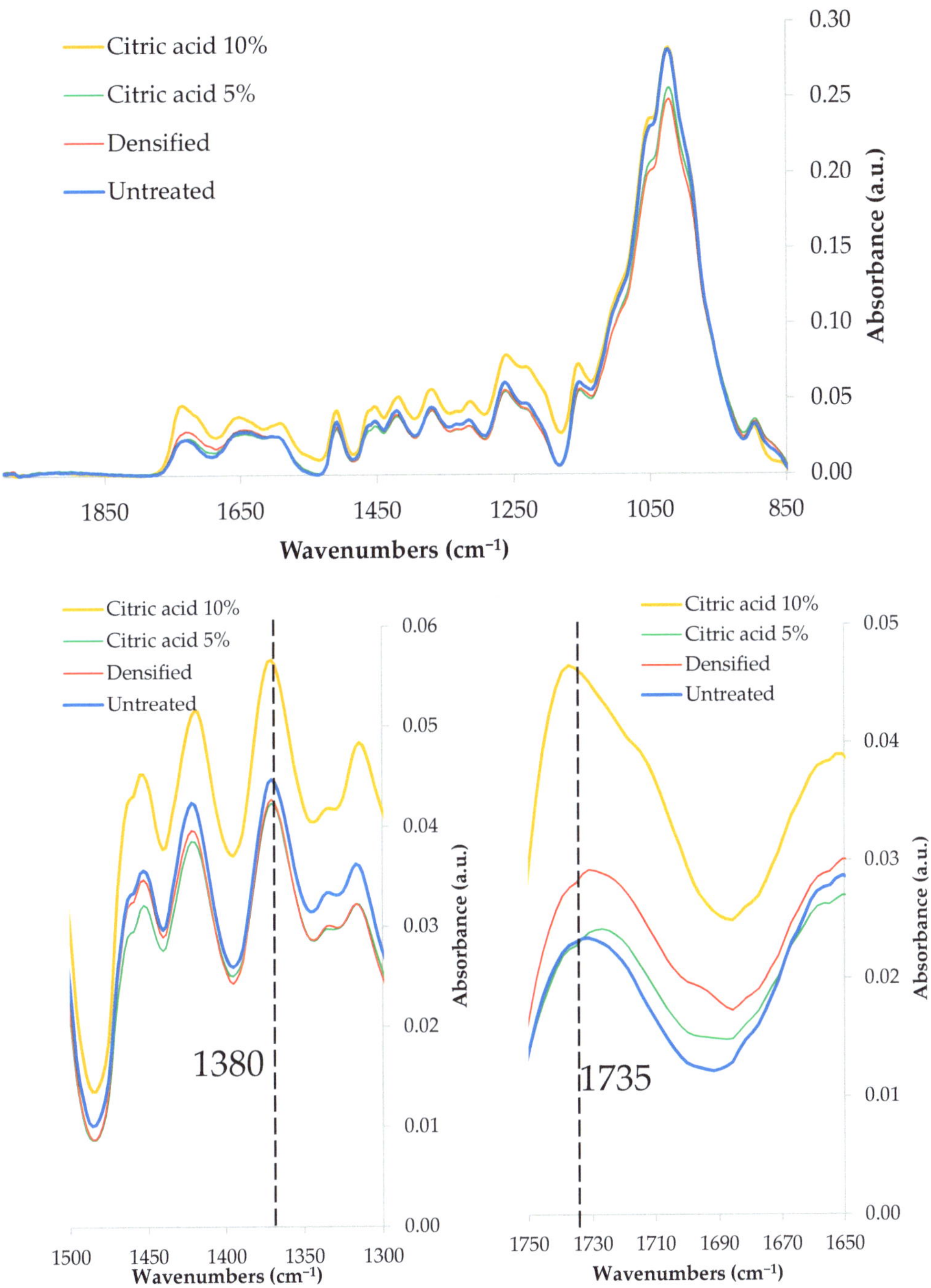

Figure 4. FTIR spectra of *Pinus sylvestris* sapwood for each treatment.

3.2.2. Scanning Electron Microscopy

Figure 5 presents the surface of *Pinus sylvestris* wood for each treatment evaluated, which were obtained using the scanning electron microscopy technique.

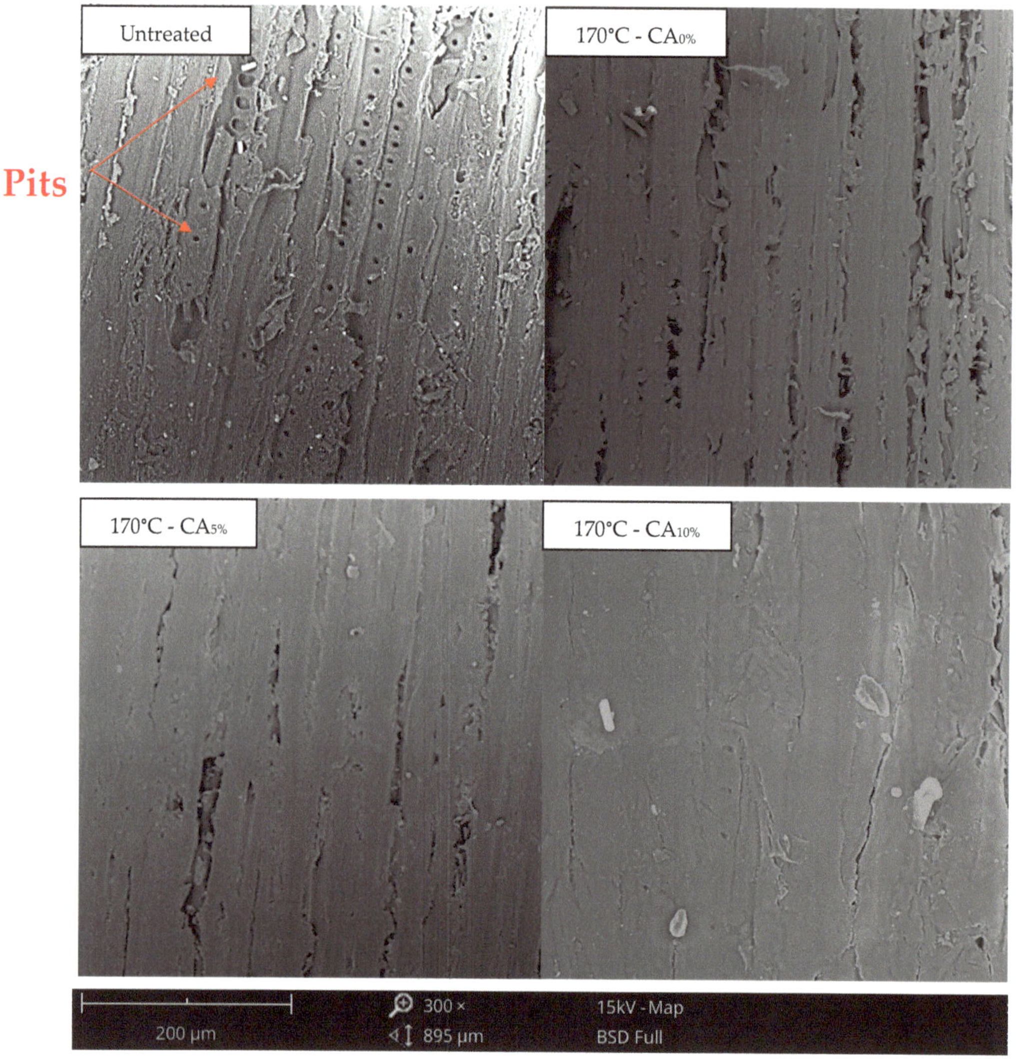

Figure 5. SEM analysis of *Pinus sylvestris* sapwood for each evaluated treatment.

From the pictures, it was possible to verify that the untreated *Pinus sylvestris* wood (control) presented a very irregular and porous surface, with many pits and valleys.

Evaluating the effect of the densification, it was possible to observe that the application of the thermomechanical treatment, as the main consequence, reduced the number of apparent pits.

Regarding the effect of citric acid, it was possible to observe that the substance acted as a film, revesting the surface and reducing considerably the number of valleys when compared to the densified wood without citric acid. Considering the amount of citric acid applied, it was possible to verify that the largest amount of citric acid (10%) presented a better coating when compared to the lowest amount of the substance (5%), so that the valleys were almost completely covered.

3.2.3. pH

Figure 6 presents the average pH of *Pinus sylvestris* for each treatment evaluated.

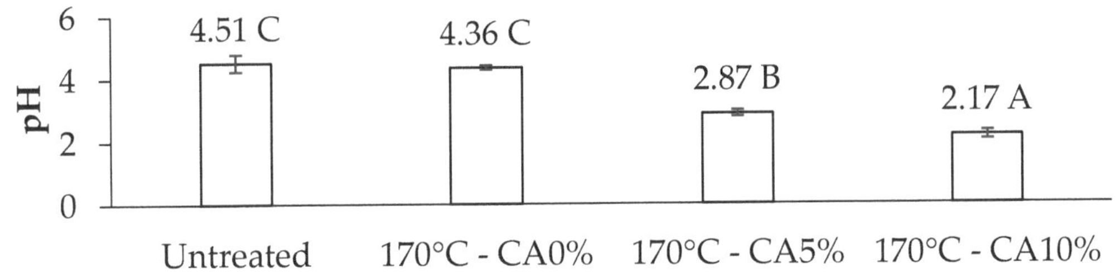

Figure 6. pH of *Pinus sylvestris* sapwood for each evaluated treatment. Values followed by the same letter are not statistically different, according to Tukey's test.

The pH showed a significant difference for all the evaluated treatments. When compared to the untreated wood (control), all the treatments applied promoted a reduction in pH and, consequently, an increase in acidity on the wood surface.

Evaluating the effect of the densification, it was possible to observe that the thermomechanical treatment caused a nonsignificant reduction of the pH (3.52%).

On the other hand, when evaluating the effect of citric acid, it was possible to observe that the combined treatment presented a significant reduction in the pH by 36.36% and 51.86% for the amounts of 5% and 10% of citric acid, respectively.

3.3. Density Profile

Figure 7 presents the density profile of *Pinus sylvestris* sapwood for each treatment evaluated.

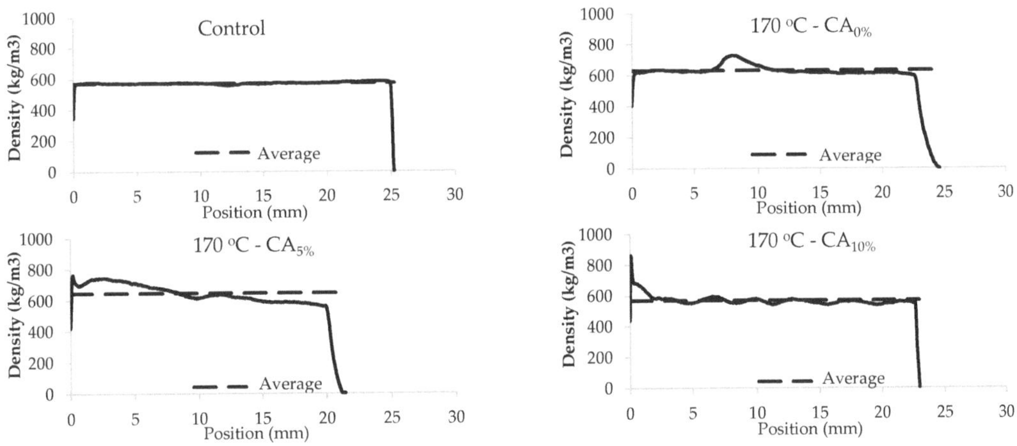

Figure 7. Density profiles of *Pinus sylvestris* sapwood for each treatment.

This way, it was possible to verify that the untreated wood presented a very homogeneous and close to the average line of 0.56 g/cm^3.

Evaluating the effect of the densification, it was possible to verify that the thermomechanical treatment provoked a peak of density (0.76 g/cm^3) close to the position 7.5 mm.

Making a visual analysis of the sample, it was possible to verify that the increase in density was a consequence of the change in the direction of the growth rings.

Lastly, for the treatments of modified citric acid, it was possible to verify a similar pattern for both concentrations (5% and 10%) where the peak of density did not occur in the middle of the sample but in the region where the substance was applied, reaching density peaks of 0.76 g/cm^3 ($CA_{5\%}$) and 0.86 g/cm^3 ($CA_{10\%}$).

3.4. Abrasion Test

Figure 8 presents the abrasion (Δt) and wear rate (WR) of *Pinus sylvestris* for each treatment.

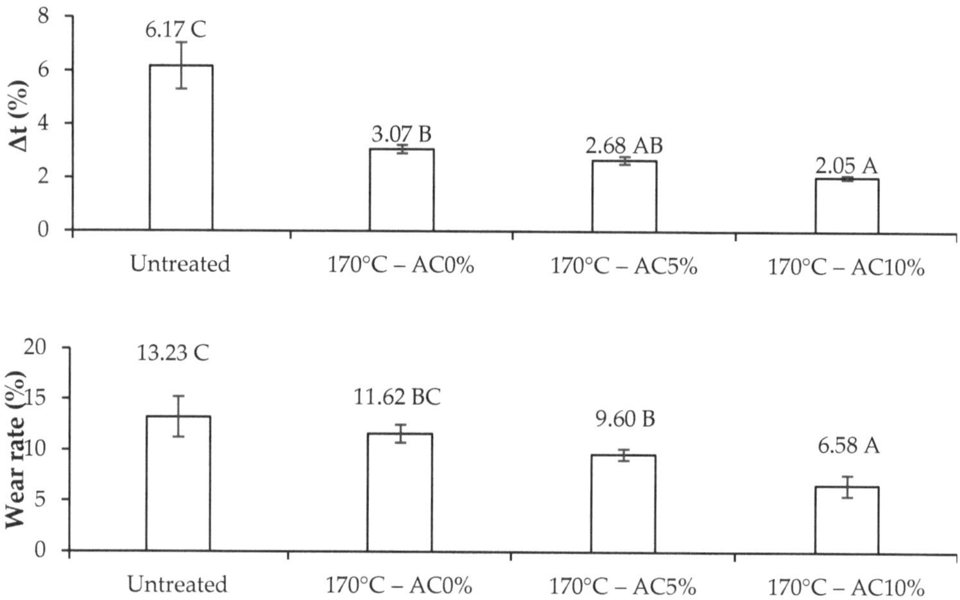

Figure 8. Abrasion and wear rate of *Pinus sylvestris* for each treatment. Values followed by the same letter are not statistically different, according to Tukey's test.

The samples of *Pinus sylvestris* without treatment presented the highest abrasion and wear rate between all the evaluated conditions.

Considering the effect of the thermomechanical treatment, the densified samples presented a significant difference in comparison with the untreated samples, with an average reduction of 50.21% for the Δt and 16.70% for the WR.

On the other hand, considering the effect of citric acid, it was possible to verify a significant increase in the performance of both properties when compared to the untreated and densified samples, with an average reduction of 61.65% for the Δt and 42.00% for the WR.

Additionally, it was possible to verify that the application of a higher amount of citric acid (10%) promoted a better performance of both properties, although significantly different only for the WR.

Lastly, when considering the visual aspect (Figure 9), it was possible to observe that the abrasion was noticeable in all the evaluated treatments. In this context, it was possible to verify that, although citric acid promoted a reduction in the abrasion through a smaller reduction in the thickness and mass loss, the technique applied was not enough efficient to penetrate the deeper layers of the wood.

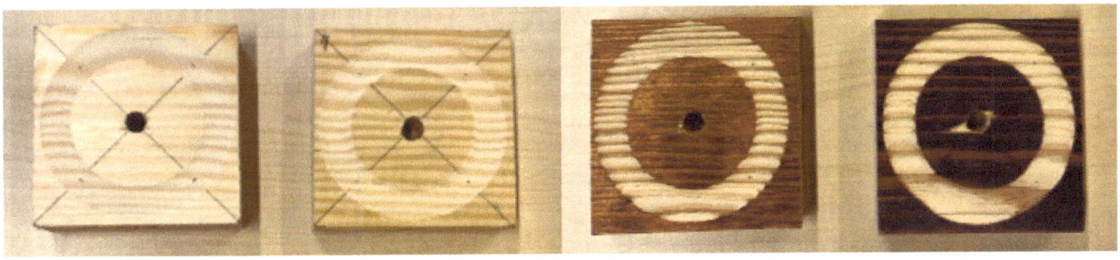

Figure 9. Visual aspect of *Pinus sylvestris* for each treatment after the abrasion test.

4. Discussion

4.1. Thermomechanical Treatment

When considering the effect of the thermomechanical treatment on the densification rate (DR), different results were found by [19,20], who applied thermomechanical treatments on boards of *Pinus sylvestris* and *Pinus elliotti* and verified a DR between 83% and 93%. In this context, ref. [21] stated that an increase in the specific mass of a wood after the application of a thermomechanical treatment may be associated withwith many factors, such as temperature, time and pressure applied.

Therefore, the divergence in the results may be associated withwith the pressure applied during the present study, since the samples were exposed to similar temperatures but with a pressure of 2.6 MPa, while the authors applied pressures of 6 MPa and 4.9 MPa, respectively.

For the effect of the thermomechanical treatment on the compression rate (CR), the results corroborated with [22], who applied a thermomechanical treatment on wood of *Pinus sylvestris* at a temperature of 150 °C for a period of 10 min. According to the same authors, the variation presented by the CR was mainly associated with the presence of internal tensions that were generated inside the cellular structure during the compression and the loss of humidity due to the exposure to high temperatures, both factors being responsible for the springback behavior of the wood after densification.

Lastly, the effect of citric acid on the mass loss matched the results obtained by [10], who verified that the application of citric acid promoted a reduction in the permeability of boards of *Pinus strobus*. Therefore, this behavior suggested that the application of the substance may have acted as a barrier, making it difficult for water to exit during the heating stage (first stage of the thermomechanical treatment), thus reducing the ML presented by the treatments.

4.2. Chemical Analysis

4.2.1. Fourier-Transform Infrared Spectroscopy

In this context, emphasis was given to two specific peaks: one between the 1725 cm^{-1} and 1740 cm^{-1} bands, related to the stretching of carbonyl groups (C=O) in the esters of wood polymers and citric acid [23], and the other one found in the 1374 cm^{-1} spectra, which, according to [24], is related to the polysaccharides (C-H bonds of CH3 groups) that are deformed in carbohydrates.

According to [25], who applied thermal treatments on wood of *Pinus pinaster*, the reduction in the peaks was associated with the thermal degradation of polysaccharides and weight loss presented in the wood as a consequence of the exposure of the material to high temperatures (170 °C).

This result corroborated with [25,26], who verified an increase in the esterification reactions after applying citric acid on wood panels of *Sorghum bicolor* L. Monech and *Vitis vinifera* L., respectively.

4.2.2. Scanning Electron Microscopy

The result of the untreated samples corroborated with [27], who evaluated the microstructure of *Pinus sylvestris* sapwood and observed that it is a species with a very porous and irregular surface, with many apparent pits, characteristics that are typical of low-density pinewood.

For the effect of the thermomechanical treatment, the results corroborated with [9], who evaluated the effect of densification on the microstructure of *Populus tomentosa* wood and verified a reduction in the number of pits per unity of area. A similar result was obtained by [28], who applied a thermomechanical treatment on *Pinus sylvestris* wood and verified a reduction in the wood porosity.

In this context, ref. [29] stated that the densification of wood through thermomechanical treatment led to the occurrence of morphological changes such as the reduction of pits and valleys, which were compressed by the pressure applied on the press.

Finally, for the effect of citric acid, the observed results corroborated with [10], who applied citric acid on *Pinus strobus* and *Pinus contorta* wood and verified a considerable reduction in the porosity of both species. According to [30], this behavior was a consequence of the capacity of citric acid to establish chemical bonds between the carboxyl groups (-COOH) available on its structure and the free hydroxyl groups (-OH) available on the wood structure, a process that is known as ester bonds.

4.2.3. pH

The *Pinus sylvestris* sapwood without treatment presented an average pH of 4.51, a result very close to that observed by [31], who verified a pH of 4.6 for untreated *Pinus roxburghii* wood.

The effect of the thermomechanical treatment differed from [11], who verified a significant reduction in the pH (3.5–4) of *Pinus radiata* wood after the application of thermal treatments due to the occurrence of reactions that led to the production of acetic and formic acids. In this context, the different results might have been associated with the temperature chosen (170 °C), which might have been insufficient to promote greater changes in the wood surface.

Lastly, the effect of citric acid corroborated with [32], who verified a reduction in the pH of *Fagus sylvatica* wood because of the application of citric acid. According to [33], such a reduction in pH observed in wood is the result of esterification reactions between the hydroxyls of the material and the carboxyls of citric acid, thus generating a new structure with a greater availability of protons.

4.3. *Density Profile*

Several studies with *Pinus sylvestris* have discovered that wood of this species can present foreign bodies such as knots [34] and resin bags [35], capable of significantly increasing the density of wood. Therefore, the obtained results suggest that the samples of *Pinus sylvestris* sapwood did not contain such bodies.

Additionally, it was not possible to verify any difference between earlywood and latewood in the obtained density profile, differing from the results obtained by [36], who were able to verify this difference in *Pinus sylvestris* wood using the X-ray tomography technique. However, this result might be associated with the resolution applied to the equipment, which might have not been sufficient to detect such differences.

For the effect of the thermomechanical treatment, it was possible to observe that the increase in density did not occur homogeneously throughout the sample but due to a peak in the density that was 20% higher than the average curve (0.63 g/cm^3). This result corroborated with [37], who applied a thermomechanical treatment on *Pinus sylvestris* and verified the formation of density peaks along the sample.

According to [38], who applied a thermomechanical treatment on boards of *Populus tomentosa*, the lower density in the surface of the sample might have been associated with the loss of moisture that occurred when the wood was in contact with the hot plates of the press.

Lastly, the effect of citric acid corroborated with [13], who evaluated the effect of citric acid applied using an impregnation chamber on the density profile of sapwood of *Pinus sylvestris* and verified the formation of peaks next to the edge of the sample, suggesting a higher accumulation of the substance on the surface of the material.

In addition, these results suggest that the citric acid might have acted as a barrier, preventing the exit of moisture present in the innermost layers of the material and, consequently, increased the density of the most superficial layer. Such behavior corroborates a study carried out by [10], who verified a reduction in the permeability of *Pinus strobus* L wood through citric acid impregnation.

4.4. Abrasion Test

The results for the abrasion test on the untreated samples of *Pinus sylvestris* corroborated with [38], who verified an abrasion rate of 6.00% for the sapwood of *Pinus sylvestris* without any treatment.

For the effect of the thermomechanical treatment, a similar result was observed by [39], who registered an increase in the resistance to abrasions after applying a thermomechanical treatment on boards of *Paulownia* spp., with a reduction in weight loss between 40% and 75% due to the pressure applied.

According to [40], the resistance to abrasions might be associated with the specific mass of the material, so that higher-density materials also tend to present a greater resistance to abrasions.

Lastly, for the effect of citric acid, it was not possible to find any results in the literature evaluating the effect of this substance on the abrasion resistance of wood. However, when considering the effect of chemical modifications on wood, ref. [41] verified an increase in the abrasion resistance after promoting an acetylation reaction on boards of *Fagus sylvatica*, and [42] obtained similar results by promoting a furfurylation reaction on boards of *Pinus radiata*.

5. Conclusions

The combination between the thermomechanical treatment and citric acid presented significant improvements in the chemical and surface properties, in the density profile and in the resistance to abrasions of *Pinus sylvestris* sapwood.

The application of a higher concentration of citric acid (10%) might be justified by the better results presented in the chemical analysis, in the density profile and in the abrasion resistance of *Pinus sylvestris* sapwood.

Lastly, when considering the visual aspect, the application of citric acid provoked an intense darkening of the surface for the same temperature applied, changing its color from a light yellow to dark brown.

Author Contributions: Conceptualization, M.C., C.D.M. and H.M.; methodology, C.D.M., H.M., L.C., J.M., K.K. and A.M.; software, M.C., K.K. and A.M.; validation, M.C., K.K. and A.M.; formal analysis, C.D.M., H.M., L.C. and J.M.; investigation, M.C.; writing—original draft preparation, M.C.; writing—review and editing, M.C.; supervision, C.D.M. and H.M. and funding acquisition, C.D.M. and H.M. All authors have read and agreed to the published version of the manuscript.

Funding: This research was funded by the Coordenação de Aperfeiçoamento de Pessoal de Nível Superior (CAPES) (Project 88882.387147/2019-01) and Deutscher Akademischer Austauschdienst (DAAD) agencies (Project 57552433).

Data Availability Statement: The data presented in this study are available on request from the corresponding author.

Conflicts of Interest: The authors declare no conflict of interest. The funders had no role in the design of the study; in the collection, analyses or interpretation of the data; in the writing of the manuscript or in the decision to publish the results.

References

1. Lee, S.H.; Tahir, M.D.P.; Lum, W.C.; Tan, L.P. A review on citric acid as green modifying agent and binder for wood. *Polymers* **2020**, *12*, 1692. [CrossRef] [PubMed]
2. Ciriminna, R.; Meneguzzo, F.; Delisi, R.; Pagliaro, M. Citric acid: Emerging applications of key biotechnology industrial product. *Chem. Cent. J.* **2017**, *11*, 22. [CrossRef] [PubMed]
3. Vukusic, S.B.; Katovic, D.; Schramm, C.; Trajkovic, J. Polycarboxylic acids as non-formaldehyde anti-swelling agents for wood. *Holzforschung* **2006**, *60*, 439–444. [CrossRef]
4. Del Menezzi, C.H.S.; Amirou, S.; Pizzi, A.; Xi, X. Reactions with wood carbohydrates and lignin of citric acid as a bond promoter of wood veneer panels. *Polymers* **2018**, *10*, 833. [CrossRef]
5. Amirou, S.; Pizzi, A.; Delmotte, L. Citric acid as waterproofing additive in butt joints linear wood welding. *Eur. J. Wood Wood Prod.* **2017**, *75*, 651–654. [CrossRef]
6. Essoua, G.G.E.; Blanchet, P.; Landry, V.; Beauregard, R. Pine wood treated with a citric acid and glycerol mixture: Biomaterial performance improved by a bio-byproduct. *Bioresources* **2016**, *11*, 3049–3072. [CrossRef]
7. Kusumah, S.S.; Umemura, K.; Yoshioka, K.; Miyafuji, H. Utilization of sweet sorghum bagasse and citric acid for manufacturing of particleboard I: Effects of pre-drying treatment and citric acid content on the board properties. *Ind. Crops Prod.* **2016**, *84*, 34–42. [CrossRef]
8. Umemura, K.; Ueda, T.; Kawai, S. Characterization of wood-based molding bonded with citric acid. *J. Wood Sci.* **2012**, *58*, 38–45. [CrossRef]
9. Bao, M.; Huang, X.; Jiang, M.; Yu, W. Effect of thermo-hydro-mechanical densification on microstructure and properties of poplar wood (*Populus tomentosa*). *J. Wood Sci.* **2017**, *63*, 591–605. [CrossRef]
10. Berube, M.A.; Schorr, D.; Ball, R.J.; Landry, V. Determination of in situ esterification parameters of citric acid-glycerol based polymers for wood impregnation. *J. Polym. Environ.* **2018**, *26*, 970–979. [CrossRef]
11. Boonstra, M.; Van Acker, J.; Kegel, E.; Stevens, M. Optimisation of a two-stage heat treatment process: Durability aspects. *Wood Sci. Technol.* **2007**, *41*, 31–57. [CrossRef]
12. Gaitán-Alvarez, J.; Berrocal, A.; Lykidis, C.; Moya, R. Furfurylation of tropical wood species with and without silver nanoparticles: Part II: Evaluation of wood properties. *Wood Mater. Sci. Eng.* **2021**, *18*, 112–119. [CrossRef]
13. Kurkowiak, K.; Emmerich, L.; Militz, H. Sorption behavior and swelling of citric acid and sorbitol (SorCA) treated wood. *Holzforschung* **2021**, *75*, 1136–1149. [CrossRef]
14. Aytin, A.; Korkut, S.; As, N.; Ünsal, Ö. Effect of heat treatment of wild cherry wood on abrasion resistance and withdrawal capacity of screws. *Drv. Ind. Znan. Časopis Za Pitanja Drv. Tehnol.* **2015**, *66*, 297–303. [CrossRef]
15. Pastore, T.C.M. *Studies of the Effect of Ultraviolet Radiation on Wood by RAMAN (FT-RAMAN), Diffuse Reflectance in the Infrared (DRIFT) and Visible (CIE-L*a*b*) Spectroscopy*; Universidade de Brasília: Brasília, Brazil, 2014.
16. Horn, B.A.; Qiu, J.; Owen, N.L.; Feist, W.C. FT-IR studies of weathering effects in western redcedar and southern pine. *Appl. Spectrosc.* **1994**, *48*, 662–668. [CrossRef]
17. Chaves, B.C.F.; Ferreira, K.M.; Dias Júnior, A.F.; Silva, V.P.C.; Silva, T.M.T.; Severo, T.S.; Valadão, D.S. Impact of pyrolysis temperature on the properties of eucalyptus wood-derived biochar. *Materials* **2020**, *13*, 5841. [CrossRef]
18. *Astm-D4060*; Standard Test Method for Abrasion Resistance of Organic Coatings by the Taber Abraser. ASTM International: West Conshohoken, PA, USA, 1995.
19. Ulker, O.; Imirzi, O.; Burdurlu, E. The effect of densification temperature on some physical and mechanical properties of Scots pine (*Pinus sylvestris* L.). *BioResources* **2012**, *7*, 5581–5592. [CrossRef]
20. Pertuzzatti, A.; Missio, A.L.; Cademartor, P.H.G.; Santini, E.J. Effect of process parameters in the thermomechanical densification of *Pinus elliottii* and *Eucalyptus grandis* fast-growing wood. *BioResources* **2018**, *13*, 1576–1590. [CrossRef]
21. Kutnar, A.; Šernek, M. Densification of wood. *Zb. Gozdarstva Lesar.* **2007**, *82*, 53–62.
22. Pelit, H.; Sönmez, A.; Budakçı, M. Effects of ThermoWood® process combined with thermo-mechanical densification on some physical properties of Scots pine (*Pinus sylvestris* L.). *BioResources* **2014**, *9*, 4552–4567. [CrossRef]
23. Brum, S.S.; Bianchi, M.L.; Silva, V.L.; Gonçalves, M. Preparação e caracterização de carvão ativado produzido a partir de resíduos do beneficiamento do café. *Química Nova* **2008**, *31*, 1048–1052. [CrossRef]
24. Michell, A.; Higgins, H. *Infrared Spectroscopy in Australian Forest Products Research*; CSIRO Forestry: Melbourne, Australia, 2002.
25. Esteves, B.; Velez Marques, A.; Domingos, I.; Pereira, H. Chemical changes of heat treated pine and eucalypt wood monitored by FTIR. *Maderas Cienc. Y Tecnol.* **2013**, *15*, 245–258. [CrossRef]
26. Santos, J.; Pereira, J.; Escobar-Avello, D.; Ferreira, I. Grape Canes (*Vitis vinifera* L.) Applications on Packaging and Particleboard Industry: New Bioadhesive Based on Grape Extracts and Citric Acid. *Polymers* **2022**, *14*, 1137. [CrossRef]
27. Barman, D.N.; Haque, M.; Hossain, M.; Paul, S.K. Deconstruction of pine wood (*Pinus sylvestris*) recalcitrant structure using alkali treatment for enhancing enzymatic saccharification evaluated by Congo red. *Waste Biomass Valoriz.* **2020**, *11*, 1755–1764. [CrossRef]
28. Laine, K.; Segerholm, K.; Wålinder, M.; Rautkari, L. Wood densification and thermal modification: Hardness, set-recovery and micromorphology. *Wood Sci. Technol.* **2016**, *50*, 883–894. [CrossRef]

29. Ahmed, S.A.; Morén, T.; Hagman, O.; Cloutier, A. Anatomical properties and process parameters affecting blister/blow formation in densified European aspen and downy birch sapwood boards by thermo-hygro-mechanical compression. *J. Mater. Sci.* **2013**, *48*, 8571–8579. [CrossRef]
30. Šefc, B.; Trajković, J.; Hasan, M.; Katović, D.; Bischof, S.; Martina, F. Dimensional stability of wood modified by citric acid using different catalysts. *Drv. Ind.* **2009**, *60*, 23–26.
31. Poonia, P.K.; Tripathi, S. Effect of microwave heating on pH and termite resistance of *Pinus roxburghii* wood. *Maderas Cienc. Y Tecnol.* **2018**, *20*, 499–504. [CrossRef]
32. Mubarok, M.; Militz, H.; Dumarçay, S.; Gérardin, P. Beech wood modification based on in situ esterification with sorbitol and citric acid. *Wood Sci. Technol.* **2020**, *54*, 479–502. [CrossRef]
33. Feng, X.; Xiao, Z.; Sui, S.; Wang, Q. Esterification of wood with citric acid: The catalytic effects of sodium hypophosphite (SHP). *Holzforschung* **2014**, *68*, 427–433. [CrossRef]
34. As, N.; Goker, Y.; Dundar, T. Effect of Knots on the physical and mechanical properties of scots pine. *Wood Res.* **2006**, *51*, 51–58.
35. Kozakiewicz, P.; Jankowska, A.; Mamiński, M.; Marciszewska, K.; Ciurzycki, W.; Tulik, M. The wood of scots pine (*Pinus sylvestris* L.) from post-agricultural lands has suitable properties for the timber industry. *Forests* **2020**, *11*, 1033. [CrossRef]
36. Campbell, R.; Mccarroll, D.; Loader, N.J.; Grudd, H. Blue intensity in *Pinus sylvestris* tree-rings: Developing a new palaeoclimate proxy. *Holocene* **2007**, *17*, 821–828. [CrossRef]
37. Belt, T.; Rautkari, L.; Laine, K.; Hill, C.A. Cupping behaviour of surface densified Scots pine wood: The effect of process parameters and correlation with density profile characteristics. *J. Mater. Sci.* **2013**, *48*, 6426–6430. [CrossRef]
38. Tu, D.; Su, X.; Zhang, T.; Fan, W. Thermo-mechanical densification of *Populus tomentosa* var. tomentosa with low moisture content. *BioResources* **2014**, *9*, 3846–3856. [CrossRef]
39. Brischke, C.; Ziegeler, N.; Bollmus, S. Abrasion resistance of thermally and chemically modified timber. *Drv. Ind. Znan. Časopis Za Pitanja Drv. Tehnol.* **2019**, *70*, 71–76. [CrossRef]
40. Li, H.; Jiang, X.; Ramaswamy, H.S.; Zhu, S. High-pressure treatment effects on density profile, surface roughness, hardness, and abrasion resistance of *Paulownia* wood boards. *Trans. ASABE* **2018**, *61*, 1181–1188. [CrossRef]
41. Swaczyna, I.; Kedzierski, A.; Tomusiak, A.; Cichy, A.; Różańska, A.N.; Policińska-Serwa, A.N. Hardness and wear resistance tests of the wood species most frequently used in flooring panels. *For. Wood Technol.* **2011**, *76*, 82–87.
42. Liu, M.; Lyu, S.; Cai, L.; Peng, L. Performance improvement of radiata pine wood by combining impregnation of furfuryl alcohol resin and densification for making fretboard materials. *Ind. Crops Prod.* **2021**, *172*, 114029. [CrossRef]

forests

Article

Improvement of the Dimensional Stability of Rubber Wood Based on the Synergies of Sucrose and Tung Oil Impregnation

Chunwang Yang [1], Susu Yang [1], Huanxin Yang [1], Buapan Puangsin [2] and Jian Qiu [3,*]

[1] Yunnan Provincial Key Laboratory of Wood Adhesives and Glued Products, Southwest Forestry University, Kunming 650224, China; yangchunwang24@163.com (C.Y.); yangsu1102@163.com (S.Y.); 15264202895@163.com (H.Y.)

[2] Department of Forest Products, Faculty of Forestry, Kasetsart University, Bangkok 10900, Thailand; fforbpp@ku.ac.th

[3] International Joint Research Center for Biomass Materials, Southwest Forestry University, Kunming 650224, China

* Correspondence: qiujian@swfu.edu.cn; Tel.: +86-137-5951-2363

Abstract: Rubber wood often exhibits dimensional instability during use, which seriously hinders its widespread application. In order to enhance the dimensional stability of rubber wood, a two-step method was employed in this study to modify rubber wood using two plant-derived compounds, namely sucrose and tung oil. Samples treated alone with sucrose or tung oil were also prepared. The water absorption, dimensional stability, and thermal stability of modified and untreated wood were evaluated. The results show that wood samples treated with 30% sucrose and tung oil had excellent water resistance and dimensional stability based on the synergistic effect of sucrose and tung oil. After 384 h of immersion, the 30% sucrose and tung oil group presented a reduction in water absorption by 76.7% compared to the control group, and the anti-swelling efficiency was 57.85%, which was 66.81% higher than that of the tung oil treatment alone. Additionally, the leaching rate of the 30% sucrose and tung oil group decreased by 81.27% compared to the sample modified with the 30% sucrose solution alone. Simultaneously, the 30% sucrose and tung oil group showed better thermal stability. Therefore, this study demonstrates that the synergistic treatment of modified rubber wood by sucrose and tung oil is an eco-friendly, economical, and highly efficient approach with the potential to expand the range of applications of rubber wood products.

Keywords: dimensional stability; sucrose; tung oil; wood modification; rubber wood

Citation: Yang, C.; Yang, S.; Yang, H.; Puangsin, B.; Qiu, J. Improvement of the Dimensional Stability of Rubber Wood Based on the Synergies of Sucrose and Tung Oil Impregnation. *Forests* **2023**, *14*, 1831. https://doi.org/10.3390/f14091831

Academic Editors: Miklós Bak and Morwenna Spear

Received: 8 August 2023
Revised: 5 September 2023
Accepted: 6 September 2023
Published: 8 September 2023

1. Introduction

The rubber tree (*Hevea brasiliensis*) is a fast-growing and versatile tree with dual economic significance, primarily cultivated for collecting rubber [1,2]. The primary regions for rubber tree cultivation and production in China are Hainan Province and Xishuangbanna, Yunnan Province [3]. Rubber wood is a by-product of rubber trees, which is taken from the trunk of rubber trees. Due to its light color, hardness, and ease of processing, rubber wood finds frequent applications in furniture manufacturing and interior decorations [3–5]. Nevertheless, one of the main drawbacks of rubber wood is its dimensional instability, which can be easily affected by environmental humidity during use, leading to dimensional deformation and seriously affecting its use [5–8]. Therefore, it is very necessary to modify rubber wood to enhance its properties and expand its range of applications.

Various methods have been researched and compared to enhance the dimensional stability and other properties of wood, mainly including impregnation treatment, chemical modification treatment, and heat treatment [9]. Among them, the methods of the impregnation modification and chemical modification of wood mainly include acetylation [10], furfurylation [11], esterification [12], silylation [13], grafting polymerization [14], paraffin modification [15], resin impregnation [16], and so on. Although these methods can enhance the properties of wood, the majority of modifiers contain hazardous chemicals or solvents that can cause serious environmental and health problems. For instance, wood treated with phenolic (PF) resins and melamine-urea-formaldehyde (MUF) resins release formaldehyde, phenols, and other harmful substances that restrict their use in indoor applications [17]. In contrast, high-temperature heat treatment avoids using chemicals, while treated wood does not release harmful substances during use. Many studies have demonstrated the efficacy of heat treatment in enhancing wood properties, including dimensional stability and resistance to biological degradation [18–21]. However, heat-treated wood may reduce both mass and volume, which can lead to a reduction in the actual recovery rate of wood and a decrease in long-term resistance to water absorption [22]. Shukla et al. discovered that the weight loss of rubber wood became more pronounced with an increasing heat treatment temperature, and heat treatment reduces of its mechanical strength [23]. With the growing awareness of sustainable development, developing an environmental-friendly, sustainable, efficient, and economical wood modifier is urgently needed.

The primary treatment agent utilized in this experiment is sucrose (S), which exists in almost the entire plant kingdom. It is mainly derived from sugarcane and sugar beets and has the advantages of easy access, low price, and high purity [24]. Sucrose is extensively utilized in various chemical and process industries such as fine chemicals, bioethanol, surfactants, etc. [25–27]. Previously, sucrose was mainly used to protect archaeological wooden artifacts [28]. Parrent used sucrose to safeguard the archaeological timber in Port Royal, Jamaica, and achieved favorable outcomes [29]. However, there has been limited research on the utilization of sucrose-modified wood for enhancing the properties of wood. In recent years, Petr et al. have discovered that vacuum impregnation of poplar wood with an aqueous mixture of sucrose and sodium chloride can increase the anti-swelling efficiency of poplar wood by up to 36%, indicating that sucrose is a highly effective and environmentally friendly agent for modifying wood [30]. Nevertheless, because sucrose is a polyhydroxy compound, it is difficult to fix in wood, and its high leachability is a major problem.

Tung oil (TO), a traditional Chinese wood oil extracted from the tung tree's seeds, consists of alpha oleic acid, oleic acid, and linoleic acid [31]. It has been extensively used for safeguarding wooden structures and furnishings for over a millennium due to its biodegradable nature and non-toxicity towards humans [32]. As a dry oil, TO can quickly dry in the air and polymerize into tough, smooth, and transparent films. Even after undergoing severe aging, tung oil still maintains excellent water resistance [33,34]. Moreover, several researchers have reported that impregnating wood with TO could enhance its dimensional stability, decay resistance, and weathering resistance [35,36]. Both S and TO treatments can improve certain properties of wood; however, currently, no research has focused on the combined effects of S and TO on the properties of rubber wood, mainly including dimensional stability and thermal stability.

In this study, a two-step method was applied to modify Xishuangbanna rubber wood using sucrose (S) and tung oil (TO) as treatment agents. In the first step, the wood was impregnated with different concentrations of an S solution. Then, impregnation of TO strengthens S fixation. Simultaneously, wood samples that were untreated and separately treated with S or TO were also prepared for comparative analysis. The effects of S and TO on the moisture absorption, water absorption, dimensional stability, color, and thermal stability of the rubber wood were evaluated. Furthermore, the microstructural and chemical structural changes of the rubber wood before and after modifications were analyzed through the utilization of scanning electron microscopy (SEM) and Fourier transform infrared spectra (FTIR). This work may provide an economical and effective approach to improving the durability of rubber wood.

2. Materials and Methods

2.1. Materials

Wood samples were collected from the sapwood of rubber wood in Xishuangbanna, Yunnan Province. After being air-dried, the lumbers were cut into pieces of 20 × 20 × 20 mm^3 (Longitudinal × Tangential × Radial, eight replicates for each test) according to the Chinese standard (GB/T 1927.2-2021, 2021) [37] for the measurement of the weight percent gain (WPG), water absorption rate (WA), equilibrium moisture content (EMC), volumetric swelling rate (VSR), anti-swelling efficiency (ASE), and leaching rate (LR). In addition, samples with dimensions of 20 × 50 × 50 mm^3 (Longitudinal × Tangential × Radial) were prepared for measuring color changes. The samples were dried to a constant weight at 80 °C and then divided into different groups. Sucrose (S) and tung oil (TO) were bought from Xilong Scientific Co., Ltd. (Shenzhen, China) and Gushi Anshan Tung Oil Sales Co., Ltd. (Xinyang, China), respectively.

2.2. Sample Preparation

The wood samples were modified using a two-step method. The samples were prepared following Table 1, resulting in twelve sets. In the first step, the wood samples were impregnated with aqueous solutions of S at concentrations of 10, 20, 30, 40, and 50% (wt/wt) and then marked S10, S20, S30, S40, and S50, respectively. The impregnation method was employed in accordance with the full cell process. Specifically, the samples were subjected to a vacuum of −0.1 MPa for 30 min, followed by the introduction of the S solution and an increased pressure to 0.6 MPa for 90 min. Afterward, the wood samples were removed and dried in an oven at 80 °C until reaching a stable weight. In the second step, the S-treated samples were impregnated by TO using an identical process. After removing the samples, the surface of the samples was wiped with a tissue to remove the excess TO on the surface. To obtain a uniformly flat cured surface, the samples were subjected to further oven drying at 80 °C until their weight remained constant. The samples treated with the S and TO two-step methods were marked as S10-TO, S20-TO, S30-TO, S40-TO, and S50-TO, respectively. Additionally, the wood specimens treated by TO only were also prepared and marked as TO. The untreated wood specimens were used as the control group and marked as C. The weight percent gain (WPG) was calculated, which included WPG_1 (for the first step), WPG_2 (for the second step), and WPG_T (for both two steps).

Table 1. Parameters of different modification treatments and weight percent gains (WPGs) of control (C), samples modified with different concentrations of sucrose (S10, S20, S30, S40, and S50), samples modified with tung oil (TO), and samples modified with both sucrose and tung oil (S10/S20/S30/S40/S50 + TO).

Groups	Treatment		WPG (%)			Density (g/cm^3)
	1st Step	2nd Step	WPG_1	WPG_2	WPG_T	
C	/	/	/	/	/	0.54 [f] (0.05)
S10	10%S	/	7.78 [e] (0.84)	/	/	0.59 [e] (0.02)
S20	20%S	/	17.98 [d] (1.99)	/	/	0.62 [e] (0.05)
S30	30%S	/	31.14 [c] (2.76)	/	/	0.75 [d] (0.09)
S40	40%S	/	44.37 [b] (2.32)	/	/	0.76 [d] (0.04)
S50	50%S	/	58.13 [a] (2.11)	/	/	0.85 [b] (0.04)
TO	/	TO	/	52.72 [a] (3.45)	/	0.80 [cd] (0.06)
S10-TO	10%S	TO	7.78 (0.84)	50.61 [ab] (1.57)	58.39 [c] (2.05)	0.84 [bc] (0.05)
S20-TO	20%S	TO	17.98 (1.99)	45.77 [ab] (2.96)	63.75 [b] (3.88)	0.88 [ab] (0.03)
S30-TO	30%S	TO	31.14 (2.76)	41.79 [b] (3.02)	72.93 [ab] (2.92)	0.89 [ab] (0.03)
S40-TO	40%S	TO	44.37 (2.32)	32.01 [c] (3.07)	76.38 [a] (5.74)	0.91 [a] (0.05)
S50-TO	50%S	TO	58.13 (2.11)	21.34 [d] (3.31)	79.47 [a] (6.07)	0.92 [a] (0.05)

Data are provided as the average (standard deviation) from replicates; different small letters represent a significant difference ($p < 0.05$) for different treatments.

2.3. Density

The density (specific gravity) of the samples was measured using the common water displacement method [38]. The density was measured by an electronic balance densitometer (XFMD-12001A, Lichen, Shanghai, China). Density can be calculated by Equation (1):

$$\text{Density} = \text{Mass}/\text{Volume} \tag{1}$$

2.4. Dimensional Stability Test

2.4.1. Moisture Sorption and Dimensional Stability Test

The samples were placed in a constant temperature and humidity chamber at 25 °C and 65% relative humidity (RH) for two weeks. The equilibrium moisture content (EMC), volumetric swelling rate (VSR), and anti-swelling efficiency (ASE) were calculated according to Equations (2)–(4).

$$\text{EMC}(\%) = (m_r - m_1)/m_1 \times 100 \tag{2}$$

$$\text{VSR}(\%) = (V_r - V_1)/V_1 \times 100 \tag{3}$$

$$\text{ASE}(\%) = (\text{VSR}_c - \text{VSR}_m)/\text{VSR}_c \times 100 \tag{4}$$

where m_r (g) is the weight of the sample at the hygroscopic equilibrium condition, m_1 (g) is the weight of the sample in the oven-dry condition before hygroscopic absorption, V_r (cm^3) is the volume of the sample at the hygroscopic equilibrium condition, and V_1 (cm^3) is the volume of the sample in the oven-dry condition before hygroscopic absorption. VSR_m (%) and VSR_c (%) are the mean VSRs in the modified and control groups, respectively.

2.4.2. Water Absorption and Dimensional Stability Test

The oven-dried wood specimens were submerged in deionized water at room temperature, with the liquid level maintained 50 mm above the surface of the sample. After being immersed for 1, 3, 6, 12, 24, 48, 96,192, and 384 h respectively, these specimens were taken out, and the water was wiped off the surface. The water absorption rate (WA), water absorption volumetric swelling rate (VSR_{WA}), and water absorption anti-swelling efficiency (ASE_{WA}) were calculated according to Equations (5)–(7).

$$\text{WA}_n(\%) = (m_n - m_2)/m_2 \times 100 \tag{5}$$

$$\text{VSR}_{WA}(\%) = (V_n - V_2)/V_2 \times 100 \tag{6}$$

$$\text{ASE}_{WA}(\%) = (\text{VSRW}_{Am} - \text{VSR}_{WAc})/\text{VSR}_{WAc} \times 100 \tag{7}$$

where m_n (g) and m_2 (g) represent the weight of the sample after n hours of immersion and the drying weight of the sample before the immersion test, respectively. V_n (cm^3) and V_2 (cm^3) represent the volume of the sample after n hours of immersion and the volume of the sample before the immersion test, respectively. The VSR_{WA} values of the modified and control samples are represented by VSR_{WAm} (%) and VSR_{WAc} (%), respectively.

2.5. Leachability Test

After being dried, the modified samples were immersed in deionized water for 384 h (with the water changed every 48 h) and were subsequently removed and dried at a temperature of 103 °C until the weight was constant. The leaching rate (LR) was calculated according to Equation (8).

$$\text{LR}(\%) = (m_m - m_3)/(m_m - m_0) \times 100 \tag{8}$$

where m_0 (g) and m_m (g) are the mass of the sample in the oven before and after modifications, respectively. m_3 (g) is the oven-dry weight of the sample after 384 h of immersion.

2.6. Color Changes

The International Commission on Illumination CIE's (1976) [39] L*a*b* standard chromaticity system was used to characterize the colors. The precision colorimeter (NH145, 3 nh, Shenzhen, China) was used to measure the L*, a*, and b* of all samples. Three measurements were taken at different locations on each surface for two tangential sections of the test samples. The measured data were used to calculate ΔL* (luminance difference), Δa* (color index difference on the red and green axes), Δb* (color index difference on the yellow and blue axes), and ΔE* (total color difference) according to the formula of the L*a*b* color system (CIE 1976) according to Equations (9)–(12).

$$\Delta L^* = {L^*}_m - {L^*}_c \tag{9}$$

$$\Delta a^* = {a^*}_m - {a^*}_c \tag{10}$$

$$\Delta b^* = {b^*}_m - {b^*}_c \tag{11}$$

$$\Delta E = \sqrt{(\Delta L^*)^2 + (\Delta a^*)^2 + (\Delta b^*)^2} \tag{12}$$

where ${L^*}_c$, ${a^*}_c$, and ${b^*}_c$ are the values measured for the control group; ${L^*}_m$, ${a^*}_m$, and ${b^*}_m$ are the values measured after the modification, respectively.

2.7. Characterizations and Microscopic Observations

2.7.1. Scanning Electron Microscope (SEM) Analysis

The surface morphologies of the samples were evaluated using a scanning electron microscope (SEM) (TESCAN MIRA LMS, Czech Republic) to investigate changes in physical structure. The samples were affixed directly onto the conductive adhesive and coated with Au using a Quorum SC7620 sputter coater at 10 mA. Subsequently, imaging of the samples was photographed using a TESCAN MIRA LMS scanning electron microscope with an accelerating voltage of 3 kV.

2.7.2. Fourier Transform Infrared (FTIR) Analysis

Fourier transform infrared (FTIR) spectra (Thermo Scientific Nicolet iS50, USA) was employed to characterize the changes in the chemical structures of the wood samples. A powder sample of 1.5 mg and pure KBr of 200 mg were finely ground and pressed into transparent sheets using a hydraulic press. The samples were then subjected to infrared spectroscopy with a wave number range of 400–4000 cm^{-1}, a scan number of 32, and a resolution of 4 cm^{-1}.

2.7.3. Thermogravimetric (TG) Analysis

The thermal stability of the wood samples was evaluated using thermogravimetric analysis (TG) before and after treatment. The thermal stability of each sample was assessed using a thermogravimetric analyzer (Netzsch TG 209 F1, Germany) under a nitrogen atmosphere with a heating rate of 10 °C/min and a final temperature of 600 °C.

2.8. Statistical Analysis

An Analysis of Variance (ANOVA) test was conducted using IBM SPSS Statistics (V 0.23) software to evaluate the effects of different modification treatments on the properties of the samples. The Least Significant Difference (LSD) method was further employed to identify the significant difference between every mean value at $p < 0.05$.

3. Results and Discussion

3.1. Weight Percent Gain and Density

Weight percent gain (WPG) is a significant parameter for assessing wood modifications, which was used to evaluate the S and TO uptake. As illustrated in Table 1, the WPG_1 of the specimens treated by S at concentrations of 10, 20, 30, 40, and 50% (wt/wt) were 7.78, 17.98, 31.14, 44.37, and 58.13%, respectively. WPG_1 increases with the rise of S concentration and exhibits a positive correlation with S concentration. The statistical analysis shows that there were significant differences between different concentrations of S and WPG_1. Due to the small molecular nature of S, it easily permeates into the wood samples during impregnation. As water evaporates, S can penetrate and deposit within the wood samples. The WPG_2 of the TO-impregnated samples alone was 52.72%, indicating that TO could penetrate into the wood and occupy its voids. The WPG_2 of S10-TO, S20-TO, S30-TO, S40-TO, and S50-TO were 50.61, 45.77, 41.79, 32.01, and 21.34%, respectively. The WPG_2 value exhibits a negative correlation with the concentration of S, and the WPG_2 values of the S30-TO, S40-TO, and S50-TO groups exhibited significant differences compared to the WPG_2 of the TO group. This may be attributed to the deposition of S inside the wood during the first impregnation process, which hinders TO penetration and consequently causes the reduction of the WPG_2 value. In addition, density is also an important indicator for the evaluation of wood properties and has a significant impact on the strength and processing properties of wood. As shown in Table 1, the density of the samples treated solely with 10 to 50% of the S solution exhibited varying degrees of increase compared to the untreated samples. After the impregnation of TO alone, the density of the samples increased by 48.15% compared to the control group. And the density of the samples treated with both different concentrations of S and TO increased significantly, ranging from 57.41% to 70.37%, compared to the untreated specimens. Furthermore, statistical analyses show significant differences in density in all modified groups compared to the control group.

3.2. *Dimensional Stability Analysis*

3.2.1. Moisture Sorption and Dimensional Stability

The dimensional stability of wood is one of the most important parameters in wood applications, including its moisture absorption and water absorption state. In the former case, wood with low dimensional stability may shrink or swell as the EMC decreases or increases in the surrounding environment [40]. Therefore, the dimensional stability of wood seriously affects the quality and application of wood products.

The EMCs and ASE values are shown in Figure 1 to evaluate the hygroscopicity and hygroscopic dimensional stability of the wood in the control and modified groups. After treatment with different concentrations of the S solution alone, the EMC was reduced to varying degrees. The EMC of the S10, S20, S30, S40, and S50 groups exhibited reductions of 5.75%, 11.19%, 21.68%, 22.74%, and 25.13%, respectively, when compared to the control group (9.28%). The EMC decreased continuously as the concentration of S increased; however, the observed differences in EMC between the S30, S40, and S50 groups were not statistically significant, probably indicating that the sucrose entering the wood had already reached saturation. Wood is hygroscopic because the cellulose chains have many hydroxyl groups which are bound to atmospheric oxygen. S can establish hydrogen bonds with these exposed hydroxyl groups, thereby reducing the number of hydroxyl groups that are bound to atmospheric moisture within the wood. At relative humidity levels above 85%, S exhibits a high degree of hygroscopicity, whereas, at lower humidity levels, it diminishes the wood's hygroscopic properties [29]. After treatment with TO, the EMC of the samples was significantly reduced compared to the untreated wood. However, no significant difference in EMC was observed between the TO group and the S10-TO group. The EMC values for the S30-TO, S40-TO, and S50-TO treated samples were 4.13%, 4.15%, and 4.11% respectively, representing a remarkable decrease of approximately 55% compared to the control group. Additionally, the statistical analyses suggested significant differences in the EMC of the S30-TO, S40-TO, and S50-TO groups compared to the TO group.

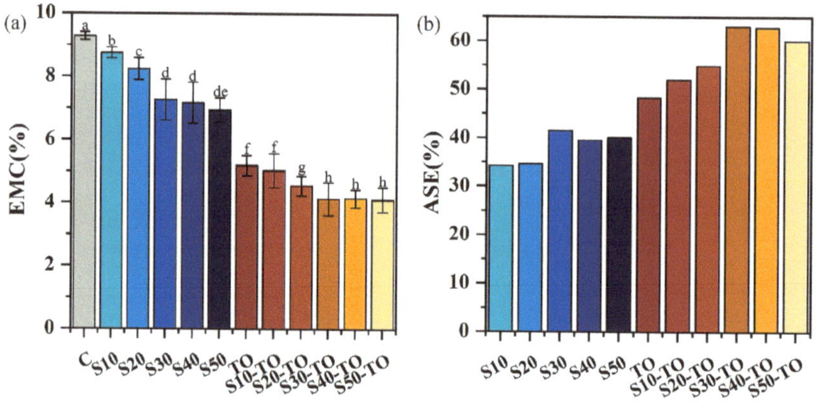

Figure 1. (**a**) Equilibrium moisture content (EMC) of control group and different treatment groups at equilibrium state of 25 °C, 65% RH. (**b**) Anti-swelling efficiency (ASE) of different treatment groups at equilibrium state of 25 °C, 65% RH. Different small letters represent significant differences ($p < 0.05$) for different treatments.

The ASEs of the modified samples are shown in Figure 1b. Both the S and TO impregnations exhibited a positive impact on enhancing dimensional stability. When impregnated with the S solution alone, S30 exhibited the best effect with an ASE of 41.57%. After the introduction of TO, the ASE values of TO, the S10-TO, S20-TO, S30-TO, S40-TO, and S50-TO groups were 48.34%, 52%, 54.98%, 63.10%, 62.90%, 60.04%, respectively. The ASE of the S30-TO group exhibited a significant increase of 30.5% compared to the sole impregnation of TO, thereby indicating a remarkable synergistic effect achieved through the combined utilization of both S and TO.

3.2.2. Water Absorption and Dimensional Stability

As a natural polymer material, wood contains a large number of free hydroxyl groups, which are very easy to combine with water molecules in the form of hydrogen bonds [41]. Simultaneously, wood has a multitude of void structures, which can store a large amount of water molecules, and this is one of the reasons why wood is very easy to absorb water. To investigate the water absorption values of wood treated with S and TO, the WA values of all the wood specimens were computed and are illustrated in Figure 2.

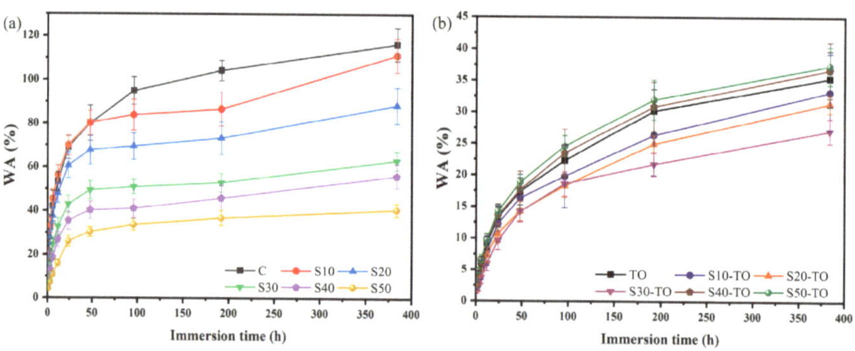

Figure 2. (**a**) Water absorption rate (WA) of control group and S10, S20, S30, S40, S50 groups. (**b**) Water absorption rate (WA) of TO group and S10-TO, S20-TO, S30-TO, S40-TO, S50-TO groups.

The WA exhibited an increase with a prolonged immersion time. The untreated wood exhibited a WA of 116.37% after being immersed for 384 h. For the S-treated samples (Figure 2a), the WA values for S10, S20, S30, S40, and S50 were 111.29, 88.52, 63.12, 56.03, and 40.62%, respectively. These results indicate that the presence of S effectively inhibits water penetration into the wood. The addition of S swelled the cell wall of the wood and blocked some of the pits [30,42], thereby impeding water movement within the cell wall and subsequently reducing WA. In comparison, the WA of the TO group exhibited a sharp decline to 35.38% after soaking for 384 h (Figure 2b), indicating its excellent water repellency. Humar and Lesar reported that TO can effectively reduce the water absorption of wood [43]. However, the samples treated by both S and TO showed different WA trends. As shown in Figure 2b, after introducing tung oil, the WA of the samples showed a significant reduction. WA reduction rates of 71.4%, 73.1%, and 76.7% were observed in the S10-TO, S20-TO, and S30-TO treatments, respectively, after 384 h of immersion compared to the control group. Noteworthily, the S30-TO group exhibited the lowest WA of 27.1% after immersing for 384 h. Based on previous studies, the WA values of wood modified with epoxidized linseed oil, carnauba wax, and rosin were 129%, 92%, and 87%, respectively [44,45]. Thus, our treatment exhibited superior water resistance properties compared to natural water repellents. However, the WA of the S40-TO group and S50-TO group were 36.72% and 37.78%, respectively, which were slightly higher than that of the TO group. This situation may be due to the lower WPG_2 values of the S40-TO and S50-TO groups because TO plays a more important role in improving water resistance performance.

Chemical modification decreases the water absorption of wood and enhances its dimensional stability. ASE_{WA} is primarily utilized for assessing the dimensional stability of wood following water immersion. The ASE_{WA} for the different modification groups is shown in Figure 3. The ASE_{WA} of the samples treated with the S solutions of different concentrations alone exhibited a significant decrease from 1 to 48 h after immersion in water. After 48 h of soaking, the ASE_{WA} tended to stabilize. The ASE_{WA} of the samples treated with S30, S40, and S50 remained stable at approximately 34% after being immersed in water for 384 h. S can easily penetrate the cell wall and forms hydrogen bonds with cellulose molecules. At the same time, S can swell the cell wall, reduce the stretching deformation of the cell wall, and ultimately enhance the dimensional stability of wood [29,42].

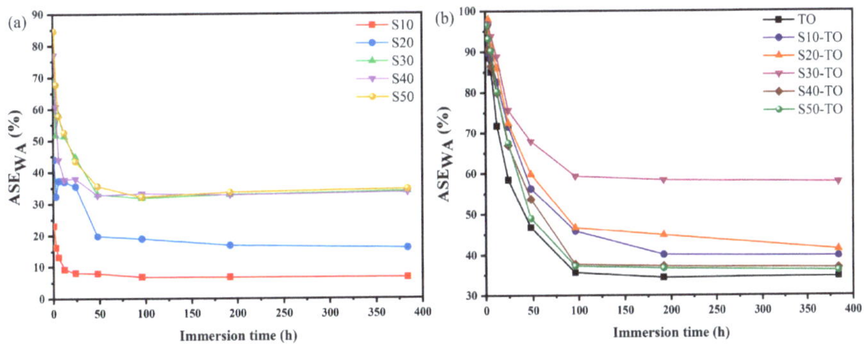

Figure 3. (**a**) Water absorption anti-swelling efficiency (ASE_{WA}) of S10, S20, S30, S40, S50 groups. (**b**) Water absorption anti-swelling efficiency (ASE_{WA}) of TO group and S10-TO, S20-TO, S30-TO, S40-TO, S50-TO groups.

Additionally, treatment with TO demonstrates a significantly greater improvement in dimensional stability. After soaking for 6 h, the ASE_{WA} of all the TO-treated samples exceeded 80%. When treated with TO alone, ASE_{WA} was 71.77% at 12 h and decreased to 34.68% at 384 h. This finding is consistent with the studies conducted by He et al. and Lee et al. [21,35], which suggest that oil uptake and the formation of a protective layer on the wood are the main factors in increasing the dimensional stability of wood. As previously mentioned, tung oil has the ability to fill cell gaps, deeply penetrate the cell wall, and reduce the number of available water-uptaking sites. Therefore, it improves the deformation resistance of wood. By comparison, the two-step S and TO treatments demonstrated superior dimensional stability due to the synergistic effect of S and TO impregnation. Particularly, for S30-TO, the ASE_{WA} still reached 75.66% after 24 h of immersion. The ASE_{WA} was 57.85% after 384 h of immersion, representing a 66.81% improvement compared to treatment with TO alone. These results suggest that the implementation of composite modifications is highly effective in reducing water absorption and enhancing dimensional stability. In conclusion, the samples treated with the 30% S solution and TO demonstrate the best dimensional stability, and the ASE_{WA} results are in agreement with WA.

3.3. Leachability Analysis

In high-humidity environments, such as bathrooms, it is essential to fix modifiers used in wood products to prevent leaching. This is due to the fact that the absence of modifiers will significantly diminish both the water resistance and dimensional stability of wood. The leaching rate (LR) of the wood samples treated with 30% S and/or TO are shown in Table 2. After 384 h of immersion, the LR of the untreated wood was 0.45%, which should be related to the extractives of the wood. This suggests that the testing process rarely affects the leaching of wood component contents. The LR of the S30 group was 94.32%. Because S is a polyhydroxy compound, it was easy to be washed away by water in the wood, and its fixation was poor. The LR of TO was 1.22%, indicating little leaching of TO during the experiment. The LR of S30-TO was 17.66%, which was 81.27% lower than S30, significantly reducing the leaching rate of the S. Furthermore, the statistical analyses demonstrated the significant efficiency of TO on the fixation of S in wood. TO has a positive fixation effect on S in wood as it forms a solidified film on the inner surface of the wood. Meanwhile, the high retention of effective modifiers contributed to the wood maintaining long-term water resistance and dimensional stability.

Table 2. Leaching rate (LR) of the C, S30, TO, and S30-TO groups.

Samples	Leaching Rate (%)
C	0.45^{a} (0.13)
S30	94.32^{b} (3.37)
TO	1.22^{c} (0.61)
S30-TO	17.66^{d} (2.68)

Data are provided as the average (standard deviation) from replicates; different small letters represent significant differences ($p < 0.05$) for different treatments.

3.4. Color Changes

The surface color of the modified wood exhibited varying degrees of change compared to the control wood. The colors of the TO and S30-TO groups were similar, as illustrated in Figure 4a. Moreover, Figure 4b illustrates the alterations in the L*, a*, and b* values of the rubber wood subjected to various treatments. For color parameters including L*, a*, and b*, the lightness of all the modified wood samples decreased, with the L* values of S30, TO, and S30-TO decreasing by 9.89, 19.89, and 20.03%, respectively, compared to the control wood. This decrease in the L* value is indicative of a darker coloration.

The a* value of S30 exhibited a 69.17% increase, while those of TO and S30-TO demonstrated respective increases of 111.85% and 110.8%. The increase in the a* value signifies an inclination for the wood surface to exhibit a red hue. The b* values of the S-30, TO, and S30-TO groups exhibited increases of 42.88, 81.28, and 77.10%, respectively, indicating a propensity for yellowing on the wood surface. The impact of modifications on the total color difference is also demonstrated in Figure 4b. The total color difference of S30, TO, and S30-TO increased from 0 to 25. The surface colors of TO and S30-TO were yellow or dark yellow, which may be due to the formation of a yellow film after TO solidification [46], resulting in a darker color of the modified sample. However, the solidified TO film does not cover the natural grain of the wood and maintains its aesthetic characteristics.

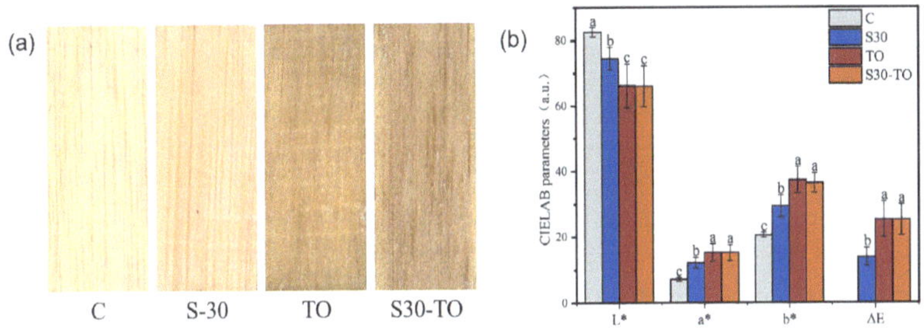

Figure 4. (**a**) Appearance of the C, S30, TO, and S30-TO groups. (**b**) Color parameters of the C, S30, TO, and S30-TO groups. Different small letters represent significant differences ($p < 0.05$) for different treatments.

3.5. Morphology and Structure

The microstructures of the control and modified wood samples are shown in Figure 5. By comparing the cross and tangential sections of the control group, all three treatment methods altered the natural cell morphology of the wood to a certain degree. Observing the cross and tangential sections of S30 reveals that sucrose can easily enter the interior of the wood and can block some pits. Many sucrose particles can be observed on S30-T. After treatment with TO, the wood cell lumens were almost completely filled with TO (TO-C), and the path of the water movement (e.g., pits and ray cells) was completely blocked in the tangential section (TO-T).

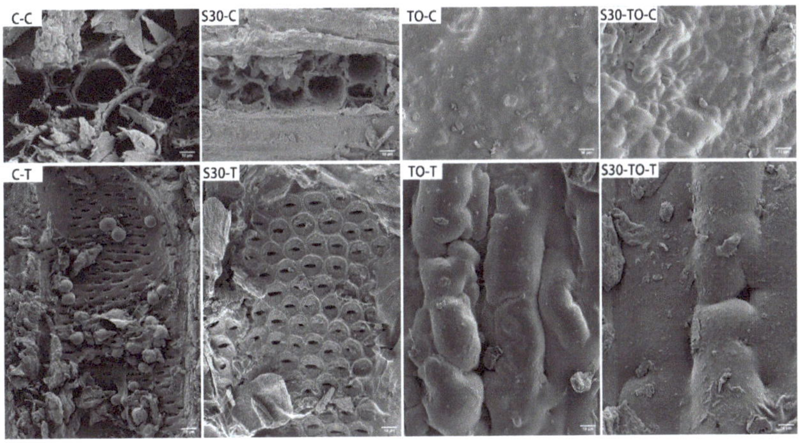

Figure 5. Microstructure images of the C, S30, TO, and S30-TO groups. (C-C, S30-C, TO-C, and S30-TO-C refer to cross sections of wood samples; C-T, S30-T, TO-T, and S30-TO-T refer to tangential sections of wood samples).

The formation of a cured film can be distinctly observed in the TO and S30-TO groups, which is due to the fact that TO is polymerized by free radicals with oxygen under sunlight and heating conditions, leading to the creation of a cured film within the wood. The microstructure of the S30-TO group was found to be similar to that of the TO group. As a consequence of blocking and covering, the number of water channels is reduced, and the hygroscopicity of the wood decreases, thus improving the dimensional stability of the wood. This result is consistent with the moisture and water absorption dimensional stability results.

3.6. FTIR Spectroscopy Analysis

Figure 6 displays the FTIR spectra of both the untreated and modified samples. It can be seen that the chemical structure of the samples treated with different modifications changed compared to the control, while the spectra remained fundamentally unaltered. The band at 3420 cm^{-1} exhibits a stretching vibration peak of hydroxyl (-OH) [47]. After modification, the intensity of the stretching vibration peak of -OH decreases, which may be attributed to the stretching vibration of free hydroxyl groups in sucrose molecules [48]. This indicates that the modification treatment reduces the relative content of hydroxyl groups in wood and improves its dimensional stability. The band at the 3420 cm^{-1} intensities of the TO group and S30-TO group decreased significantly, indicating that after TO treatment, the relative number of hydroxyl groups was reduced, which is beneficial for improving the dimensional stability of wood [49]. The bands of 2927 cm^{-1} and 2855 cm^{-1} only appear in the TO and S30-TO groups, which are attributed to the saturated C–H symmetric and antisymmetric stretching vibrations of tung oil [50]. The control group and S30 group did not find these two peaks, further confirming that tung oil successfully impregnated the wood interior. In addition, the band intensities at 1740 cm^{-1} and 1596 cm^{-1} correspond to the tensile vibration of carbonyl groups (C=O) [51,52], and their strength changes can be observed after modification treatment. The band intensities at 1374 cm^{-1} and 895 cm^{-1} correspond to C–H stretching vibrations [53,54], and the band at 1160 cm^{-1} corresponds to C–O stretching vibrations [55]. After modifications, the composition of the wood was essentially unchanged, but some of the chemical structure of the wood changed to some extent. Therefore, it can be inferred that sucrose penetration into the cell wall and crystallization there, wood oil uptake, deposition of oil in the cell wall, and the production of a hydrophobic oil film are all important factors contributing to the dimensional stability of wood.

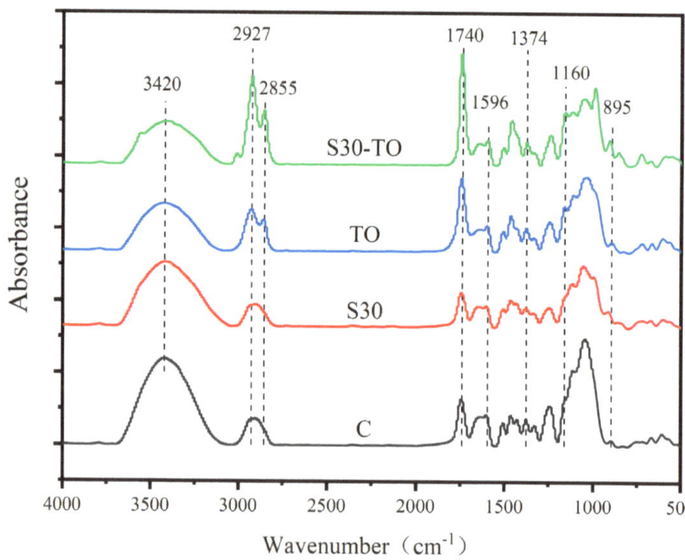

Figure 6. FTIR spectra of the C, S30, TO, and S30−TO groups.

3.7. Thermal Stability

In addition to being able to resist water, it is also crucial for wood products to possess thermal stability. In order to estimate the changes in the thermal properties of wood before and after the modification treatments, thermogravimetric (TG) and derivative thermogravimetric (DTG) curves were plotted for the modified and control samples, as shown in Figure 7.

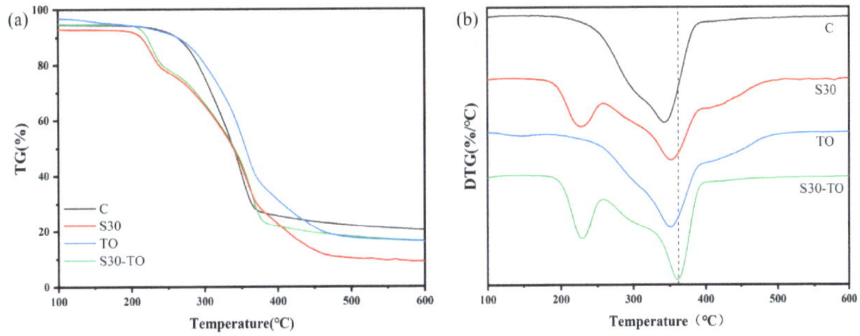

Figure 7. (**a**) Thermogravimetric and (**b**) derivative thermogravimetric thermographies of the C, S30, TO, and S30-TO groups.

For the control group, the different components of the wood degraded accordingly with temperature variations. In the first stage, TG curves in the range of 100–220 °C are considered as a degradation of unstable components such as hemicellulose and a bond breaking of complex components [56,57]. Subsequently, temperatures ranging from 220 to 450 °C result in a significant reduction in mass loss by approximately 80%, which encompasses both thermal decomposition (220–290 °C) and carbonization phases (260–450 °C) [35,58]. The control sample exhibited a peak at 342 °C, which can be ascribed to the decomposition of cellulose resulting in wood mass loss. The pyrolysis temperature of TO ranges from 420 to 450 °C. Thus, treatment with TO can effectively enhance the thermal stability of wood [59]. As depicted in Figure 7b, the pyrolysis temperature of the samples treated with TO was higher than that of the control group. The S30 and S30-TO groups exhibited a distinct peak at 228 °C, attributed to the sucrose dehydration and condensation leading to caramel formation; however, this peak was of a short duration. When the temperature rises to 256 °C, a large edge peak appears, which is mainly due to further carbonation and polymerization after the complete caramelization of sucrose; concurrently, gaseous byproducts such as CO_2, CO, acetic acid, acetone, and furfural compounds are generated [60,61]. The S30, TO, and S30-TO groups exhibited a maximum mass loss at 351 °C, 353 °C, and 361 °C, respectively, all of which were higher than that of the control group at 342 °C. This suggests that all the modified treatments enhanced the thermal stability of the wood. The levels of residual substances were observed to decrease in the TO and S30-TO groups as compared to the control group. This reduction in residual substances could be attributed to the degradation of most of the tung oil, while some wood components remained undecomposed [59]. In the same weight loss range (80 to 20%), the S30-TO group showed the highest pyrolysis temperature. It exhibited better thermal stability due to the synergistic effect of sucrose caramelization and tung oil impregnation. Thus, modified wood is suitable for heat-resistant applications like solid wood flooring with a heating system and furniture in the kitchen.

4. Conclusions

In this study, sucrose (S) and tung oil (TO) were used as treatment agents to modify rubber wood by a two-step method. The wood samples treated with 30% S and TO exhibited excellent anti-hygroscopicity, water resistance, and dimensional stability based on the synergistic effect of S and TO. After 384 h of immersion, the S30-TO group presented a 76.7% reduction in WA compared to the control group, and the ASE_{WA} was 57.85%, which was 66.81% higher than the TO treatment alone. Compared with the S30 group, the leaching rate of the S30-TO group decreased by 81.27%, indicating that TO has a remarkable fixation effect on S. The addition of S could swell the cell wall and effectively reduce the tensile deformation of the cell wall. Meanwhile, TO had the capability to obstruct wood voids and form a solidified protective film on the wood surface, thereby impeding water infiltration. Furthermore, S and TO treatments significantly enhance the thermal stability of rubber wood. These findings suggest that, as environmental-friendly natural materials, the treatment of S and TO, probably by synergistic effects, could impart rubber wood with better water resistance, dimensional stability, and thermal stability. Therefore, this work offers an eco-friendly, economical, and highly efficient approach to improving the performance of rubber wood, with broad potential applications and commercial values.

Author Contributions: Conceptualization, C.Y. and J.Q.; methodology, C.Y.; software, S.Y.; validation, H.Y.; investigation, C.Y.; resources, J.Q.; data curation, C.Y.; writing—original draft preparation, C.Y.; writing—review and editing, J.Q. and B.P.; supervision, J.Q.; project administration, J.Q. All authors have read and agreed to the published version of the manuscript.

Funding: This research was funded by the Natural Science Foundation of China (31971586) and the China Scholarship Council (CSC) scholarship.

Data Availability Statement: Data will be made available on request.

Acknowledgments: The authors gratefully thank the technicians and scientific associates from the College of Materials and Chemical engineering of the Southwest Forestry University and the Faculty of the Forestry of Kasetsart University.

Conflicts of Interest: The authors declare no conflict of interest.

References

1. Srivaro, S.; Lim, H.; Li, M.; Pasztory, Z. Properties of Mixed Species/Density Cross Laminated Timber Made of Rubberwood and Coconut Wood. *Structures* **2022**, *40*, 237–246. [CrossRef]
2. Meethaworn, B.; Khongtong, S. The Tuneable Rubberwood: Roles of Impregnated Polymer Level. *Wood Mater. Sci. Eng.* **2020**, *16*, 397–406. [CrossRef]
3. Yang, S.; Wu, X.; Liu, L.; Yan, Y.; Qiu, J.; Qin, L. Analysis of Fungal Diversity before and after Discoloration of Rubberwood in Xishuangbanna. *Diversity* **2023**, *15*, 471. [CrossRef]
4. Teoh, Y.P.; Don, M.M.; Ujang, S. Assessment of the Properties, Utilization, and Preservation of Rubberwood (*Hevea brasiliensis*): A Case Study in Malaysia. *J. Wood Sci.* **2011**, *57*, 255–266. [CrossRef]
5. Chen, C.; Tu, D.; Zhao, X.; Zhou, Q.; Cherdchim, B.; Hu, C. Influence of Cooling Rate on the Physical Properties, Chemical Composition, and Mechanical Properties of Heat-Treated Rubberwood. *Holzforschung* **2020**, *74*, 1033–1042. [CrossRef]
6. de Jesus Eufrade Junior, H.; Ohto, J.M.; da Silva, L.L.; Lara Palma, H.A.; Ballarin, A.W. Potential of Rubberwood (*Hevea brasiliensis*) for Structural Use after the Period of Latex Extraction: 'A Case Study in Brazil. *J. Wood Sci.* **2015**, *61*, 384–390. [CrossRef]
7. Long, X.; Fang, Y.; Qin, Y.; Yang, J.; Xiao, X. Latex-Specific Transcriptome Analysis Reveals Mechanisms for Latex Metabolism and Natural Rubber Biosynthesis in Laticifers of *Hevea brasiliensis*. *Ind. Crops Prod.* **2021**, *171*, 113835. [CrossRef]
8. Umar, I.; Zaidon, A.; Lee, S.; Halis, R. Oil-Heat Treatment of Rubberwood for Optimum Changes in Chemical Constituents and Decay Resistance. *J. Trop. For. Sci.* **2016**, *28*, 88–96.
9. Dong, Y.; Wang, K.; Li, J.; Zhang, S.; Shi, S.Q. Environmentally Benign Wood Modifications: A Review. *ACS Sustain. Chem. Eng.* **2020**, *8*, 3532–3540. [CrossRef]
10. Bryne, L.E.; Wålinder, M.E.P. Ageing of Modified Wood. Part 1: Wetting Properties of Acetylated, Furfurylated, and Thermally Modified Wood. *Holzforschung* **2010**, *64*, 295–304. [CrossRef]
11. Shen, X.; Yang, S.; Li, G.; Liu, S.; Chu, F. The Contribution Mechanism of Furfuryl Alcohol Treatment on the Dimensional Stability of Plantation Wood. *Ind. Crops Prod.* **2022**, *186*, 115143. [CrossRef]
12. Chabert, A.J.; Fredon, E.; Rémond, R. Improving the Stability of Beech Wood with Polyester Treatment Based on Malic Acid. *Holzforschung* **2022**, *76*, 268–275. [CrossRef]
13. Ghorbani, M.; Poorzahed, N.; Amininasab, S.M. Morphological, Physical, and Mechanical Properties of Silanized Wood-Polymer Composite. *J. Compos. Mater.* **2019**, *54*, 1403–1412. [CrossRef]
14. Wang, K.; Dong, Y.; Yan, Y.; Zhang, S.; Li, J. Improving Dimensional Stability and Durability of Wood Polymer Composites by Grafting Polystyrene onto Wood Cell Walls. *Polym. Compos.* **2016**, *39*, 119–125. [CrossRef]
15. Jiang, J.; Chen, Y.; Cao, J.; Mei, C. Improved Hydrophobicity and Dimensional Stability of Wood Treated with Paraffin/Acrylate Compound Emulsion through Response Surface Methodology Optimization. *Polymers* **2020**, *12*, 86. [CrossRef] [PubMed]
16. Schwarzkopf, M. Densified Wood Impregnated with Phenol Resin for Reduced Set-Recovery. *Wood Mater. Sci. Eng.* **2020**, *16*, 35–41. [CrossRef]

17. Shi, J.; Li, J.; Zhou, W.; Zhang, D. Improvement of Wood Properties by Urea-Formaldehyde Resin and Nano-SiO_2. *Front. For. China* **2007**, *2*, 104–109. [CrossRef]
18. Hill, C.; Altgen, M.; Rautkari, L. Thermal Modification of Wood—A Review: Chemical Changes and Hygroscopicity. *J. Mater. Sci.* **2021**, *56*, 6581–6614. [CrossRef]
19. Ditommaso, G.; Gaff, M.; Kačík, F.; Sikora, A.; Sethy, A.; Corleto, R.; Razaei, F.; Kaplan, L.; Kubš, J.; Das, S.; et al. Interaction of Technical and Technological Factors on Qualitative and Energy/Ecological/Economic Indicators in the Production and Processing of Thermally Modified Merbau Wood. *J. Clean. Prod.* **2019**, *252*, 119793. [CrossRef]
20. Candelier, K.; Thevenon, M.-F.; Petrissans, A.; Dumarcay, S.; Gerardin, P.; Petrissans, M. Control of Wood Thermal Treatment and Its Effects on Decay Resistance: A Review. *Ann. For. Sci.* **2016**, *73*, 571–583. [CrossRef]
21. Lee, S.H.; Ashaari, Z.; Lum, W.C.; Abdul Halip, J.; Ang, A.F.; Tan, L.P.; Chin, K.L.; Md Tahir, P. Thermal Treatment of Wood Using Vegetable Oils: A Review. *Constr. Build. Mater.* **2018**, *181*, 408–419. [CrossRef]
22. Allegretti, O.; Cuccui, I.; Terziev, N.; Sorini, L. A Model to Predict the Kinetics of Mass Loss in Wood during Thermo-Vacuum Modification. *Holzforschung* **2021**, *75*, 474–479. [CrossRef]
23. Shukla, S.R.; Sharma, S.K. Effect of High Temperature Processing under Different Environments on Physical and Surface Properties of Rubberwood (*Hevea brasiliensis*). *J. Indian Acad. Wood Sci.* **2014**, *11*, 182–189. [CrossRef]
24. Kusumah, S.S.; Umemura, K.; Guswenrivo, I.; Yoshimura, T.; Kanayama, K. Utilization of Sweet Sorghum Bagasse and Citric Acid for Manufacturing of Particleboard II: Influences of Pressing Temperature and Time on Particleboard Properties. *J. Wood Sci.* **2017**, *63*, 161–172. [CrossRef]
25. Sheldon, R.A. Green and Sustainable Manufacture of Chemicals from Biomass: State of the Art. *Green Chem.* **2013**, *16*, 950–963. [CrossRef]
26. Kobayashi, H.; Fukuoka, A. Synthesis and Utilisation of Sugar Compounds Derived from Lignocellulosic Biomass. *Green Chem.* **2013**, *15*, 1740–1763. [CrossRef]
27. Plat, T.; Linhardt, R.J. Syntheses and Applications of Sucrose-Based Esters. *J. Surfactants Deterg.* **2001**, *4*, 415–421. [CrossRef]
28. Morgós, A.; Imazu, S.; Ito, K. *Conservation and Digitalization: Conference Proceedings*; Piotrowska, K., Konieczny, P., Gdańsku, N.M.M.W., Eds.; National Maritime Museum: Gdańsk, Poland, 2015; ISBN 978-83-64150-10-4.
29. Parrent, J.M. The Conservation of Waterlogged Wood Using Sucrose. *Stud. Conserv.* **1985**, *30*, 63–72. [CrossRef]
30. Petr, P.; Aleš, D. Moisture Absorption and Dimensional Stability of Poplar Wood Impregnated with Sucrose and Sodium Chloride. *Maderas Cienc. Tecnol.* **2014**, *16*, 299–311. [CrossRef]
31. Ribeiro, B.O.; Valério, V.S.; Gandini, A.; Lacerda, T.M. Copolymers of Xylan-Derived Furfuryl Alcohol and Natural Oligomeric Tung Oil Derivatives. *Int. J. Biol. Macromol.* **2020**, *164*, 2497–2511. [CrossRef]
32. Xiong, X.Z.; Liu, Y.; Huang, X.H.; Wang, X.X.; Chen, Y.; Yin, X.H. Current Situation and Development Prospect of Tung Oil Tree (*Vernicia fordii*) in Chongqing Three Gorges Reservoir Area. *Adv. Mater. Res.* **2012**, *518–523*, 5385–5389. [CrossRef]
33. Samadzadeh, M.; Boura, S.H.; Peikari, M.; Ashrafi, A.; Kasiriha, M. Tung Oil: An Autonomous Repairing Agent for Self-Healing Epoxy Coatings. *Prog. Org. Coat.* **2011**, *70*, 383–387. [CrossRef]
34. Li, H.; Cui, Y.; Li, Z.; Zhu, Y.; Wang, H. Fabrication of Microcapsules Containing Dual-Functional Tung Oil and Properties Suitable for Self-Healing and Self-Lubricating Coatings. *Prog. Org. Coat.* **2018**, *115*, 164–171. [CrossRef]
35. He, Z.; Qian, J.; Qu, L.; Yan, N.; Yi, S. Effects of Tung Oil Treatment on Wood Hygroscopicity, Dimensional Stability and Thermostability. *Ind. Crops Prod.* **2019**, *140*, 111647. [CrossRef]
36. Peng, Y.; Wang, Y.; Zhang, R.; Wang, W.; Cao, J. Improvement of Wood against UV Weathering and Decay by Using Plant Origin Substances: Tannin Acid and Tung Oil. *Ind. Crops Prod.* **2021**, *168*, 113606. [CrossRef]
37. *GB/T 1927.2-2021*; Test Methods for Physical and Mechanical Properties of Small Clear Wood Specimens—Part 2: Sampling Methods and General Requirements. Standard Press: Beijing, China, 2021.
38. Wang, J.P.; Matthews, M.L.; Williams, C.M.; Shi, R.; Yang, C.; Tunlaya-Anukit, S.; Chen, H.-C.; Li, Q.; Liu, J.; Lin, C.-Y.; et al. Improving Wood Properties for Wood Utilization through Multi-Omics Integration in Lignin Biosynthesis. *Nat. Commun.* **2018**, *9*, 1579. [CrossRef]
39. Brocco, V.F.; Paes, J.B.; Costa, L.G. da; Kirker, G.T.; Brazolin, S. Wood Color Changes and Termiticidal Properties of Teak Heartwood Extract Used as a Wood Preservative. *Holzforschung* **2020**, *74*, 233–245. [CrossRef]
40. Arends, T.; Pel, L.; Smeulders, D. Moisture Penetration in Oak during Sinusoidal Humidity Fluctuations Studied by NMR. *Constr. Build. Mater.* **2018**, *166*, 196–203. [CrossRef]
41. Banks, W.B. Water Uptake by Scots Pine Sapwood, and Its Restriction by the Use of Water Repellents. *Wood Sci. Technol.* **1973**, *7*, 271–284. [CrossRef]
42. Tahira, A.; Howard, W.; Pennington, E.R.; Kennedy, A. Mechanical Strength Studies on Degraded Waterlogged Wood Treated with Sugars. *Stud. Conserv.* **2017**, *62*, 223–228. [CrossRef]
43. Humar, M.; Lesar, B. Efficacy of Linseed- and Tung-Oil-Treated Wood against Wood-Decay Fungi and Water Uptake. *Int. Biodeterior. Biodegrad.* **2013**, *85*, 223–227. [CrossRef]
44. Chen, J.; Wang, Y.; Cao, J.; Wang, W. Improved Water Repellency and Dimensional Stability of Wood via Impregnation with an Epoxidized Linseed Oil and Carnauba Wax Complex Emulsion. *Forests* **2020**, *11*, 271. [CrossRef]
45. Dong, Y.; Yan, Y.; Wang, K.; Li, J.; Zhang, S.; Xia, C.; Shi, S.Q.; Cai, L. Improvement of Water Resistance, Dimensional Stability, and Mechanical Properties of Poplar Wood by Rosin Impregnation. *Eur. J. Wood Wood Prod.* **2016**, *74*, 177–184. [CrossRef]

46. Liu, M.; Lyu, S.; Peng, L.; Lyu, J.; Huang, Z. Radiata Pine Fretboard Material of String Instruments Treated with Furfuryl Alcohol Followed by Tung Oil. *Holzforschung* **2021**, *75*, 480–493. [CrossRef]
47. Kubovský, I.; Kačíková, D.; Kačík, F. Structural Changes of Oak Wood Main Components Caused by Thermal Modification. *Polymers* **2020**, *12*, 485. [CrossRef] [PubMed]
48. Lei, H.; Du, G.; Wu, Z.; Xi, X.; Dong, Z. Cross-Linked Soy-Based Wood Adhesives for Plywood. *Int. J. Adhes. Adhes.* **2014**, *50*, 199–203. [CrossRef]
49. Kumar, R.; Mago, G.; Balan, V.; Wyman, C.E. Physical and Chemical Characterizations of Corn Stover and Poplar Solids Resulting from Leading Pretreatment Technologies. *Bioresour. Technol.* **2009**, *100*, 3948–3962. [CrossRef]
50. Feng, Y.; Cui, Y.; Zhang, M.; Li, M.; Li, H. Preparation of Tung Oil-Loaded PU/PANI Microcapsules and Synergetic Anti-Corrosion Properties of Self-Healing Epoxy Coatings. *Macromol. Mater. Eng.* **2020**, *306*, 2000581. [CrossRef]
51. Li, M.-F.; Shen, Y.; Sun, J.-K.; Bian, J.; Chen, C.-Z.; Sun, R.-C. Wet Torrefaction of Bamboo in Hydrochloric Acid Solution by Microwave Heating. *ACS Sustain. Chem. Eng.* **2015**, *3*, 2022–2029. [CrossRef]
52. Guo, J.; Song, K.; Salmén, L.; Yin, Y. Changes of Wood Cell Walls in Response to Hygro-Mechanical Steam Treatment. *Carbohydr. Polym.* **2014**, *115*, 207–214. [CrossRef]
53. Jin, T.; Zeng, H.; Huang, Y.; Liu, L.; Ji, D.; Guo, H.; Shi, S.; Du, G.; Zhang, L. Synthesis of Fully Biomass High-Performance Wood Adhesives from Xylitol and Maleic Anhydride. *ACS Sustain. Chem. Eng.* **2023**, *11*, 11781–11789. [CrossRef]
54. Yin, Y.; Berglund, L.; Salmén, L. Effect of Steam Treatment on the Properties of Wood Cell Walls. *Biomacromolecules* **2010**, *12*, 194–202. [CrossRef]
55. Pandey, K.K.; Pitman, A.J. Examination of the Lignin Content in a Softwood and a Hardwood Decayed by a Brown-Rot Fungus with the Acetyl Bromide Method and Fourier Transform Infrared Spectroscopy. *J. Polym. Sci. Part Polym. Chem.* **2004**, *42*, 2340–2346. [CrossRef]
56. Lin, B.-J.; Colin, B.; Chen, W.-H.; Pétrissans, A.; Rousset, P.; Pétrissans, M. Thermal Degradation and Compositional Changes of Wood Treated in a Semi-Industrial Scale Reactor in Vacuum. *J. Anal. Appl. Pyrolysis* **2018**, *130*, 8–18. [CrossRef]
57. Kesik, H.I.; Korkut, S.; Hiziroglu, S.; Sevik, H. An Evaluation of Properties of Four Heat Treated Wood Species. *Ind. Crops Prod.* **2014**, *60*, 60–65. [CrossRef]
58. Wang, Y.; Zhang, R.; Yang, M.; Peng, Y.; Cao, J. Improvement on Dimensional Stability and Mold Resistance of Wood Modified by Tannin Acid and Tung Oil. *Holzforschung* **2022**, *76*, 929–940. [CrossRef]
59. Yang, J.J.; Guai, W.S.; Che, H.R. Study on Thermogravimetry and Pyrolysis Dynamics of Tung Oil. *Adv. Mater. Res.* **2012**, *524–527*, 1719–1722. [CrossRef]
60. Kim, S. Environment-Friendly Adhesives for Surface Bonding of Wood-Based Flooring Using Natural Tannin to Reduce Formaldehyde and TVOC Emission. *Bioresour. Technol.* **2008**, *100*, 744–748. [CrossRef] [PubMed]
61. Wu, Z.; Zhang, B.; Zhou, X.; Li, L.; Yu, L.; Liao, J.; Du, G. Influence of Single/Collective Use of Curing Agents on the Curing Behavior and Bond Strength of Soy Protein-Melamine-Urea-Formaldehyde (SMUF) Resin for Plywood Assembly. *Polymers* **2019**, *11*, 1995. [CrossRef]

forests

MDPI

Article

Prediction of Thermally Modified Wood Color Change after Artificial Weathering Based on IPSO-SVM Model

Juncheng Li, Ning Li, Jinze Li, Wei Wang * and Haolin Wang

College of Mechanical and Electrical Engineering, Northeast Forestry University, Harbin 150040, China; lijuncheng165134@nefu.edu.cn (J.L.)
* Correspondence: vickywong@nefu.edu.cn; Tel.: +86-133-1361-3588

Abstract: The support vector machine (SVM) model was applied to predict the color change of heat-modified wood after artificial weathering. In order to improve the prediction performance, the improved particle swarm optimization (IPSO) algorithm was used to optimize the parameters of the SVM model, and an improved particle swarm optimized support vector machine (IPSO-SVM) model was established on the basis of the nonlinear descending weight strategy to improve the particle swarm optimization. To verify the performance of the established model, the MAE, RMSE, and R^2 of the test set and training set were compared with the PSO-SVM model and the SVM model. Analysis of the results showed that compared to the PSO-SVM model and the SVM model, the IPSO-SVM model reduced the RMSE of the training set data by 49% and 72%, the MAE by 52% and 78%, the STD by 14% and 68%, the test set data by 6% and 24%, the MAE by 2% and 25%, and the STD by 22% and 29%, respectively. The results show that modeling studies using the IPSO-SVM model provide results showing that color changes in heat-modified wood after artificial weathering can be successfully predicted without expensive and time-consuming experimental studies.

Keywords: prediction model; particle swarm optimization; support vector machine; heat treatment of wood

check for updates

Citation: Li, J.; Li, N.; Li, J.; Wang, W.; Wang, H. Prediction of Thermally Modified Wood Color Change after Artificial Weathering Based on IPSO-SVM Model. *Forests* **2023**, *14*, 948. https://doi.org/10.3390/f14050948

Academic Editors: Morwenna Spear and Miklós Bak

Received: 6 April 2023
Revised: 28 April 2023
Accepted: 28 April 2023
Published: 4 May 2023

1. Introduction

Wood has a wide variety of uses. Since ancient times, wood has been widely used in various industries such as architectural decoration, wood furniture manufacturing, and cultural and educational office supplies, and the demand for wood is growing [1]. However, it is well known that in the natural environment, due to such environmental factors as sunlight, oxygen, water, and temperature [2], the surface of the wood will change color, especially under the weathering, and the color change will be apparent to the eye [3]. Due to the photodegradation of lignin and wood extracts, the color of wood changes [4]. Moreover, after heat treatment of wood, its color will change greatly, and the demand for wood after heat treatment is increasing in various fields. The color of the wood after heat treatment is generally brown, which is similar to the color of some precious wood, and the color of the heat-treated material is uniform inside and outside, making it aesthetically pleasing [5]. In addition, compared to wood without heat treatment, heat-treated wood exposed to ultraviolet light has better color stability, which may be related to the phenol content in the wood [6]. Tomak et al. [7] believed that heat treatment could improve the color stability and surface quality of weathering samples. Studies have shown that heat-treated wood of different colors can be obtained by adjusting the temperature and time of heat treatment [8,9], and the gloss value of the wood after heat treatment is low. Due to the heat treatment, the uniformity of the color and gloss of the exposed surface of the wood improves [10], so as to realize the diversification of product varieties. Most studies have shown that the color change in heat-treated wood is highly dependent on the wood species [11]. Sikora et al. [12] studied the effect of thermal modification temperature on the color change of spruce and oak and found that the surface brightness of the wood (L*) decreased with the increase in the treatment temperature and maximum color difference (ΔE*), and the total color difference reached its maximum value at 210 °C. The change in wood color is due mainly to the change in chemical composition. The decrease in hemicellulose content will lead to a decrease in the brightness of the wood and a change in the total color difference [13]. The reduction in the wood's brightness is also influenced by the lignin content [14].

In general, to obtain data on wood color change under different heat treatment processes it is necessary to conduct a large number of difficult and time-consuming comprehensive experiments. To obtain data more quickly and efficiently, it is necessary to explore appropriate modeling methods for the study of color change in wood after heat treatment. The principle of SVM is to minimize structural risks, so SVM is particularly effective in small-sample and nonlinear problems. Liping Sun et al. [15] proposed the least squares support vector machine (LSSVM) method to establish an online model of a wood drying system. Based on data for the deceleration drying stage obtained through a drying experiment as samples, the online prediction model of the wood drying system was established and forecasted according to the requirement of actual predictive control. Jicheng Li et al. [16] proposed a modeling method using an improved ant colony algorithm (MACA) to optimize the least squares support vector machine (LSSVM). Accurately and reliably assessing changes in wood moisture content is the key to improving wood drying quality.

In this paper, a support vector machine model is established to predict change in wood color under different artificial weathering times. The prediction model can effectively reduce the experimental cost. In this study, the prediction model is used to learn and predict the data to verify the accuracy of the model. The measurement is not carried out through real experiments, which is a limitation of this study. Therefore, the proposed model is compared with the original model and other models to verify its advantages and practicability.

2. Experimental Part

2.1. Change in Wood Color

Table 1 shows the experimental data published by Nguyen et al. [17], which can be queried on the European Journal of Wood and Wood Products website, specifically at [17] https://doi.org/10.1007/s00107-019-01449-0 (accessed 29 October 2022).

Table 1. Experimental measurements of color changes in heat-treated wood during artificial weathering.

Heat-Treated Wood	Artificial Weathering Time (h)	Average of the Sample.							
		Heat Treated Larch				Heat Treated Polar			
		ΔL	Δa	Δb	ΔE	ΔL	Δa	Δb	ΔE
180 °C	240	−1.01	0.55	−0.03	1.15	8.15	−2.97	−0.94	8.72
	480	−0.86	0.25	−0.28	0.94	6.52	−2.13	−0.53	6.88
	720	−0.56	−0.19	−0.54	0.79	8.02	−2.23	−0.27	8.33
	960	−3.22	0.70	−0.41	3.32	6.19	−0.92	2.33	6.68
	1200	−3.88	1.08	−0.4	4.05	5.92	−0.52	2.05	6.29
	1440	−3.19	1.20	−0.53	3.44	4.15	0.19	2.89	5.06
	1680	−3.58	1.22	−0.38	4.06	4.34	−0.07	2.88	5.20
	1920	−5.09	1.88	0.41	5.44	5.48	−0.6	2.86	6.21
	2160	−5.19	2.51	−0.25	5.77	5.57	−0.74	2.46	6.13
	2400	−5.00	2.47	−0.45	5.59	5.90	−1.14	0.98	6.09
	2700	−4.68	1.73	−0.57	5.02	5.47	−2.55	−0.48	6.05
	3000	−3.76	1.45	−1.28	4.23	5.24	−2.75	−0.78	5.97
190 °C	240	−2.73	1.02	0.71	3.00	7.73	−2.21	0.15	8.04
	480	−2.12	0.73	0.80	2.38	14.30	−1.69	2.10	6.58
	720	−2.03	0.38	0.80	2.22	14.63	−1.37	3.05	7.89
	960	−3.33	1.04	0.83	3.58	5.00	0.01	6.06	7.85

Table 1. *Cont.*

Heat-Treated Wood	Artificial Weathering Time (h)	Average of the Sample.							
		Heat Treated Larch				Heat Treated Polar			
		ΔL	Δa	Δb	ΔE	ΔL	Δa	Δb	ΔE
	1200	−3.81	1.21	0.71	4.05	4.70	0.57	6.34	7.91
	1440	−3.24	1.26	0.85	3.57	5.72	0.57	6.93	9.01
	1680	−3.63	1.74	0.87	4.12	5.77	0.58	7.67	9.62
	1920	−4.28	1.78	1.27	4.81	5.81	0.64	9.17	10.88
	2160	−4.44	2.03	1.43	5.09	6.05	1.61	9.39	11.28
	2400	−3.94	1.90	1.37	4.58	6.57	1.27	8.60	10.90
	2700	−3.66	1.77	1.05	4.20	6.86	0.61	8.09	10.62
	3000	−3.09	1.35	0.82	3.47	6.33	0.17	6.96	9.41
200 °C	240	0.54	1.12	0.17	1.26	13.40	−3.67	0.10	8.36
	480	1.00	1.05	0.35	1.49	5.09	−0.07	4.10	8.53
	720	1.95	0.38	0.31	2.01	7.52	−1.07	3.69	8.44
	960	0.30	1.09	0.79	1.38	8.60	−0.92	5.06	10.02
	1200	0.50	1.46	1.53	2.17	6.45	0.27	5.57	8.52
	1440	0.14	1.29	1.02	1.65	7.50	0.31	5.81	9.49
	1680	0.07	2.02	1.87	2.75	8.44	−0.02	6.24	10.50
	1920	−0.04	2.17	2.96	3.67	9.79	−0.35	6.78	11.91
	2160	0.11	2.22	3.63	4.26	9.90	−0.42	6.82	12.03
	2400	0.27	2.20	3.31	3.98	10.12	−0.65	6.03	11.80
	2700	0.29	1.75	2.62	3.17	10.35	−0.86	5.34	11.68
	3000	0.85	1.28	1.52	2.16	9.91	−0.97	4.68	11.00
210 °C	240	0.72	2.04	6.07	6.44	7.74	−0.62	2.19	8.07
	480	4.04	2.04	7.89	9.10	3.85	1.33	3.17	5.16
	720	2.98	2.69	8.66	9.54	5.44	0.95	3.79	6.70
	960	−1.09	3.16	7.04	7.80	3.48	2.19	6.00	7.28
	1200	−0.05	3.23	6.74	7.48	2.92	2.83	6.52	7.68
	1440	0.55	3.16	7.20	7.88	4.77	2.88	7.17	9.08
	1680	0.27	3.47	7.20	8.00	4.08	2.92	7.65	9.15
	1920	−0.59	3.73	7.62	8.51	3.96	3.39	8.64	10.09
	2160	−0.57	3.77	7.90	8.77	3.18	3.48	9.09	12.23
	2400	−0.35	3.66	7.77	8.60	3.69	3.01	8.41	9.66
	2700	−0.25	3.56	7.65	8.44	4.00	2.86	8.22	9.58
	3000	0.27	3.23	7.07	7.78	3.36	2.83	7.97	9.10
220 °C	240	−3.42	2.19	4.31	5.92	6.13	−0.39	3.33	6.99
	480	−2.52	3.64	5.38	6.50	−0.22	2.04	3.87	4.38
	720	−1.43	2.93	5.85	6.69	2.65	1.82	4.92	5.88
	960	−3.45	2.73	5.00	6.66	−0.13	2.71	5.82	6.42
	1200	−2.37	2.59	4.55	5.75	1.19	2.74	5.54	6.29
	1440	−2.26	2.64	4.32	5.55	2.47	2.86	5.77	6.90
	1680	−2.17	2.90	4.70	5.94	2.43	2.92	5.84	6.96
	1920	−2.02	3.39	5.30	6.60	1.99	3.38	7.87	8.79
	2160	−1.71	3.53	5.64	6.87	1.71	3.66	11.38	12.08
	2400	−1.64	3.41	5.41	6.60	1.93	3.54	10.94	11.66
	2700	−1.47	3.23	5.30	6.38	3.07	3.38	10.50	11.45
	3000	−0.93	3.00	4.96	5.87	2.71	3.22	9.98	10.83

In this experiment, two different types of wood, cork and hardwood, were selected for heat treatment processing; the specific species were heat-treated larches *(Larix gmelini)* and heat-treated poplar (*Populus alba*), and they were provided by the College of Materials Science and Engineering of Northeast Forestry University. The heat treatment time for both wood types was four hours, and the heat treatment temperature was 180, 190, 200, 210, and 220 °C, respectively. One-hundred-five pieces of each of the two types of wood were randomly divided into ten groups of 21 samples each. The sample size was 80 × 30 × 3 mm^3 (l × t × r). The sample was placed at a room temperature of 20 ± 2 °C and a relative humidity of 65 ± 5% until the moisture content of the sample was about 12%. The samples were subjected to an accelerated weathering test in a weathering tester to simulate the damage to the wood that would normally take months or years. The weathering time was 0, 240, 480, 720, 960, 1200, 1440, 1680, 1920, 2160, 2400, 2700, and 3000 h, respectively. Color variations on the sample area were evaluated. The color change was measured using the CIE 1976 L*a*b* color measurement system [18], and the total color change (ΔE) was calculated as follows:

$$\Delta L^* = L_2{}^* - L_1{}^* \quad (1)$$

$$\Delta a^* = a_2{}^* - a_1{}^* \quad (2)$$

$$\Delta b^* = b_2{}^* - b_1{}^* \quad (3)$$

$$\Delta E = \sqrt{\Delta L^{*2} + \Delta a^{*2} + \Delta b^{*2}} \quad (4)$$

where ΔL*, Δa*, and Δb* represent color coordinate changes; $L_1{}^*$, $a_1{}^*$, $b_1{}^*$ indicates the brightness, red and green coordinates, and yellow and blue coordinates of the untreated sample; and $L_2{}^*$, $a_2{}^*$, $b_2{}^*$ represents the brightness, red and green coordinates, and yellow and blue coordinates of the treated sample.

As Figure 1a shows, the untreated larch ΔE increased before 1680 h and decreased from 1680 h to 3000 h. According to Figure 1b, the ΔE of the untreated poplars showed an upward trend before 1200 h, and then a downward trend, which was consistent with the results of Xing et al. [19]. The ΔE of the two untreated woods was greater than that of the heat-treated woods, suggesting that the heat-treated artificially weathered woods had better color stability.

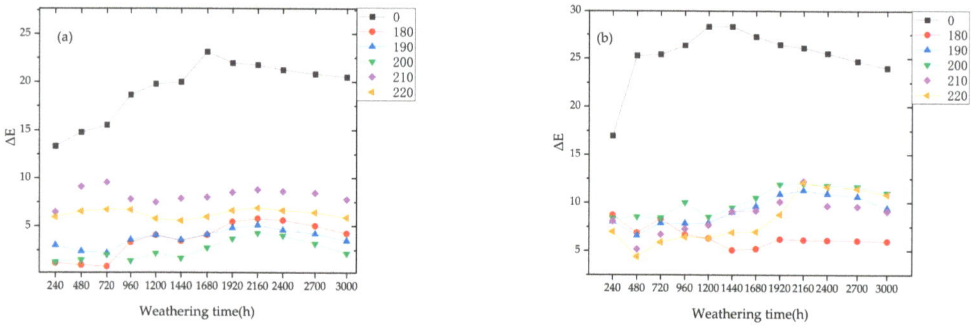

Figure 1. (**a**) ΔE of heat-treated larch and (**b**) ΔE of heat-treated poplar.

2.2. Prediction Model

2.2.1. Support Vector Machine Model

The Support Vector Machine (SVM) was initially devised by Vapnik, a former Soviet professor [20]. The SVM regression model has good adaptability in solving linear and nonlinear prediction and can better solve the complicated problems of influencing factors in the prediction of wood color change in the process of wood heat treatment. Support vector machines have performed well for small-sample, nonlinear, and large dimensional problems and are widely used to treat classification and regression problems. Finally, support vector machines can be transformed into a dual optimization form [21]. This is given in Formulas (5) and (6):

Objective function:

$$min\frac{1}{2}\sum_{ij=1}^{1}(\alpha_i^*-\alpha_i)\left(\alpha_j^*-\alpha_j\right)K(x_i,x_j)+\varepsilon\sum_{i=1}^{1}(\alpha_i^*-\alpha_i)-\sum_{i=1}^{1}y_i(\alpha_i^*-\alpha_i) \quad (5)$$

Conditions:

$$\begin{cases}\sum\limits_{i=1}^{1}y_i(\alpha_i-\alpha_i^*)=0\\ \alpha_i,\alpha_i^*\geq 0, i=1,2,\ldots l\end{cases} \quad (6)$$

In Equations (5) and (6), α_i and α_i^* are Lagrange multipliers, and $K(x_i,x_j)$ is the kernel function. Commonly used kernel functions include the linear kernel function (LinearKernel, LK), the polynomial kernel function (PolynomialKernel, PK), and the radial basis kernel function (RBF), as shown in Table 2. For nonlinear problems, the kernel function is introduced, the input space is transformed into a high-dimensional space through nonlinear mapping, and the nonlinear problem becomes a linear problem. Because RBF has strong nonlinear mapping ability, it can map the original features to infinite dimensions, which has been widely used at present. RBF is adopted as the kernel function in this paper. RBF can be expressed by Formula (7):

$$K(x_i,x_j)=\exp\left(-\delta\cdot|x_i-x_j|^2\right) \quad (7)$$

where δ is the kernel function parameter representing the space range that a specific training sample can reach.

Table 2. Model parameters.

Algorithm	Parameter
SVM	c = 4.0; g = 0.8
PSO-SVM	c1 = 1.5; c2 = 1.7; maxgen = 50; sizepop = 5; popcmax = 100; popcmin = 0.1 popgmax = 100; popgmin = 0.1
IPSO-SVM	c1 = 1.5; c2 = 1.7; maxgen = 50; sizepop = 5; popcmax = 100; popcmin = 0.1 popgmax = 100; popgmin = 0.1; w_star = 2; w_end = 0.4

After the kernel function is determined, in order to optimize the predictive performance of the SVM model, two optimal parameters, namely penalty factor c and kernel parameter g, should be found. In order to determine the c and g of the SVM reasonably quickly, this paper introduces an improved PSO to optimize the c and g of SVM and presents an improved particle swarm optimization support vector machine (IPSO-SVM) algorithm.

2.2.2. Improved Particle Swarm Optimization

Based on simulated social behaviors such as bird feeding and human cognition, J. Kennedy et al. [22] proposed a global optimization algorithm–particle swarm optimization (PSO), which is an evolutionary computing technology based on swarm intelligence. The particle swarm optimization algorithm has the advantages of simplicity, easy implementation, few parameters to be adjusted, and no gradient information. However, since all the particles fly toward the optimal region, the diversity will be lost and the particles will tend to be homogeneous, which causes the convergence speed to be obviously slow in the late stage of convergence. At the same time, when the algorithm converges to a certain degree, it cannot continue to optimize, and it is easy to fall into local optimal, so the convergence accuracy is not high. Therefore, many scholars are committed to improving the convergence accuracy of the PSO algorithm. In view of the above limitations, Shi Y et al. [23] in 1998 proposed linear decreasing weight strategy (LDIW), an improved particle swarm optimization algorithm with inertial weights. At the same time, aiming at the problems of premature convergence and non-convergence in the particle swarm optimization algorithm, this paper proposes an improved particle swarm optimization algorithm with nonlinear decreasing inertia weight. In the algorithm iteration process, the particle boundary velocity is limited by the maximum velocity nonlinear decreasing change strategy. Nonlinear decreasing variation in inertia weight is used to balance the global research capacity at the early stage and the local optimization at the late stage. It can be expressed by Formula (8):

$$w = w_s - (w_s - w_e)\left|f_1\sqrt{\frac{t}{T_{max}}}\right| \quad (8)$$

where w_s is the initial inertia weight; w_e is the end value of the maximum number of iterations allowed; T_{max} is the maximum number of evolutions; t is the number of current iterations; and f_1 is the regulating factor, in order to control the rate of change of w. The root sign in the formula is to generate the square to form the nonlinear effect, and the absolute value is added to prevent the negative sign of the open square root. As the iteration progresses, the inertia weight decreases nonlinearly. The inertia weight w takes a larger value in the algorithm of early iteration so as to avoid the algorithm falling into the local extreme value and to maintain a strong global search ability. At the same time, a small inertia weight is selected in the late iteration to enhance the local search ability and accelerate the convergence speed.

2.2.3. IPSO-SVM Model

This paper presents an improved particle swarm optimization (IPSO) algorithm to optimize the support vector machine model. As a machine learning algorithm, SVM has unique advantages in dealing with nonlinear relation. Through nonlinear mapping, complex nonlinear problems can be transformed into linear regression problems in high dimensional space, and the complex relationship between multiple factors can be quickly learned and analyzed. The essence of the support vector machine (SVM) regression prediction model is to find the optimal relationship between each input and output sample on the basis of the limited sample data and give a reasonable output for the input samples in the test set. In order to optimize the predictive performance of the SVM model, IPSO is used to optimize and improve the SVM model. The process is as follows: First, determine the input and output of the model, divide the sample data into two categories—training and testing—and normalize all the sample data. Then the improved particle swarm optimization algorithm is used to optimize the population parameters and set the population size, nonlinear decreasing weight factor, iteration times, and other parameters. Next, the particle fitness value is calculated and the particle velocity, position, and other parameters are updated to determine the end conditions of the optimal parameters. Finally, the IPSO-SVM regression prediction performance is tested through the test set data. If the predicted value meets the set requirements, the predicted value is output. Otherwise, continue optimizing the parameters until the end condition is met. This process is shown in Figure 2:

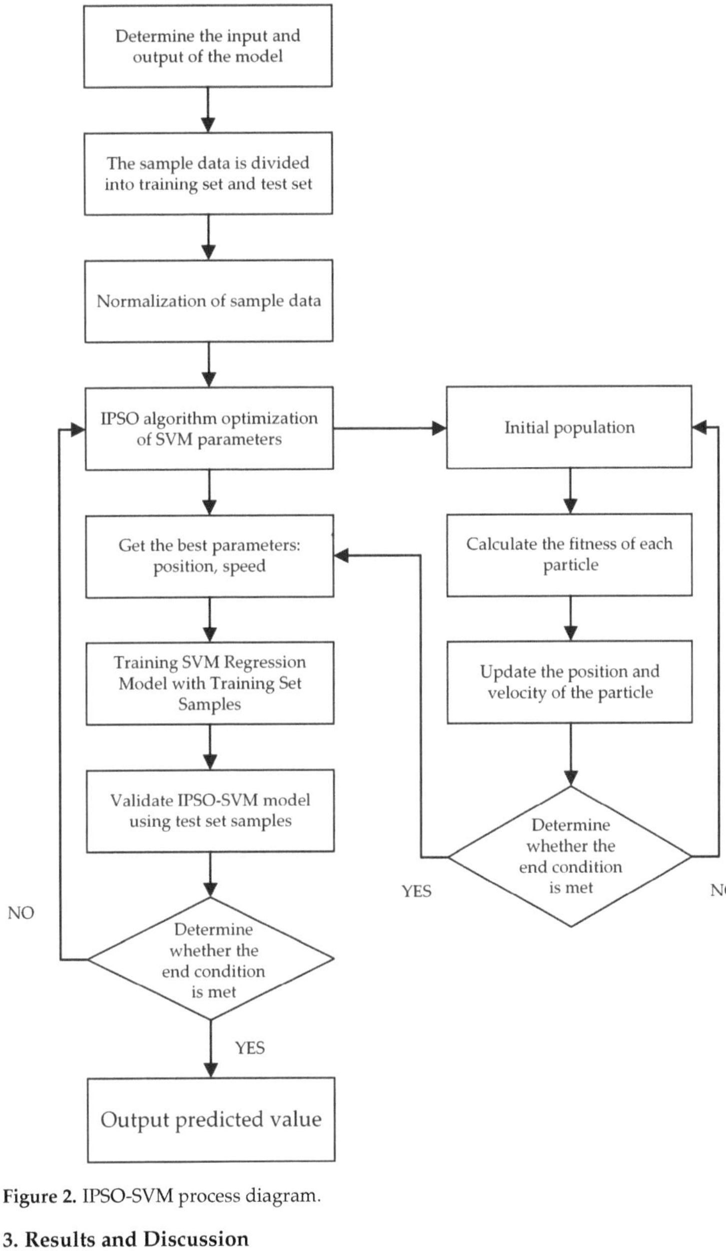

Figure 2. IPSO-SVM process diagram.

3. Results and Discussion

3.1. Model Parameter Settings

In order to verify the predictive performance of the proposed model, the improved particle swarm optimization support vector machine model (IPSO-SVM) was compared with the particle swarm optimization support vector machine model (PSO-SVM) and the support vector machine model (SVM). The parameters of each model are shown in Table 2. In the SVM model, c is the penalty factor and g is the radial basis function parameter. In the IPSO-SVM model, c1 represents the local search capability of the PSO parameters, c2 represents the global search capability of the PSO parameters, maxgen is the maximum number of evolution, sizepop is the maximum number of population, popcmax is the maximum change value of SVM parameter c, popcmin is the minimum change value of SVM parameter c, popgmax is the maximum change value of SVM parameter g, popgmin is the minimum change value of SVM parameter g, and w_star and w_end are the initial and end values of the nonlinear decreasing weight factors. In the PSO-SVM model, parameter representation is the same as that in the IPSO-SVM model.

3.2. Model Evaluation Criteria

Common regression evaluation indexes include mean absolute error (MAE), mean square error (MSE), root mean square error (RMSE), goodness of fit (R^2), and standard deviation (STD), where RMSE is the square root of the MSE and the order of magnitude is the same as the true value. MSE and RMSE are essentially the same, but RMSE is used for better data description. Therefore, RMSE only is used as the evaluation index in this paper. In order to illustrate the degree of dispersion of the samples, the smaller the RMSE, the better. R^2 reflects the accuracy of the model fitting data. Generally, R^2 varies in the range of 0 to 1. The closer the value is to 1, the stronger the fitting ability is and the better the fitting effect is; otherwise, the worse the fitting effect is. Formulas (9)–(12) are shown as follows:

$$\mathrm{RMSE} = \sqrt{\frac{\sum_{i=1}^{N}(A_i - F_i)^2}{N}} \tag{9}$$

$$\mathrm{MAE} = \frac{1}{N}\sum_{i=1}^{N}|A_i - F_i| \tag{10}$$

$$\mathrm{R}^2 = 1 - \frac{\sum(F_i - A_i)^2}{\sum(F_i - A_i)^2} \tag{11}$$

$$\mathrm{STD} = \sqrt{\frac{\sum_{i=1}^{N}(x_i - \overline{x})^2}{N-1}} \tag{12}$$

where A_i and F_i denote the actual and predicted values, respectively, x_i denotes the error between the predicted and actual values, and $\overline{x}$ denotes the average of all the errors.

3.3. Comparative Analysis of Model Performance

This paper compares the IPSO-SVM model with the PSO-SVM model and the SVM model. The evaluation results of each model are listed in Table 3.

Table 3. Model evaluation results.

Model		IPSO-SVM	PSO-SVM	SVM
RMSE	Training	0.37363	0.73098	1.3534
	Testing	0.9166	0.97504	1.2056
MAE	Training	0.21446	0.44525	0.95681
	Testing	0.73084	0.7491	0.9691
R^2	Training	0.98288	0.9369	0.77346
	Testing	0.90127	0.87627	0.83279
STD	Training	0.43294	0.64479	1.35062
	Testing	0.8536	1.09641	1.20145

It can be seen from Table 3 that the RMSE of the training set and the test set of the IPSO-SVM model are 0.37363 and 0.9166, the MAE values are 0.21446 and 0.73084, and the STD values are 0.43294 and 0.8536, respectively. In addition, the R^2 of the training set and of the test set is 0.98288 and 0.90127, respectively. It is proved that the measured results have a good fit with the model prediction. Compared to the PSO-SVM model and the SVM model, the IPSO-SVM model reduced the RMSE of the training set data by 49% and 72%, the MAE by 52% and 78%, the STD by 14% and 68%, the test set data by 6% and

24%, the MAE by 2% and 25%, and the STD by 22% and 29%, respectively. This indicates that the optimization effect of the IPSO-SVM model is obvious, and the prediction effect of IPSO-SVM model is better than that of the other two models.

Figure 3 shows the fitness convergence curves of the IPSO-SVM model and the PSO-SVM model. It can be seen from Figure 3 that the IPSO-SVM model reaches the optimum in the 27th generation while the PSO-SVM model reaches the optimum in the 34th generation. Moreover, the fitness value of the IPSO-SVM model is always smaller than that of the PSO-SVM model. It can be proved that the IPSO-SVM model is superior to the PSO-SVM model in terms of convergence speed and convergence accuracy.

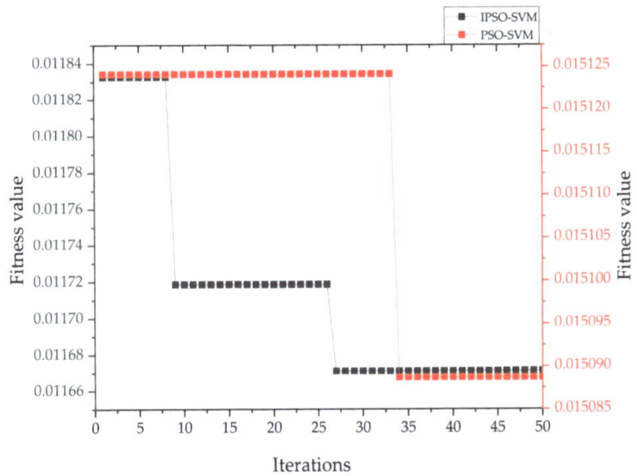

Figure 3. Fitness convergence curves of IPSO-SVM and PSO-SVM models.

The predicted effect of the model can also be observed in the line graph. Figure 4 shows the predicted results of the three prediction models. Compared with the other three models, the RMSE and MAE of the IPSO-SVM model are significantly lower than those of the other three models, and the R^2 of the IPSO-SVM model is closer to 1 than that of the other three models, indicating the highest goodness of fit. From Figure 4 combined with Table 3, it can be clearly seen that the predicted effect of the IPSO-SVM model is significantly better than that of the other two prediction models, and the predicted effect from good to bad is represented by the sequence of IPSO-SVM model, PSO-SVM model, and SVM model.

Concerning the thermal modification of artificial weathering wood color change to R^2 of the values listed in Table 3, we confirmed the IPSO-SVM model between the prediction results and the real value as having a very good fitting effect. The results show that the IPSO-SVM model for predicting thermal modification after artificial weathering wood color change is very accurate. a and b in Figure 4 show the comparison between the actual value and the predicted value of the color change under heat treatment by the IPSO-SVM model. Obviously, compared with other models, the actual value matches the predicted value better. Therefore, after the IPSO-SVM model is trained by the training set, the prediction effect of this model is satisfactory, and the prediction result is also within the acceptable range and can be used to predict the change in wood color after artificial weathering.

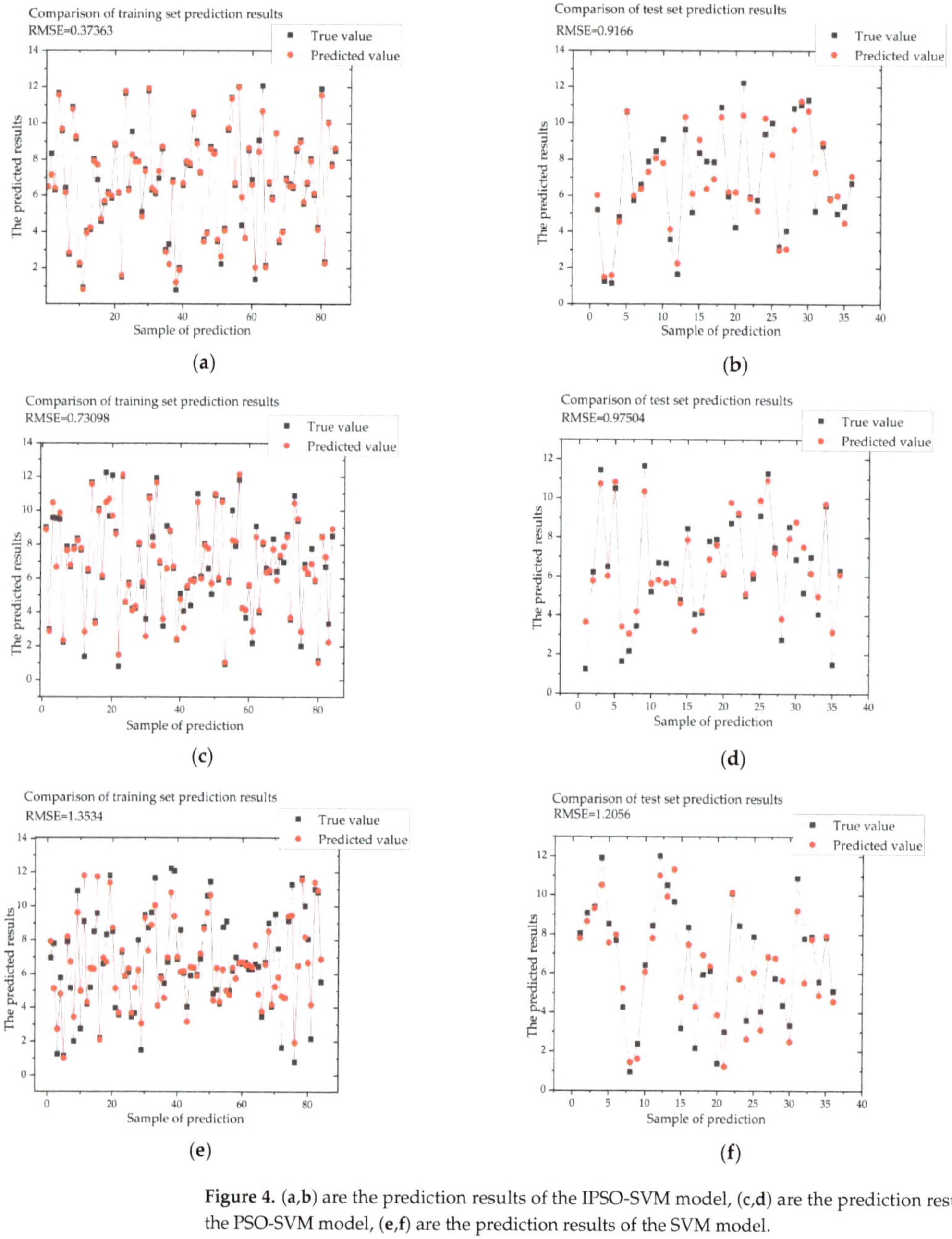

Figure 4. (**a**,**b**) are the prediction results of the IPSO-SVM model, (**c**,**d**) are the prediction results of the PSO-SVM model, (**e**,**f**) are the prediction results of the SVM model.

4. Conclusions

In order to solve the problems of wood color prediction, an SVM prediction model based on improved particle swarm optimization was proposed. In this study, the color change of heat-modified wood after artificial weathering was predicted in larch (*Larix gmelini*) and poplar (*Populus alba*). The heat treatment temperature, artificial weathering time, and wood type were used as the input variables, and the total color difference (ΔE) was used as the output variable to predict the color change. In order to verify the performance of the IPSO-SVM model in predicting wood color change, the mean absolute error (MAE), root mean square error (RMSE), goodness of fit (R^2), and standard deviation (STD) of the SVM model and of the PSO-SVM model were respectively compared. It was found that the IPSO-SVM model had the highest correlation coefficient and the lowest root mean square error, mean absolute error, and standard deviation. This shows that the IPSO-SVM model has the best predictive effect. Modeling studies using the IPSO-SVM model have provided results showing that the color changes of heat-modified wood after artificial weathering can be successfully predicted.

Author Contributions: Conceptualization, J.L. (Juncheng Li) and W.W.; methodology, J.L. (Juncheng Li); software, N.L.; validation, J.L. (Juncheng Li), W.W. and H.W.; formal analysis, J.L. (Juncheng Li); investigation, H.W.; resources, J.L. (Juncheng Li); data curation, J.L. (Juncheng Li); writing—original draft preparation, J.L. (Juncheng Li); writing—review and editing, J.L. (Juncheng Li); visualization, J.L. (Jinze Li); supervision, H.W.; project administration, W.W.; funding acquisition, W.W. All authors have read and agreed to the published version of the manuscript.

Funding: This study was funded by the Fundamental Research Funds for the Central Universities, grant number 2572019BL04, and the Scientific Research Foundation for the Returned Overseas Chinese Scholars of Heilongjiang Province, grant number LC201407.

Institutional Review Board Statement: Not applicable.

Informed Consent Statement: Not applicable.

Data Availability Statement: The data from this paper are openly available in a public repository that issues datasets with DOIs. The data that support the findings of this study are openly available in the European Journal of Wood and Wood Products at https://doi.org/10.1007/s00107-019-01449-0, reference number [17] (accessed on 29 October 2022). The data presented in this study are available in the article.

Conflicts of Interest: The authors declare no conflict of interest.

References

1. Rahman, M.O.; Hussain, A.; Basri, H. A critical review on waste paper sorting techniques. *Int. J. Environ. Sci. Technol.* **2014**, *11*, 551–564. [CrossRef]
2. Can, A.; Sivrikaya, H. Surface characterization of wood treated with boron compounds combined with water repellents. *Color Res. Appl.* **2019**, *44*, 462–472. [CrossRef]
3. Pratiwi, L.A.; Darmawan, W.; Priadi, T.; George, B.; Gérardin, P. Characterization of thermally modified short and long rotation teaks and the effects on coatings performance. *Maderas Cienc. Y Tecnol.* **2019**, *21*, 209–222. [CrossRef]
4. Calienno, L.; Pelosi, C.; Picchio, R.; Agresti, G.; Santamaria, U.; Balletti, F.; Monaco, A. Light-induced color changes and chemical modification of treated and untreated chestnut wood surface. *Stud. Conserv.* **2015**, *60*, 131–139. [CrossRef]
5. Borůvka, V.; Šedivka, P.; Novák, D.; Holeček, T.; Turek, J. Haptic and Aesthetic Properties of Heat-Treated Modified Birch Wood. *Forests* **2021**, *12*, 1081. [CrossRef]
6. Ayadi, N.; Lejeune, F.; Charrier, F.; Charrier, B. Color stability of heat-treated wood during artificial weathering. *Holz Als Roh-Und Werkst.* **2003**, *61*, 221–226. [CrossRef]
7. Tomak, E.D.; Ustaomer, D.; Yildiz, S.; Pesman, E. Changes in surface and mechanical properties of heat treated wood during natural weathering. *Measurement* **2014**, *53*, 30–39. [CrossRef]
8. Sivrikaya, H.; Tesařová, D.; Jeřábková, E.; Can, A. Color change and emission of volatile organic compounds from Scots pine exposed to heat and vacuum-heat treatment. *J. Build. Eng.* **2019**, *26*, 100918. [CrossRef]
9. Jiang, H.; Quanji, L.U.; Guanjun, L.I.; Min, L.I.; Jianing, L.I. Effect of heat treatment on the surface color of rubber wood (*Hevea brasiliensis*). *Wood Res.* **2020**, *65*, 633–644. [CrossRef]
10. Zhou, F.; Zhou, Y.; Fu, Z.; Gao, X. Effects of density on colour and gloss variability changes of wood induced by heat treatment. *Color Res. Appl.* **2021**, *46*, 1151–1160. [CrossRef]
11. Varga, D.; Van der Zee, M.E. Influence of steaming on selected wood properties of four hardwood species. *Holz Als Roh-Und Werkst.* **2008**, *66*, 11–18. [CrossRef]
12. Sikora, A.; Kačík, F.; Gaff, M.; Vondrová, V.; Bubeníková, T.; Kubovský, I. Impact of thermal modification on color and chemical changes of spruce and oak wood. *J. Wood Sci.* **2018**, *64*, 406–416. [CrossRef]
13. Bourgois, P.J.; Janin, G.; Guyonnet, R. La mesure de couleur. Une methode d'etude et d'optimisation des transformations chimiques du bois thermolyse. *Holzforschung* **1991**, *45*, 377–382. [CrossRef]

14. Tofani, G.; Cornet, I.; Tavernier, S. Multiple linear regression to predict the brightness of waste fibres mixtures before bleaching. *Chem. Pap.* **2022**, *76*, 4351–4365. [CrossRef]
15. Sun, L.; Fan, Y.; Zhang, D.; Cao, J. Online modeling for wood drying based on least squares support vector machine. *Chin. J. Sci. Instrum.* **2009**, *30*, 1991–1995. [CrossRef]
16. Li, J.; Sun, L. Forecasting of Wood Moisture Content Based on Modified Ant Colony Algorithm to Optimize LSSVM Parameters. *IEEE Access* **2020**, *99*, 85116–85127. [CrossRef]
17. Nguyen, T.T.; Nguyen, T.H.V.; Ji, X.D.; Yuan, B.N.; Trinh, H.M.; Do, K.T.L.; Guo, M.H. Prediction of the color change of heat-treated wood during artificial weathering by artificial neural network. *Eur. J. Wood Wood Prod.* **2019**, *77*, 1107–1116. [CrossRef]
18. Mitsui, K.; Takada, H.; Sugiyama, M.; Hasegawa, R. Changes in the properties of light-irradiated wood with heat treatment. part 1. effect of treatment conditions on the change in color. *Holzforschung* **2001**, *55*, 601–605. [CrossRef]
19. Xing, D.; Wang, S.; Li, J. Effect of Artificial Weathering on the Properties of Industrial-Scale Thermally Modified Wood. *Bioresources* **2015**, *10*, 8238–8252. [CrossRef]
20. Cortes, C.; Vapnik, V.N. Support Vector Networks. *Mach. Learn.* **1995**, *20*, 273–297. [CrossRef]
21. Vapnik, V.N. *The Nature of Statistical Learning Theory*; Springer: Berlin/Heidelberg, Germany, 1995. [CrossRef]
22. Kennedy, J.; Eberhart, R. Particle Swarm Optimization. In Proceedings of the ICNN'95-International Conference on Neural Networks, Perth, WA, Australia, 27 November–1 December 1995. [CrossRef]
23. Shi, Y.-H.; Eberhart, R.-C. A Modified Particle Swarm Optimizer. In Proceedings of the IEEE World Congress on Computational Intelligence, Anchorage, AK, USA, 4–9 May 1998. [CrossRef]

forests

MDPI

Article

Predicting the Mechanical Properties of Heat-Treated Woods Using Optimization-Algorithm-Based BPNN

Runze Zhang and Yujie Zhu *

College of Engineering and Technology, Northeast Forestry University, Harbin 150040, China; cherish9z@nefu.edu.cn
* Correspondence: zhuyujie004@126.com; Tel.: +86-136-2460-2246

Abstract: This paper aims to enhance the accuracy of predicting the mechanical behavior of wood subjected to thermal modification using an improved dung beetle optimization (IDBO) model. The IDBO algorithm improves the original DBO algorithm via three main steps: (1) using piecewise linear chaotic mapping (PWLCM) to generate the initial dung beetle species and increase its heterogeneity; (2) adopting an adaptive nonlinear decreasing producer ratio model to control the number of producers and boost the algorithm's convergence rate; and (3) applying a dimensional learning-enhanced foraging (DLF) search strategy that optimizes the algorithm's ability to explore and exploit the search space. The IDBO algorithm is evaluated on 14 benchmark functions and outperforms other algorithms. The IDBO algorithm is then applied to optimize a back-propagation (BP) neural network for predicting five mechanical property parameters of heat-treated larch-sawn timber. The results indicate that the IDBO-BP model significantly reduces the error compared with the BP, tent-sparrow search algorithm (TSSA)-BP, grey wolf optimizer (GWO)-BP, nonlinear adaptive grouping grey wolf optimizer (IGWO)-BP and DBO-BP models, demonstrating its superiority in predicting the physical characteristics of lumber after heat treatment.

Keywords: dung beetle optimization; BP neural network; wood heat treatment; timber mechanical performance forecast

Citation: Zhang, R.; Zhu, Y. Predicting the Mechanical Properties of Heat-Treated Woods Using Optimization-Algorithm-Based BPNN. *Forests* **2023**, *14*, 935. https://doi.org/10.3390/f14050935

Academic Editors: Morwenna Spear and Miklós Bak

Received: 29 March 2023
Revised: 21 April 2023
Accepted: 27 April 2023
Published: 2 May 2023

1. Introduction

Timber is a widely utilized material in the construction and furniture industries because it has numerous benefits, such as environmental sustainability, aesthetic appeal and ease of processing. However, its limited stability and durability hinder its application [1,2]. These limitations have prompted the development of various wood modification techniques, such as chemical, physical and biological methods [3]. Heat treatment is a prevalent technique that enhances wood properties by altering its chemical, physical and structural characteristics through exposure to specific temperature and humidity conditions [4]. This treatment increases wood stability, durability and resistance to corrosion and hydrolysis while improving mechanical properties such as strength, stiffness and hardness [5,6]. Common heat treatment methods include vacuum, dry and moist treatments [7].

The improvement of timber properties via heat treatment has been demonstrated by studies. Korkut et al. [8] examined how the thermal process affects red bud maple's surface roughness and mechanical behaviors. The results indicated that increasing temperatures reduce density and moisture content but increase bending strength and surface roughness. Icel et al. [9] demonstrated that heat treatment significantly improves the physical properties, chemical composition and microstructure of spruce and pine, resulting in enhanced stability, durability performance and service life. Xue et al. [10] investigated how high-temperature heat treatment and impregnation modification techniques affect aspen lumber's physical and mechanical characteristics and found significant improvements in mechanical strength and preservation.

Despite its effectiveness in improving wood mechanical properties, heat treatment has certain limitations. Boonstra et al. [11] reported the decomposition of natural wood components during heat treatment, resulting in reduced wood quality. Hill [12] noted that the efficacy of heat treatment is influenced by various factors such as treatment time, temperature, humidity and wood species, making it challenging to control and optimize the process. Goli et al. [1] investigated the impact of heat treatment on the physical and mechanical properties of birch plywood, revealing an increase in density and hardness but a decrease in moisture content and bending strength.

To address these limitations, researchers have explored the use of neural network models to predict wood mechanical properties. Kohonen [13] introduced self-organizing mapping (SOM) as one of the earliest prototypes for applying neural networks to nonlinear prediction problems. C.G.O. [14] highlighted the potential for neural networks to model complex nonlinear relationships for predicting mechanical properties such as strength and stiffness. Adamopoulos et al. [15] investigated the relationship between the fiber properties of recycled pulp and the mechanical properties of corrugated base paper. Multiple linear regression and artificial neural network models were used to predict the tensile strength and compressive strength of corrugated base paper with different fiber sources, and the results showed that the artificial neural network model was more accurate and stable than the multiple linear regression model. You et al. [16] demonstrated that an artificial neural network (ANN) model based on nondestructive vibration testing can successfully predict the MOE of bamboo–wood composites with high accuracy.

Although employing the BP neural network models to forecast the physical characteristics of heat-treated lumber reduces experimental costs, it presents certain challenges, such as susceptibility to local minima during the learning process and a poor generalization ability, resulting in the inaccurate prediction of new data. To address these limitations, some researchers have explored combining BP neural networks with meta-heuristic algorithms to improve prediction accuracy and model robustness. Chen et al. [17] integrated the Aquila Optimization Algorithm (AOA) [18] with BP neural networks to accurately predict the balance water rate and weight ratio of thermal processing timber, and Wang et al. [19] utilized the Carnivorous Plant Algorithm (CPA) [20] to ameliorate BP neural networks for predicting the adhesion intensity and coarseness of the surfaces of heat-treated wood. Their results indicated that both the AOA-BP and CPA-BP models outperform traditional BP neural network models.

Meta-heuristic algorithms can effectively avoid local optima and improve prediction accuracy when combined with BP neural networks. However, local optima may still occur due to inappropriate algorithm parameters or unreasonable algorithm combinations, resulting in poor model performance. To address this issue, some researchers have suggested improving the original meta-heuristic algorithms before applying them to optimize BP neural networks, aiming to increase the model's generalization ability and reliability. For example, Li et al. [21] enhanced the sparrow search algorithm (SSA) [22] with tent chaotic mapping and applied it to optimize BP neural networks for predicting the mechanical characteristics of heat-treated timber. They found that the TSSA-BP model performs well. Ma et al. [23] proposed a nonlinear adaptive grouping strategy for the Gray Wolf Optimization (GWO) [24] algorithm and used it to optimize BP neural networks for timber mechanical performance forecasts. They demonstrated that the proposed IGWO-BP model has much higher prediction accuracy than that of conventional models.

Similarly, the original Dung Beetle Optimization (DBO) [25] algorithm has drawbacks in avoiding local optima and achieving satisfactory algorithmic accuracy for practical engineering applications. To address these flaws, this article proposes an Improved Dung Beetle Optimizer (IDBO) for optimizing BP neural networks. The IDBO algorithm incorporates three main improvements: first, utilizing piece-wise linear chaotic mapping (PWLCM) to initialize the dung beetle population to increase diversity; second, introducing an adaptive parameter adjustment strategy to enhance the early-stage best-finding ability and improve algorithmic search efficiency; and finally, balancing local and global search capabilities by incorporating a dimensional learning-enhanced foraging strategy (DLF).

The rest of this article is structured as follows: Section 2 introduces the basic theory of BP and DBO; Section 3 presents the IDBO algorithm model; Section 4 verifies the performance of the IDBO algorithm using benchmark functions; Section 5 evaluates the reliability of the suggested IDBO model for wood mechanical property predictions; and Section 6 concludes.

2. Theoretical Analysis of the Algorithm

2.1. Back-Propagation (BP) Neural Network Models

The BP neural network is a multi-layered feedforward model primarily utilized for supervised learning tasks. It typically includes input, hidden and output layers [26]. Neurons receive inputs from preceding layers and compute a weighted sum that is transformed by an activation function before being output to subsequent layers. The error between the desired and actual outputs is calculated and propagated backward through the network via a back-propagation algorithm. Weights are updated according to each neuron's contribution to the error using the chain rule. Multiple iterations minimize error and enable the network to approximate desired outputs.

This paper used MATLAB's machine learning toolbox (2019a) to create a BP network. Input data included heat treatment temperature, time and relative humidity, and output data comprised Longitudinal Compressive Strength (LCS), Transverse Rupture Strength (TRS), Transverse Modulus of Elasticity (TME), Radial Hardness (RH) and Tangential Hardness (TH). Five separate prediction models were developed using trainlm with a learning rate of 0.01. Figure 1 shows the structure of the BP neural network with a single hidden layer.

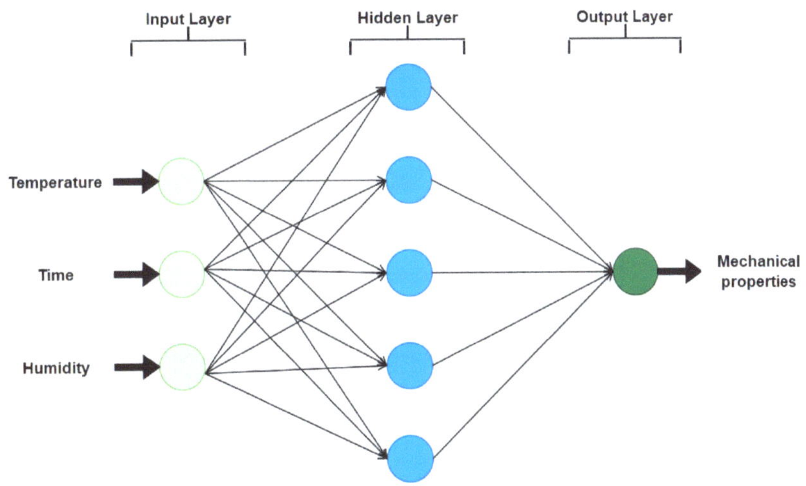

Figure 1. BP Network Structure.

2.2. The Traditional DBO Algorithm

The DBO is a novel swarm intelligence algorithm that simulates dung beetle habits, such as ball rolling, dancing, foraging, stealing, breeding and other behaviors, and the DBO algorithm comprises four optimization processes: rolling balls, breeding, foraging and stealing [25].

2.2.1. Dung Beetle Ball Rolling

Dung beetle rolling behavior is divided into an obstructed mode and an unobstructed mode.

Obstacle-Free Mode

When the dung beetle is moving forward without obstacles, the dung beetle uses the sun for navigation during dung ball rolling. In this model, the place of the dung beetle alters as the light intensity changes, and the position is renewed, as follows:

$$x_i^{t+1} = x_i^t + a \times k \times x_i^{t-1} + b \times \left|x_i^t - x_{wrost}^t\right| \quad (1)$$

where t indicates the count of the current iterations, and x_i^t is in terms of the position of the ith dung beetle in the population at the tth permutation. $k\epsilon(0,\ 0.2]$ shows a fixed parameter representing the flexure coefficient, b is an invariant quantity in the range of (0, 1), and α represents a natural coefficient with values of either -1 or 1, with 1 indicating no deviation and -1 indicating deviation from the original direction. x_{wrost}^t means the worst location in the present specie, and the change in light intensity is simulated by $\left|x_i^t - x_{wrost}^t\right|$.

Barrier Mode

The dung beetle, when it encounters an obstacle that prevents it from moving forward, needs to dance to regain a new forward direction. The authors use a tangent function to mimic the dancing behavior as a way to obtain a new rolling direction, which is only considered to be in the range of $[0, \pi]$, and the beetle continues rolling the dung ball once it has determined a new direction. The equation for updating the position at this point is as follows:

$$x_i^{t+1} = x_i^t + tan\theta\left|x_i^t - x_i^{t-1}\right| \quad (2)$$

When $\theta = 0, \frac{\pi}{2}, \pi$, no change occurs in the dung beetle's position.

2.2.2. Dung Beetle Breeding

In nature, female dung beetles roll their dung balls to a safe place suitable for egg laying and hide them as a way to provide a suitable habitat for their progeny. Inspired by this, the authors propose a frontier option strategy to model the brood ball location of female dung beetles:

$$\begin{cases} Lf^* = max\left\{x_{gbest}^t \times (1-R), Lf\right\} \\ Uf^* = min\left\{x_{gbest}^t \times (1+R), Uf\right\} \end{cases} \quad (3)$$

where $R = \frac{1-t}{T_{max}}$, and T_{max} is the upper limit of iterations. The lower and upper limits of the optimization problem are Lf and Uf, respectively. The current population attains the global optimum at x_{gbest}^t. The authors define the spawning's lower and upper edges region with Lf and Uf, which means that the region where the dung beetles spawn is dynamically adjusted with the number of iterations.

When a female dung beetle determines the spawning area, she lays her eggs in that area. Each female dung beetle generates a single brood ball per cycle. The area where oviposition occurs is dynamically adjusted with the count of iterations, so the position of the nestling sphere is also dynamic during the iterations, as defined below:

$$B_i^{t+1} = x_{gbest}^t + b_1 \times \left(B_i^t - Lf^*\right) + b_2 \times \left(B_i^t - Uf^*\right) \quad (4)$$

where B_i^{t+1} is the location of the ith brood ball at the tth iteration, b_1, b_2 represent two random and independent vectors that have D components each, and D is the number of parameters in the optimization problem. The position of the nestling ball must be restricted to the spawning area.

2.2.3. Dung Beetle Foraging

This behavior is mainly aimed at small dung beetles. Some mature dung beetles emerge from the ground in search of food, and the optimal foraging area for small dung beetles is dynamically updated, as indicated below:

$$\begin{cases} Lf^l = max\{x^t_{lbest} \times (1-R), Lf\} \\ Uf^l = min\{x^t_{lbest} \times (1+R), Uf\} \end{cases} \tag{5}$$

where R is the same as the previous definition, and x^t_{lbest} represents the best local position for the current population. The authors use Lf^l and Uf^l to define the bottom and top boundaries of the foraging region of the small dung beetle, respectively. The equation for updating the position at this point is as follows:

$$x_i^{t+1} = x_i^t + C_1 \times \left(x_i^t - Lf^l\right) + C_2 \times \left(x_i^t - Uf^l\right) \tag{6}$$

where C_1 is a number that follows a normal distribution when chosen randomly, namely C_1~N(0, 1), and C_2 is a random vector belonging to a range of $(0,1)$ of $1 \times D$.

2.2.4. Dung Beetle Stealing

In the population, there are some dung beetles that steal dung balls from other dung beetles, and the authors update the location of the thieving dung beetles as follows:

$$x_i^{t+1} = x^t_{lbest} + S \times g \times \left(\left|x_i^t - x^t_{gbest}\right| + \left|x_i^t - x^t_{lbest}\right|\right) \tag{7}$$

where g is a vector of dimension D that is randomly chosen, obeying a normal distribution, and S indicates a constant value.

The diagram of the DBO algorithm's process is presented in Figure 2. The algorithm first generates a random initial population of dung beetles in the search space and defines its relevant parameters and then calculates the value of each agent's fitness to adjust their positions based on the objective function, and it finally repeats the above steps until the termination criteria are met, showing the globally optimal solution and its corresponding value of suitability.

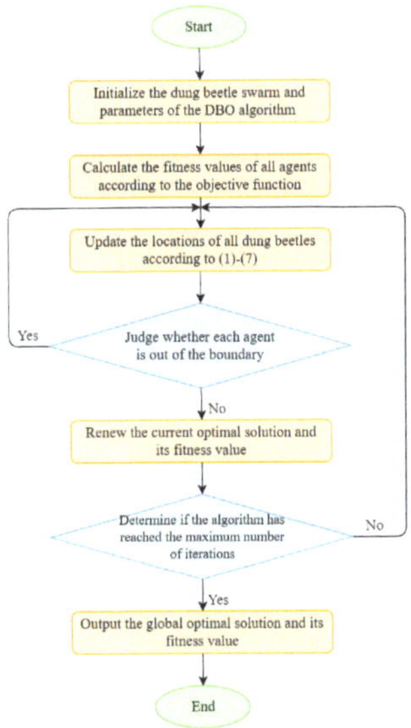

Figure 2. Diagram of the DBO algorithm.

3. Proposed Method

3.1. Improved Dung Beetle Optimizer

Despite its simplicity and successful application to several engineering design problems, the DBO algorithm exhibits limitations, such as poor global searchability and premature convergence to local optima. To address these deficiencies, this paper proposes an improved dung beetle optimizer with specific enhancement strategies.

3.1.1. Piece-Wise Linear Chaotic Mapping

When tackling sophisticated optimization projects, the simple random generation of initial populations by the DBO can result in rapid declines in population diversity and excessive convergence during later iterations. Chaotic sequences have recently been adopted for improving population diversity in meta-heuristic algorithms due to their randomness and ergodicity [27]. The basic approach involves mapping chaotic sequences into individual search spaces using chaos models such as Tent [21], Logistic [28] or Kent [29] chaos mapping.

When selecting a chaotic mapping, two important characteristics—simplicity and ergodicity—must be considered. Segmented linear chaotic mapping satisfies these criteria with its relatively uniform phase distribution and simple equations compared to those of other one-dimensional chaotic systems. This paper uses PWLCM mapping to generate a random sequence with dynamical equations [30] defined as follows:

$$x_{i+1} = F_p(x_i) = \begin{cases} \frac{x_i}{p}, \ 0 \leq x_i < p \\ x_i - \frac{p}{0.5} - p, \ p \leq x_i < 0.5 \\ F_p(1 - x_i), \ 0.5 \leq x_i < 1 \end{cases} \quad (8)$$

With the control parameter $p\epsilon(0, 0.5)$, the $x_i\epsilon(0, 1)$ system is in a chaotic state. Assigning initial values to the control function p and x_0, after circular iterations, a random sequence in the interval (0, 1) can be obtained, which has excellent statistical properties and is commonly applied to generate the initial solution of the algorithm to increase the diversity of the species. When $p = 0.4$, the initial overall (one-dimensional) distribution is as shown in Figure 3.

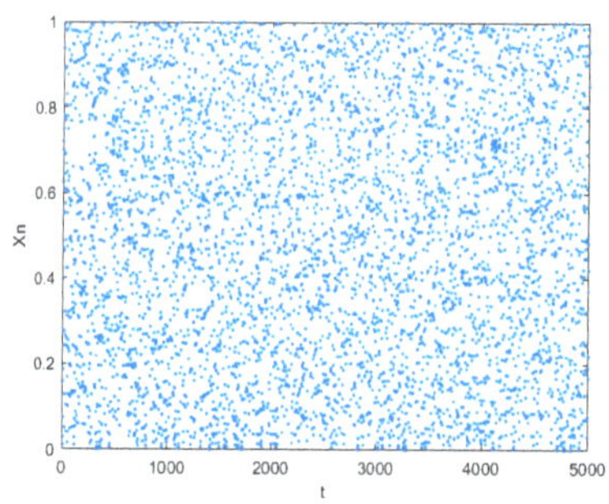

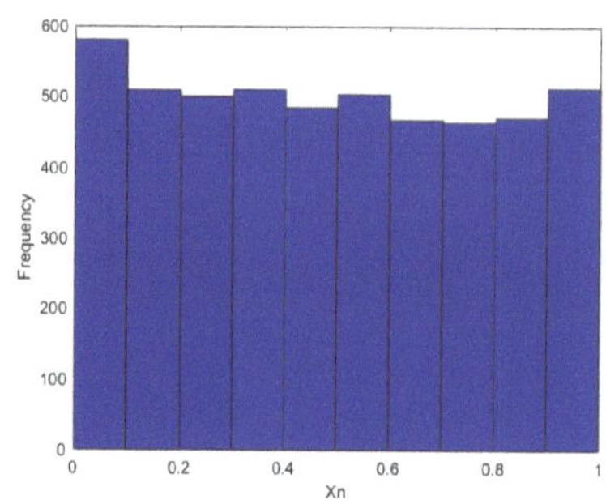

Figure 3. Population initialization in PWLCM: (**a**) scatter map; (**b**) frequency distribution histogram.

3.1.2. Self-Adaptive Parameter Adjustment Tactics

The DBO algorithm comprises four main components: the global development of producers, egg-laying by female dung beetles, foraging by small dung beetles and stealing behavior by stealing dung beetles. The number of producers determines the explored scope and convergence rate of the algorithm. However, in the original algorithm, the authors do not specify the distribution ratio of these four agents, which may result in incomplete coverage of the search space or slow convergence. To address this issue, this article suggests an adaptively derived non-linear decreasing producer ratio model (Equation (9)) with an initial producer ratio set to 0.4. With a sufficient number of producers, the algorithm can conduct a more extensive global search during early iterations and fully exploit potential solutions. As iterations progress and the demand for producers decreases, their proportion decreases from 0.4 to 0.2, and exploitation is minimized during the middle and late stages to facilitate rapid convergence.

$$P_{percent} = 0.4 - \frac{t(0.4 - 0.2)}{M} \tag{9}$$

This model enhances algorithm diversity and robustness by dynamically adjusting the number of producers and controlling competition and cooperation among them according to certain strategies. This maintains algorithm versatility while gradually reducing producer numbers during the search process to effectively balance convergence speed and exploration ability. As a result, global exploration capability and convergence speed are improved.

3.1.3. Dimension Learning-Enhanced Foraging Search Strategy

During the search process, the DBO algorithm may select a locally optimal solution while ignoring a more optimal global solution due to its random strategy and lack of effective evasion methods. To address this issue, we introduce the Dimension Learning-enhanced Foraging (DLF) search strategy.

In the original DBO algorithm, position updates are obtained according to objective functions corresponding to different agents. This can contribute to slow convergence, trapping in local optima, and the premature loss of population diversity due to random agent selection. In contrast, our proposed DLF search strategy enables agents to update their locations by learning from their neighbors and completing their behaviors accordingly.

In the DLF search strategy, the new location of the dung beetles $X_i(t)$ is obtained from Equation (12), in which the beetle gains information from various neighbors and a randomly chosen agent from the population. Then, in addition to $X_{i-DBO}(t+1)$, the DLF search strategy generates another agent for the new location of beetle $X_i(t)$, named $X_{i-DLF}(t+1)$. To this purpose, first, using Formula (10), the radius $R_i(t)$ is obtained with the magnitude of the displacement vector between the X_i current position $X_i(t)$ and the agent position $X_{i-DBO}(t+1)$.

$$R_i(t) = \|X_i(t) - X_{i-DBO}(t+1)\| \tag{10}$$

Next, the neighborhood of $X_i(t)$ expressed by $N_i(t)$ is derived using Equation (11), which is related to the radius $R_i(t)$, where D_i is the length of the line segment joining $X_i(t)$ and $X_j(t)$.

$$N_i(t) = \{X_j(t) | D_i(X_i(t), X_j(t)) \leq R_i(t), X_j(t) \epsilon Pop\} \tag{11}$$

Then, multi-domain learning is performed using Equation (12), where d denotes dimensionality.

$$X_{i-DLF,d}(t+1) = X_{i,d}(t) + rand \times (X_{n,d}(t) - X_{r,d}(t)) \tag{12}$$

Finally, the locations are updated with Formula (13), and the above steps are repeated until a predefined maximum number of iterations is reached and returns the global optimal solution and its corresponding fitness value.

$$X_i(t+1) = \begin{cases} X_{i-DBO}(t+1), \; if \; f(X_{i-DBO}) < f(X_{i-DLH}) \\ X_{i-DLF}(t+1) \qquad\qquad\qquad\qquad otherwise \end{cases} \tag{13}$$

3.2. The IDBO-BP Algorithm

The BP neural network models are assigned random weights and thresholds with numerous variable parameters that can cause instability in model computation [31]. The predictive performance of these models can be enhanced by optimizing BP neural networks using DBO. However, the DBO algorithm has issues such as an uneven initial population distribution, susceptibility to local optima and a slow convergence speed. To address these problems, this paper proposes the IDBO algorithm.

First, PWLCM is introduced to initialize the population to produce a more uniform initial solution distribution and high-quality initial solutions while augmenting population richness. Second, using an adaptive parameter adjustment strategy to dynamically tune producer numbers according to the search process accelerates the convergence rate and enhances global exploration capability. Finally, employing the DLF search strategy balances the exploration and exploitation abilities of the algorithm.

The main idea of the IDBO-BP algorithm is to update the weights and thresholds of the BP neural network by continuously updating the positions of the dung beetle swarm until the global best position is found, i.e., the optimal solution.

The diagram of the IDBO-BP algorithm's process is presented in Figure 4. Data are first normalized using Equation (15) before the proportion of dung beetles pushing the ball is dynamically adjusted according to Formula (9). Dung beetle population locations are then initialized using PWLCM mapping, as shown in Equation (8). Fitness values for all dung beetles are calculated, and their locations are updated according to Formulas (1)–(7). The current optimal solution is updated in combination with the DLF strategy before the positions of all dung beetles are updated again in combination with Equation (13). When the iteration limit is reached, the best solution is output along with the optimal parameters of the BP neural network.

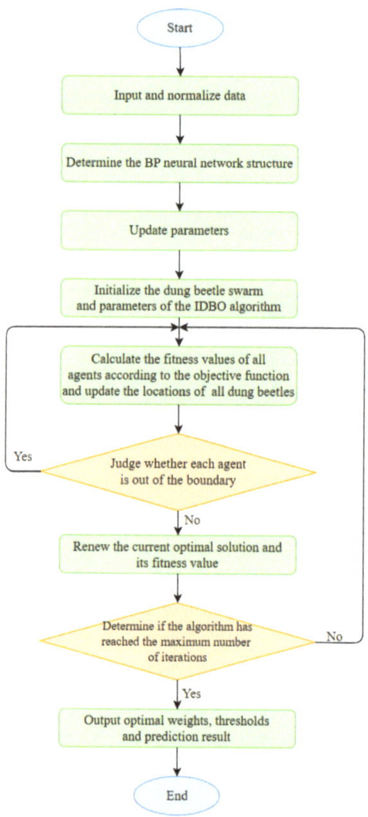

Figure 4. Diagram of the IDBO-BP algorithm.

4. Evaluate the Effectiveness of the Suggested IDBO Model

The efficacy of the suggested IDBO method is assessed through a series of experiments utilizing various benchmark functions in this section.

4.1. Benchmark Functions

To objectively appraise the effectiveness of various meta-heuristic algorithms and to validate the usefulness of the IDBO amelioration strategy, 14 standard test functions were selected from the literature [32], and the CEC2017 test function was utilized to evaluate the capability of the IDBO algorithm. Functions F1–F8 are unimodal with a single global optimal solution and were employed to gauge the velocity and exactness of convergence of the algorithm. Functions F9–F14 are multimodal with a single global optimum and several local optima and were used to estimate the global search and excavation capabilities of the algorithm. The details of these benchmark functions, including their expressions, dimensions, search ranges and theoretical optimal solutions, are given in Tables 1 and 2. To provide a more intuitive understanding of these benchmark functions and their optimal values, Figures 5 and 6 depict 3D views (30 dimensions) of some of these functions.

Table 1. Unimodal benchmark functions.

Function	Dim	Range	F_{min}
$f_1(x) = \sum_{i=1}^{n} x_i^2$	30/50/100	[−100, 100]	0
$f_2(x) = \sum_{i=1}^{n} \lvert x_i \rvert + \prod_{i=1}^{n} \lvert x_i \rvert$	30/50/100	[−10, 10]	0
$f_3(x)=\sum_{i=1}^{n}\left(\sum_{j=1}^{i} x_j\right)^2$	30/50/100	[−100, 100]	0
$f_4(x) = max\{\lvert x_i \rvert, 1 \le i \le n\}$	30/50/100	[−100, 100]	0
$f_5(x) = \sum_{i=1}^{n-1}\left[100(x_{i+1} - x_{i^2})^2 + (x_i - 1)^2\right]$	30/50/100	[−30, 30]	0
$f_6(x)=\sum_{i=1}^{n}([x_i + 0.5])^2$	30/50/100	[−100, 100]	0
$f_7(x)=x_1{}^2+10^6\sum_{i=2}^{n} x_i{}^2$	30/50/100	[−100, 100]	0
$f_8(x)=\sum_{i=1}^{n} x_i{}^2 + (\sum_{i=1}^{n} 0.5ix_i)^2 + (\sum_{i=1}^{n} 0.5ix_i)^4$	30/50/100	[−5, 10]	0

Table 2. Multimodal benchmark functions.

Function	Dim	Range	F_{min}
$f_9(x)=\sum_{i=1}^{n}\left[x_i{}^2 - 10cos(2\pi x_i) + 10\right]$	30/50/100	[−5.12, 5.12]	0
$f_{10}(x)=\sum_{i=1}^{n}\lvert x_i sin(x_i) + 0.1x_i \rvert$	30/50/100	[−10, 10]	0
$f_{11}(x)=\frac{\pi}{n}\left\{10sin(\pi y_1) + \sum_{i=1}^{n-1}(y_i - 1)^2\left[1 + 10sin^2(\pi y_{i+1})\right] + (y_n - 1)^2\right\} + \sum_{i=1}^{n} u(x_i, 10, 100, 4)$, where $y_i = 1+\frac{x_i+1}{4}$, for all $i = 1, \ldots, n$; $u(x_i,a,k,m)=\begin{cases} k(x_i - a)^m & x_i > a \\ 0 & -a < x_i < a \\ k(-x_i - a)^m & x_i < -a \end{cases}$	30/50/100	[−50, 50]	0
$f_{12}(x) = 0.1\{sin^2(3\pi x_1) + \sum_{i=1}^{n}(x_i - 1)^2[1 + sin^2(3\pi x_i + 1)] + (x_n - 1)^2[1+ sin^2(2\pi x_n)]\} + \sum_{i=1}^{n} u(x_i, 5, 100, 4)$	30/50/100	[−50, 50]	0
$f_{13}(x)=\left[\frac{1}{n-1}\sum_{i=1}^{n-1}(\sqrt{s_i} \times (sin(50s_i^{0.2}) + 1)\right]^2$, $s_i = \sqrt{x_i^2 + x_{i+1}^2}$	30/50/100	[−100, 100]	0
$f_{14}(x) = sin^2(\pi y_1) + \sum_{i=1}^{n-1}(y_i - 1)^2\left[1 + 10sin^2(\pi y_i + 1)\right] + (y_n - 1)^2\left[1 + sin^2(2\pi y_n)\right]$, where $y_i = 1 + \frac{x_i-1}{4}$, for all $i = 1, \ldots, n$	30/50/100	[−10, 10]	0

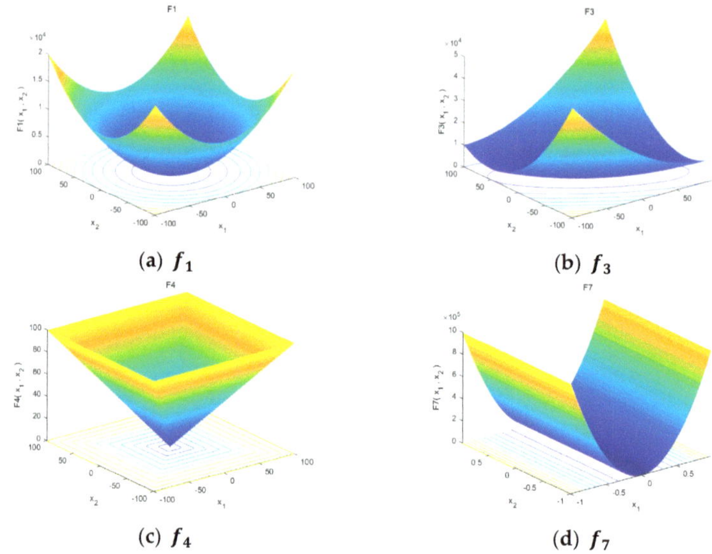

Figure 5. Three-dimensional view of partial unimodal test functions.

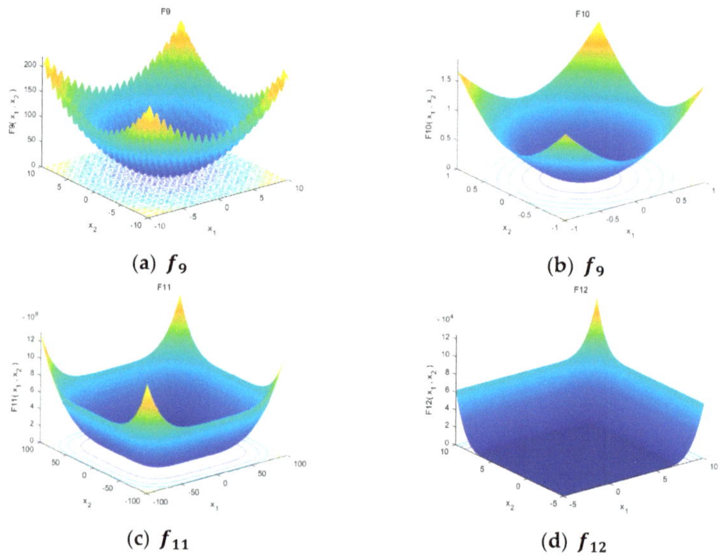

Figure 6. Three-dimensional view of partial multimodal test functions.

4.2. Contrast Algorithm and Experimental Parameter Settings

To fully validate the reliability of the presented IDBO model, its results were compared with those of four widely used basic metaheuristics: PSO (Eberhart et al., 1995) [33], GWO (Mirjalli et al., 2014), WOA (Mirjalili et al., 2016) [34] and DBO (Xue et al., 2022) [25]. As indicated in Table 3, the parameter settings that were recommended in their respective original works were adopted for the experiments involving these comparison algorithms.

Table 3. Algorithms' parameters setting.

Algorithm	Parameter	Setting
WOA	a	Gradually reduced from 2 to 0
GWO	a	Uniformly lowered from 2 to 0
PSO	C_1 and C_2	2
	Inertia weight	Linearly decreased from 0.9 to 0.1
DBO	α and β	0.1
	a and b	0.3 and 0.5
IDBO	α and β	0.1
	a and b	0.3 and 0.5
	The proportions of the ball-rolling dung beetle, the brood ball, the small dung beetle and the thief were [0.4 0.2], 0.2, 0.2 and [0.2 0.4]	

To more accurately evaluate the efficacy of the IDBO algorithm and its comparative algorithms, a population size of $N = 30$ was uniformly set, and the upper limit of iterations was fixed at 500. Each model was executed separately 30 times. The dimension D was set to 30, 50 and 100 to examine the effectiveness of the suggested approach in searching for merits across different dimensions. To minimize the influence of randomness in the simulation results, the optimal values, means and standard deviations of the optimization results (fitness) were recorded separately to appraise the exploration performance, accuracy and reliability of the models.

The experiment was implemented on a Windows 11 operating system with an 11th Gen Intel® Core™ i7-11700 processor with 2.5 GHz and 16 GB RAM using MATLAB 2019a for simulation. The optimal fitness, mean fitness and standard error of fitness for the IDBO algorithm and its comparative algorithms are presented in Table 4, where bold values indicate the best consequences. Additionally, the bottom three lines of each table show the 'w/t/l' for the wins (w), ties (t) and losses (l) of each algorithm.

Table 4. Unimodal benchmark function optimization results.

F	D	Index	WOA	GWO	PSO	DBO	IDBO
F1	30	Best	3.94×10^{-83}	7.29×10^{-30}	6.50×10^{2}	1.03×10^{-176}	5.01×10^{-201}
		Mean	7.03×10^{-74}	9.20×10^{-28}	1.77×10^{3}	1.77×10^{-89}	8.31×10^{-151}
		STD	3.41×10^{-73}	9.69×10^{-28}	6.99×10^{2}	9.68×10^{-89}	4.46×10^{-150}
	50	Best	4.13×10^{-89}	1.29×10^{-29}	6.20×10^{2}	1.71×10^{-163}	1.64×10^{-199}
		Mean	6.15×10^{-72}	9.52×10^{-28}	1.93×10^{3}	1.71×10^{-110}	8.00×10^{-151}
		STD	4.33×10^{-71}	1.13×10^{-27}	8.76×10^{2}	1.21×10^{-109}	5.66×10^{-150}
	100	Best	3.31×10^{-88}	2.43×10^{-29}	7.52×10^{2}	2.11×10^{-182}	1.89×10^{-208}
		Mean	2.61×10^{-72}	1.20×10^{-27}	2.09×10^{3}	4.94×10^{-105}	6.48×10^{-152}
		STD	2.36×10^{-71}	2.64×10^{-27}	7.95×10^{2}	4.90×10^{-104}	6.48×10^{-151}
F2	30	Best	1.71×10^{-57}	1.67×10^{-17}	1.41×10^{1}	7.24×10^{-81}	3.30×10^{-111}
		Mean	8.35×10^{-50}	9.15×10^{-17}	2.03×10^{1}	6.05×10^{-55}	1.98×10^{-83}
		STD	4.56×10^{-49}	5.34×10^{-17}	3.98×10^{0}	3.31×10^{-54}	9.03×10^{-83}
	50	Best	2.53×10^{-58}	2.48×10^{-17}	9.95×10^{0}	1.67×10^{-83}	1.64×10^{-103}
		Mean	9.04×10^{-51}	9.55×10^{-17}	1.95×10^{1}	2.02×10^{-48}	1.46×10^{-80}
		STD	4.88×10^{-50}	6.54×10^{-17}	4.77×10^{0}	1.43×10^{-47}	1.04×10^{-79}
	100	Best	5.90×10^{-59}	1.06×10^{-17}	1.05×10^{1}	5.00×10^{-84}	8.80×10^{-107}
		Mean	3.65×10^{-50}	9.57×10^{-17}	1.94×10^{1}	2.17×10^{-56}	1.80×10^{-80}
		STD	3.22×10^{-49}	7.80×10^{-17}	4.20×10^{0}	1.71×10^{-55}	1.45×10^{-79}

Table 4. *Cont.*

F	D	Index	WOA	GWO	PSO	DBO	IDBO
F3	30	Best	8.29×10^{3}	5.82×10^{-9}	2.23×10^{3}	1.61×10^{-148}	2.25×10^{-187}
		Mean	4.90×10^{4}	1.64×10^{-5}	4.87×10^{3}	5.61×10^{-79}	3.79×10^{-95}
		STD	1.40×10^{4}	3.50×10^{-5}	1.71×10^{3}	3.07×10^{-78}	2.08×10^{-94}
	50	Best	1.48×10^{4}	2.04×10^{-9}	1.34×10^{3}	9.33×10^{-144}	6.40×10^{-178}
		Mean	4.36×10^{4}	3.33×10^{-5}	5.71×10^{3}	5.63×10^{-81}	4.74×10^{-99}
		STD	1.31×10^{4}	1.23×10^{-4}	2.02×10^{3}	2.82×10^{-80}	3.35×10^{-98}
	100	Best	1.47×10^{4}	3.82×10^{-9}	2.16×10^{3}	2.31×10^{-157}	9.16×10^{-183}
		Mean	4.27×10^{4}	3.43×10^{-5}	5.39×10^{3}	9.29×10^{-56}	1.63×10^{-85}
		STD	1.26×10^{4}	1.57×10^{-4}	1.60×10^{3}	9.29×10^{-55}	1.63×10^{-84}
F4	30	Best	2.60×10^{-1}	4.49×10^{-8}	1.79×10^{1}	2.07×10^{-77}	1.13×10^{-93}
		Mean	5.48×10^{1}	6.54×10^{-7}	2.85×10^{1}	3.47×10^{-50}	3.07×10^{-63}
		STD	2.85×10^{1}	5.12×10^{-7}	6.42×10^{0}	1.90×10^{-49}	1.68×10^{-62}
	50	Best	8.90×10^{-1}	2.11×10^{-8}	1.60×10^{1}	2.22×10^{-85}	5.72×10^{-100}
		Mean	4.97×10^{1}	7.76×10^{-7}	2.74×10^{1}	1.01×10^{-51}	1.35×10^{-68}
		STD	2.53×10^{1}	1.22×10^{-6}	5.21×10^{0}	7.18×10^{-51}	9.52×10^{-68}
	100	Best	2.73×10^{-1}	5.70×10^{-8}	1.57×10^{1}	3.24×10^{-81}	1.78×10^{-100}
		Mean	5.02×10^{1}	6.89×10^{-7}	2.84×10^{1}	4.82×10^{-48}	4.69×10^{-66}
		STD	2.65×10^{1}	8.72×10^{-7}	4.94×10^{0}	4.38×10^{-47}	4.42×10^{-65}
F5	30	Best	2.71×10^{1}	2.59×10^{1}	3.58×10^{4}	2.54×10^{1}	2.48×10^{1}
		Mean	2.79×10^{1}	2.69×10^{1}	4.52×10^{5}	2.58×10^{1}	2.52×10^{1}
		STD	4.45×10^{-1}	7.15×10^{-1}	4.03×10^{5}	1.83×10^{-1}	3.06×10^{-1}
	50	Best	2.72×10^{1}	2.59×10^{1}	4.35×10^{4}	2.52×10^{1}	2.47×10^{1}
		Mean	2.82×10^{1}	2.72×10^{1}	4.51×10^{5}	2.58×10^{1}	2.52×10^{1}
		STD	4.52×10^{-1}	7.16×10^{-1}	3.86×10^{5}	2.68×10^{-1}	3.10×10^{-1}
	100	Best	2.69×10^{1}	2.53×10^{1}	3.65×10^{4}	2.53×10^{1}	2.47×10^{1}
		Mean	2.80×10^{1}	2.70×10^{1}	4.21×10^{5}	2.58×10^{1}	2.52×10^{1}
		STD	4.45×10^{-1}	7.56×10^{-1}	3.26×10^{5}	2.17×10^{-1}	2.55×10^{-1}
	30	Best	0.00×10^{0}	0.00×10^{0}	8.05×10^{2}	0.00×10^{0}	0.00×10^{0}
		Mean	0.00×10^{0}	0.00×10^{0}	2.56×10^{3}	0.00×10^{0}	0.00×10^{0}
		STD	0.00×10^{0}	0.00×10^{0}	8.87×10^{2}	0.00×10^{0}	0.00×10^{0}
	50	Best	0.00×10^{0}	0.00×10^{0}	9.63×10^{2}	0.00×10^{0}	0.00×10^{0}
		Mean	0.00×10^{0}	0.00×10^{0}	2.60×10^{3}	0.00×10^{0}	0.00×10^{0}
		STD	0.00×10^{0}	0.00×10^{0}	8.53×10^{2}	0.00×10^{0}	0.00×10^{0}
	100	Best	0.00×10^{0}	0.00×10^{0}	8.35×10^{2}	0.00×10^{0}	0.00×10^{0}
		Mean	0.00×10^{0}	0.00×10^{0}	2.37×10^{3}	0.00×10^{0}	0.00×10^{0}
		STD	0.00×10^{0}	0.00×10^{0}	1.09×10^{3}	0.00×10^{0}	0.00×10^{0}
F7	30	Best	3.68×10^{-77}	4.83×10^{-23}	8.52×10^{8}	1.93×10^{-159}	9.32×10^{-196}
		Mean	1.85×10^{-66}	1.04×10^{-21}	1.75×10^{9}	6.52×10^{-108}	9.14×10^{-140}
		STD	9.98×10^{-66}	1.42×10^{-21}	5.93×10^{8}	2.48×10^{-107}	5.01×10^{-139}
	50	Best	1.87×10^{-81}	8.86×10^{-24}	6.67×10^{8}	1.04×10^{-158}	8.72×10^{-212}
		Mean	3.32×10^{-66}	7.31×10^{-22}	1.78×10^{9}	1.48×10^{-105}	3.17×10^{-148}
		STD	2.35×10^{-65}	1.05×10^{-21}	8.39×10^{8}	9.64×10^{-105}	2.23×10^{-147}
	100	Best	1.50×10^{-82}	2.06×10^{-23}	3.42×10^{8}	2.07×10^{-173}	1.84×10^{-200}
		Mean	4.67×10^{-68}	9.61×10^{-22}	1.83×10^{9}	4.86×10^{-93}	2.34×10^{-141}
		STD	3.28×10^{-67}	1.86×10^{-21}	8.66×10^{8}	4.86×10^{-92}	1.68×10^{-140}

Table 4. *Cont.*

F	D	Index	WOA	GWO	PSO	DBO	IDBO
F8	30	Best	7.79×10^{-86}	3.58×10^{-31}	6.83×10^{0}	2.68×10^{-173}	2.53×10^{-198}
		Mean	8.40×10^{-77}	2.89×10^{-29}	3.41×10^{1}	3.19×10^{-109}	6.83×10^{-161}
		STD	4.17×10^{-76}	5.64×10^{-29}	1.84×10^{1}	1.68×10^{-108}	2.95×10^{-160}
	50	Best	1.90×10^{-95}	3.62×10^{-31}	7.56×10^{0}	1.65×10^{-180}	1.59×10^{-215}
		Mean	1.34×10^{-75}	1.93×10^{-29}	2.92×10^{1}	1.24×10^{-116}	3.43×10^{-153}
		STD	5.27×10^{-75}	2.53×10^{-29}	2.23×10^{1}	6.24×10^{-116}	2.42×10^{-152}
	100	Best	7.99×10^{-93}	2.83×10^{-31}	8.12×10^{0}	2.98×10^{-177}	3.62×10^{-212}
		Mean	1.79×10^{-74}	2.53×10^{-29}	2.90×10^{1}	1.13×10^{-103}	8.64×10^{-153}
		STD	1.29×10^{-73}	4.61×10^{-29}	1.60×10^{1}	1.06×10^{-102}	8.64×10^{-152}
Rank	30	w/t/l	0/1/7	0/1/7	0/0/8	0/1/7	**7/1/0**
	50	w/t/l	0/1/7	0/1/7	0/0/8	0/1/7	**7/1/0**
	100	w/t/l	0/1/7	0/1/7	0/0/8	0/1/7	**7/1/0**

4.3. Evaluation of Exploration and Exploitation

The single-peak functions are well-suited to verify the development capability of algorithms in finding optimal solutions. Multimodal functions with numerous locally optimal solutions can assess the ability of IDBO to evade local optima during exploration.

As indicated in Table 4, the IDBO algorithm demonstrates significant improvement for all seven test functions except F6 across all dimensions. Table 5 reveals that the IDBO algorithm outperforms other algorithms in three different dimensions for all five test functions except F13 and that its optimal value, average and standard error are optimal. Thus, it can be inferred that the IDBO algorithm is more effective than DBO in evaluating optimal solutions, which proves that the modification tactic presented in this article can feasibly enhance the original algorithm's ability to explore.

Table 5. Multimodal benchmark function optimization results.

F	D	Index	WOA	GWO	PSO	DBO	IDBO
F9	30	Best	0.00×10^{0}	0.00×10^{0}	7.52×10^{1}	0.00×10^{0}	0.00×10^{0}
		Mean	0.00×10^{0}	2.79×10^{0}	1.09×10^{2}	9.62×10^{-1}	0.00×10^{0}
		STD	0.00×10^{0}	3.25×10^{0}	1.72×10^{1}	3.66×10^{0}	0.00×10^{0}
	50	Best	0.00×10^{0}	0.00×10^{0}	7.23×10^{1}	0.00×10^{0}	0.00×10^{0}
		Mean	3.41×10^{-15}	7.50×10^{0}	1.13×10^{2}	2.79×10^{-1}	0.00×10^{0}
		STD	1.78×10^{-14}	2.92×10^{1}	1.82×10^{1}	1.27×10^{0}	0.00×10^{0}
	100	Best	0.00×10^{0}	0.00×10^{0}	6.65×10^{1}	0.00×10^{0}	0.00×10^{0}
		Mean	1.14×10^{-15}	2.47×10^{0}	1.07×10^{2}	3.38×10^{0}	0.00×10^{0}
		STD	8.00×10^{-15}	4.00×10^{0}	1.83×10^{1}	1.68×10^{1}	0.00×10^{0}
F10	30	Best	4.32×10^{-58}	1.97×10^{-17}	6.43×10^{0}	4.37×10^{-89}	9.47×10^{-108}
		Mean	3.25×10^{-32}	5.71×10^{-4}	1.08×10^{1}	1.26×10^{-4}	2.44×10^{-81}
		STD	1.78×10^{-31}	7.99×10^{-4}	2.60×10^{0}	3.44×10^{-4}	1.34×10^{-80}
	50	Best	4.00×10^{-58}	1.75×10^{-16}	3.44×10^{0}	1.99×10^{-88}	2.32×10^{-101}
		Mean	3.83×10^{-1}	4.95×10^{-4}	1.01×10^{1}	1.32×10^{-1}	8.49×10^{-81}
		STD	2.71×10^{0}	5.42×10^{-4}	2.85×10^{0}	9.17×10^{-1}	6.01×10^{-80}
	100	Best	2.83×10^{-60}	3.32×10^{-17}	4.88×10^{0}	3.72×10^{-87}	1.69×10^{-109}
		Mean	2.28×10^{-1}	4.47×10^{-4}	1.08×10^{1}	1.99×10^{-3}	3.31×10^{-80}
		STD	2.28×10^{0}	5.21×10^{-4}	2.61×10^{0}	1.57×10^{-2}	2.91×10^{-79}

Table 5. *Cont.*

F	D	Index	WOA	GWO	PSO	DBO	IDBO
F11	30	Best	6.35×10^{-3}	1.31×10^{-2}	1.15×10^{1}	1.11×10^{-7}	4.56×10^{-6}
		Mean	2.76×10^{-2}	5.00×10^{-2}	1.07×10^{3}	3.57×10^{-3}	5.73×10^{-5}
		STD	2.02×10^{-2}	2.95×10^{-2}	4.83×10^{3}	1.89×10^{-2}	1.03×10^{-4}
	50	Best	2.47×10^{-3}	1.22×10^{-2}	1.26×10^{1}	7.74×10^{-8}	3.29×10^{-6}
		Mean	2.34×10^{-2}	4.54×10^{-2}	1.10×10^{3}	2.26×10^{-3}	3.79×10^{-5}
		STD	1.81×10^{-2}	2.56×10^{-2}	3.36×10^{3}	1.47×10^{-2}	4.79×10^{-5}
	100	Best	3.42×10^{-3}	1.32×10^{-2}	7.66×10^{0}	5.39×10^{-8}	2.17×10^{-6}
		Mean	2.26×10^{-2}	4.46×10^{-2}	1.61×10^{3}	9.76×10^{-5}	7.55×10^{-5}
		STD	1.92×10^{-2}	2.36×10^{-2}	6.56×10^{3}	7.08×10^{-4}	1.96×10^{-4}
F12	30	Best	9.44×10^{-2}	3.15×10^{-1}	7.93×10^{2}	1.79×10^{-4}	1.50×10^{-4}
		Mean	5.49×10^{-1}	6.39×10^{-1}	2.14×10^{5}	5.44×10^{-1}	3.32×10^{-2}
		STD	3.20×10^{-1}	1.90×10^{-1}	2.87×10^{5}	4.09×10^{-1}	4.58×10^{-2}
	50	Best	1.81×10^{-1}	1.00×10^{-1}	6.48×10^{1}	7.70×10^{-4}	5.05×10^{-5}
		Mean	6.09×10^{-1}	6.13×10^{-1}	1.89×10^{5}	6.14×10^{-1}	2.66×10^{-2}
		STD	2.75×10^{-1}	2.44×10^{-1}	3.94×10^{5}	4.19×10^{-1}	4.10×10^{-2}
	100	Best	1.17×10^{-1}	1.02×10^{-1}	7.83×10^{1}	1.35×10^{-3}	6.25×10^{-5}
		Mean	4.87×10^{-1}	6.46×10^{-1}	3.87×10^{5}	7.15×10^{-1}	3.72×10^{-2}
		STD	2.78×10^{-1}	2.30×10^{-1}	4.63×10^{5}	4.89×10^{-1}	6.20×10^{-2}
F13	30	Best	0.00×10^{0}	0.00×10^{0}	0.00×10^{0}	0.00×10^{0}	0.00×10^{0}
		Mean	8.23×10^{-5}	0.00×10^{0}	0.00×10^{0}	0.00×10^{0}	0.00×10^{0}
		STD	3.25×10^{-4}	0.00×10^{0}	0.00×10^{0}	0.00×10^{0}	0.00×10^{0}
	50	Best	0.00×10^{0}	0.00×10^{0}	0.00×10^{0}	0.00×10^{0}	0.00×10^{0}
		Mean	5.65×10^{-5}	0.00×10^{0}	0.00×10^{0}	0.00×10^{0}	0.00×10^{0}
		STD	3.11×10^{-4}	0.00×10^{0}	0.00×10^{0}	0.00×10^{0}	0.00×10^{0}
	100	Best	0.00×10^{0}	0.00×10^{0}	0.00×10^{0}	0.00×10^{0}	0.00×10^{0}
		Mean	7.81×10^{-5}	0.00×10^{0}	0.00×10^{0}	0.00×10^{0}	0.00×10^{0}
		STD	3.89×10^{-4}	0.00×10^{0}	0.00×10^{0}	0.00×10^{0}	0.00×10^{0}
F14	30	Best	3.69×10^{-1}	8.22×10^{-1}	4.86×10^{2}	8.97×10^{-2}	2.88×10^{-4}
		Mean	9.42×10^{-1}	1.33×10^{0}	1.03×10^{3}	5.50×10^{-1}	8.37×10^{-2}
		STD	4.24×10^{-1}	2.92×10^{-1}	3.21×10^{2}	4.22×10^{-1}	9.28×10^{-2}
	50	Best	2.55×10^{-1}	6.37×10^{-1}	3.73×10^{2}	2.69×10^{-4}	4.52×10^{-4}
		Mean	9.42×10^{-1}	1.23×10^{0}	9.25×10^{2}	4.83×10^{-1}	1.05×10^{-1}
		STD	3.68×10^{-1}	2.13×10^{-1}	3.46×10^{2}	2.10×10^{-1}	1.11×10^{-1}
	100	Best	1.93×10^{-1}	8.13×10^{-1}	4.09×10^{2}	1.40×10^{-3}	2.75×10^{-4}
		Mean	8.70×10^{-1}	1.25×10^{0}	9.66×10^{2}	5.16×10^{-1}	7.45×10^{-2}
		STD	3.82×10^{-1}	2.29×10^{-1}	3.04×10^{2}	2.94×10^{-1}	7.85×10^{-2}
Rank	30	w/t/l	0/1/5	0/1/5	0/1/5	0/1/5	4/2/0
	50	w/t/l	0/0/6	0/1/5	0/1/5	0/1/5	4/2/0
	100	w/t/l	0/0/6	0/1/5	0/1/5	0/1/5	4/2/0

4.4. Evaluation of Convergence Curves

To more intuitively observe and compare the convergence rate, accuracy and ability of each algorithm to evade local optima, the convergence curves for IDBO and four basic meta-heuristic algorithms f_1~f_{14} (30 dimensions) are presented in Figure 7. The transverse axis represents the number of iterations, whereas the longitudinal axis denotes the order of magnitude of fitness values. Fitness values are expressed as logarithms to base 10 to better illustrate convergence trends.

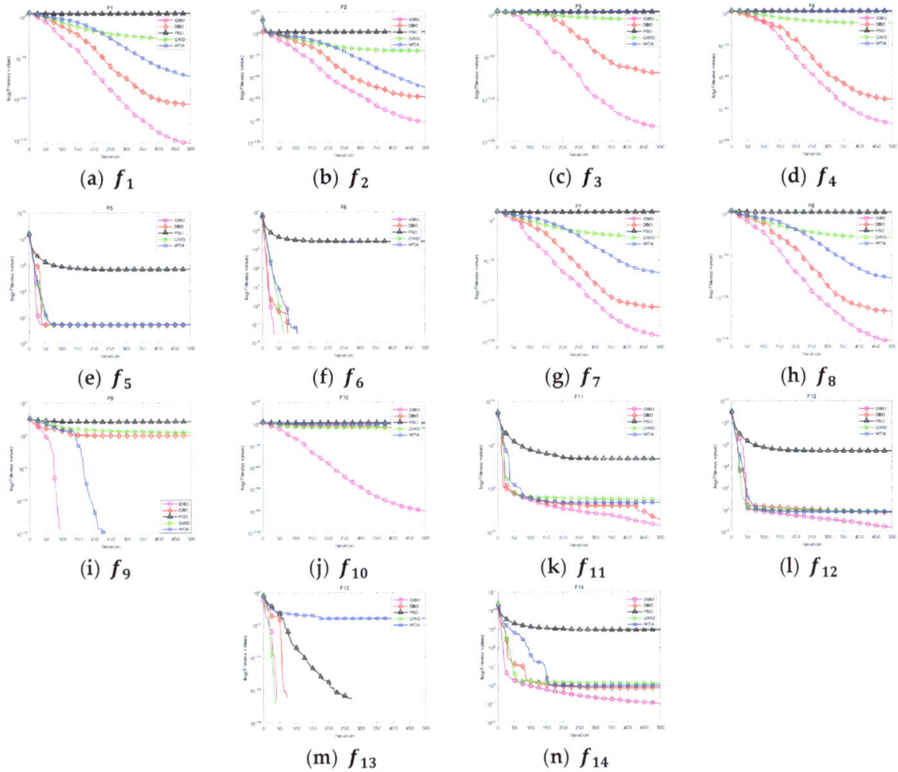

Figure 7. Convergence curves of unimodal and multimodal functions.

As shown in Figure 7, IDBO exhibits the fastest convergence and highest accuracy in convergence curves for functions F1–F5, F7–F10 and F14 with a near-linear decrease to theoretical optimal values without stagnation. DBO performs second only to IDBO for these functions and outperforms other algorithms. DBO, GWO and WOA converge to optimal values with minimal stagnation for function F6 but at a slower rate than that of IDBO, and PSO exhibits stagnation at local optima. In the convergence curves for functions F11–F13, IDBO converges rapidly with a quick inflection point to achieve optimal accuracy. This demonstrates that the amelioration method recommended in this paper effectively enhances the original algorithm in terms of convergence rate and accuracy.

4.5. Local Optimal Circumvention Evaluation

As previously mentioned, multimodal functions can be used to examine the search behavior of algorithms. As indicated in Table 5, IDBO achieves the best (fitness) optimal values across three different dimensions of 30, 50 and 100 and outperforms other algorithms. This demonstrates that IDBO effectively equilibrates local and global searches to evade local optima. The improvement approach suggested in this article dramatically augments the exploratory potential of the original model.

4.6. High-Dimensional Robustness Evaluation

General algorithms may not exhibit robustness and stability when solving complex problem functions in high dimensions, and their ability to find optimal solutions may decrease abruptly. To assess the performance of IDBO in high dimensions, results for IDBO and other algorithms were compared in 50 and 100 dimensions. As presented in Table 4, for unimodal functions other than F3 and F6, PSO, WOA and GWO all exhibit decreased convergence accuracy in higher dimensions, and DBO and IDBO show decreased accuracy in 50 dimensions but little change in 100 dimensions, indicating stability for both DBO and IDBO at higher dimensions. For function F3, the convergence accuracy of IDBO is 18 orders of scale above that of DBO in 50 dimensions but increases to 29 orders of magnitude higher than that of DBO in 100 dimensions, indicating slightly inferior performance for DBO at higher dimensions.

According to Table 5, for high-dimensional multimodal function F9, the convergence accuracy for WOA and DBO decreases from theoretical optimal values as dimensionality increases, and only IDBO consistently converges to theoretical optimal values with a mean and standard deviation of zero, indicating stable performance for IDBO when seeking high-dimensional multimodal functions. For four test functions, excluding F9 and F13, IDBO's performance at high dimensions is comparable to that at 30 dimensions, achieving optimal mean and standard deviation values. Overall, IDBO exhibits a strong performance when finding optimal solutions for high-dimensional optimization problems, demonstrating its stability and robustness at high dimensions.

Table 6 presents a summary of the performance outcomes for IDBO and other algorithms, as shown in Tables 4 and 5. The total performance metric is employed to calculate TP for each algorithm using Equation (14), in which each algorithm has Q trials and M failed tests.

$$\mathrm{TP} = \left(\frac{Q - M}{Q}\right) \times 100\% \tag{14}$$

Table 6. Total performance of IDBO and other basic prevalent meta-heuristics algorithms.

	WOA	GWO	PSO	DBO	IDBO
	w/t/l	w/t/l	w/t/l	w/t/l	w/t/l
D = 30	0/2/12	0/2/12	0/1/13	0/2/12	11/3/0
D = 50	0/1/13	0/2/12	0/1/13	0/2/12	11/3/0
D = 100	0/1/13	0/2/12	0/1/13	0/2/12	11/3/0
Total	0/4/38	0/6/36	0/3/39	0/6/36	33/9/0
TP	9.52%	14.29%	7.14%	14.29%	100.00%

4.7. Statistical Analysis

To further evaluate the validity of the suggested enhancement tactics, this paper used a Wilcoxon signed-rank test to compare IDBO with four meta-heuristic algorithms and applied the Friedman test (Equation (15)) to calculate each algorithm's ranking. The number of populations $N = 30$ was set, each test function was subjected to 30 independent runs of each algorithm, dimension $D = 30$, and the Wilcoxon signed-rank test with $\alpha = 0.05$ was implemented for IDBO and other algorithms on 14 test functions. The α-values are presented in along with statistics for "}}+", "}}−" and "}} =". "}}+" shows that IDBO clearly outperforms other comparison algorithms, "}}−" indicates inferiority, and "}} =" denotes no significant difference. N/A represents not applicable when both searches for superiority result in 0, indicating a comparable performance. Bold text indicates insignificant or comparable differences.

Table 7 shows that WOA, GWO and DBO have comparable search performances with IDBO for F6, and PSO differs significantly from IDBO. WOA and IDBO have a comparable search performances for F9, and DBO differs insignificantly from IDBO. GWO, PSO and DBO have equivalent search behavior with IDBO for F13, and WOA differs insignificantly from IDBO. GWO, PSO and DBO differ significantly from IDBO for all functions except F6, F9 and F13.

Table 7. α-values of Wilcoxon signed-rank test.

Function	WOA	GWO	PSO	DBO
F1	3.02×10^{-11}	3.02×10^{-11}	3.02×10^{-11}	6.07×10^{-11}
F2	3.02×10^{-11}	3.02×10^{-11}	3.02×10^{-11}	4.20×10^{-10}
F3	3.02×10^{-11}	3.02×10^{-11}	3.02×10^{-11}	2.57×10^{-7}
F4	3.02×10^{-11}	3.02×10^{-11}	3.02×10^{-11}	1.61×10^{-10}
F5	3.02×10^{-11}	3.02×10^{-11}	3.02×10^{-11}	4.18×10^{-9}
F6	N/A	N/A	1.21×10^{-12}	N/A
F7	3.02×10^{-11}	3.02×10^{-11}	3.02×10^{-11}	1.61×10^{-10}
F8	3.02×10^{-11}	3.02×10^{-11}	3.02×10^{-11}	2.61×10^{-10}
F9	N/A	1.16×10^{-12}	1.21×10^{-12}	3.34×10^{-1}
F10	3.02×10^{-11}	3.02×10^{-11}	3.02×10^{-11}	3.69×10^{-11}
F11	3.02×10^{-11}	3.02×10^{-11}	3.02×10^{-11}	1.99×10^{-2}
F12	3.69×10^{-11}	3.02×10^{-11}	3.02×10^{-11}	1.33×10^{-10}
F13	8.15×10^{-2}	N/A	N/A	N/A
F14	3.02×10^{-11}	3.02×10^{-11}	3.02×10^{-11}	1.09×10^{-10}
+/=/−	11/3/0	12/2/0	13/1/0	12/2/0

Table A2 in Appendix A shows the results of Friedman's test. The IDBO algorithm has a lower average ranking value than that of the other algorithms in all three dimensions, indicating its superior performance. Moreover, Table A2 reveals that the IDBO algorithm's mean value decreases relative to DBO as dimensionality increases. This shows that IDBO is more robust in higher dimensions than DBO and further verifies the effectiveness of our optimization strategy.

$$F_f = \frac{12n}{k(k+1)}\left[\sum_j R_j^2 - \frac{k(k+1)^2}{4}\right] \quad (15)$$

where n is the count of case tests, k is the quantity of algorithms, and R_j is the mean ranking of the jth algorithm.

The IDBO algorithm shows significant improvements in both local and global exploration abilities based on a comprehensive analysis of benchmark function test results, convergence curves, Wilcoxon signed-rank test results, and Friedman test results for each algorithm. It exceeds the original DBO and WOA algorithms and other optimization algorithms that we compare it with in terms of convergence velocity, accuracy and stability. This verifies the performance of the optimization scheme this paper recommends.

5. Experimental Research

5.1. Data Preprocessing

To ensure an accurate comparison of algorithm results, this paper uses the same data as those used in [35]. The authors used larch-sawn timber of 22 mm thickness from Northeast China. Samples were heat treated at atmospheric pressure with temperature, time and relative humidity as the process parameters. The temperature was divided into five levels (120 °C to 210 °C), time was divided into four levels (0.5 to 3 h), and relative humidity was divided into four levels (0 to 100%). After treatment, specimens were placed at an ambient temperature of (20 ± 2) °C and relative humidity of (65 ± 3)% until reaching a balanced moisture level. Mechanical properties were then measured by GB/T1935-2009 to GB/T1941-2009 standards. For each test, the mean of five replicates was computed, yielding 88 sets of data in total.

To guarantee equity in model comparisons, this article uses the same training and testing samples as those used in [23]. The first 58 samples in Table A1 in Appendix A formed the training set, and the last 30 samples constituted the testing set. Input data were normalized using Equation (16) to avoid effects on training speed and prediction accuracy.

$$Y_{norm} = \frac{(y - y_{min})}{(y_{max} - y_{min})} \quad (16)$$

Y_{norm} denotes the value of y after scaling it to a unit interval, and y is the original value. The range of y is bounded by y_{min} and y_{max} as the lower and upper limits, respectively.

5.2. Model Parameter Setting

The IDBO-BP model was used to forecast the mechanical features and compare the results with those of BP, TSSA-BP, GWO-BP, IGWO-BP and DBO-BP neural networks to demonstrate its prediction capability. The upper limit of the iterations of the BP neural network was set to 1000 with a target error of 0.0001 and a population size of 50.

5.2.1. Selection of Activation Functions

The activation function is a vital component of a neural network that determines how the neurons produce the output from the input. The activation function gives neural networks nonlinear modeling capabilities, allowing them to approximate complex data and functions. The performance and convergence of neural networks depend on the choice of activation function, so selecting a suitable activation function is a critical step in neural network design. Four common activation functions for BP neural network models in MATLAB are LOGSIG, TANSIG, POSLIN and PURELIN.

Table A3 in Appendix A shows the activation function combinations that minimize the error of different models. Table A3 indicates that the optimal activation function combination for the IDBO-BP model is LOGSIG-PURELIN for LCS, obtained via the exhaustive method. Similarly, the optimal activation functions for other models can be derived.

5.2.2. Determination of the Topology

The number of neurons in each layer and the connection between two adjacent layers constitute the topology of a BP neural network model. The topology affects the neural network's complexity and expressiveness, which in turn influence the neural network's performance and convergence. Hence, selecting an appropriate topology is a crucial step in neural network design.

Determination of the Number of Neurons in the Hidden Layer

The BP neural network model's structure and performance depend on the number of hidden layer neurons, a key parameter that affects the model's fit to the data. The optimal number of hidden layer neurons should avoid both underfitting and overfitting. Underfitting occurs when the network has too few hidden layer neurons to capture the data's complex features; overfitting or gradient vanishing occurs when the network has too many hidden layer neurons that fit the training data too closely. The number of hidden layer neurons is not fixed but varies according to the problem's complexity and the data's size. Selecting the appropriate number of hidden layer neurons is essential to enhance the model's generalization ability and prediction accuracy.

This paper proposes an empirical formula (Formula (17)) for estimating the number of neurons in the hidden layer as a reference. The number of neurons in the hidden layer varies from 2 to 7. Table A3 in Appendix A shows the neuron configurations that minimize the error of different models. For example, Table A3 indicates that the optimal neuron configuration for the IDBO-BP model for LCS is 2 (single hidden layer), obtained by the trial-and-error method. Similarly, the optimal neuron configuration for other models can be derived.

$$N_h = \frac{N_s}{(\alpha \times (N_i + N_o))}, \alpha \epsilon [2, 7] \quad (17)$$

where N_h, N_i and N_o are the number of neurons in the hidden, input and output layers, respectively, and N_s is the number of samples in the training set.

Determination of the Number of Hidden Layers

The hidden layer of a neural network enables it to process non-linearly separable data. Without hidden layers, neural networks can only represent linearly separable functions or decisions. The number of hidden layers and the activation function influence the neural network's representational power and fit. Generally, more hidden layers reduce the error but also increase the network's complexity and training difficulty, and they may cause overfitting. This paper uses neural network models with single and double hidden layers and employs different activation functions to determine the optimal network structure.

Table A3 in Appendix A shows the topologies of the different models at the error minimum. For example, Table A3 indicates that the optimal topology for the IDBO-BP model for TRS is 3-4-6-1, obtained via iterative attempts. Figure 8 shows the corresponding topology schematic diagram. Similarly, the optimal topology for other models can be derived.

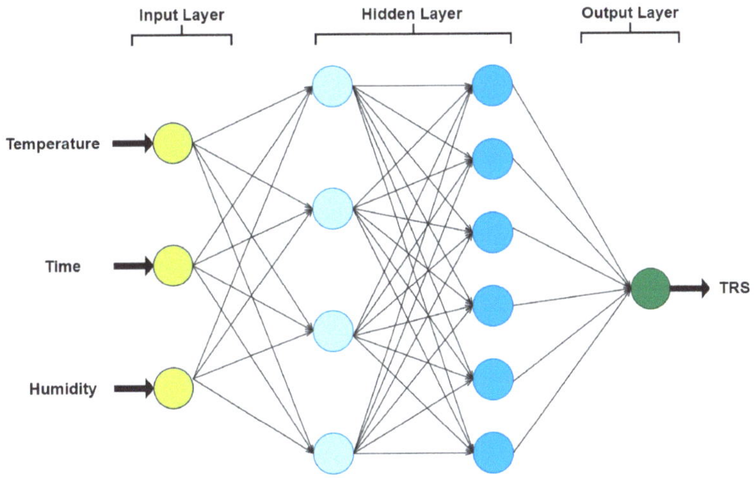

Figure 8. Schematic diagram of the IDBO-BP model for the mechanical properties of transverse rupture strength.

5.3. Model Assessment Standards

Statistical error is commonly used to evaluate model prediction properties. Common regression evaluation metrics include the mean absolute error (MAE), mean squared error (MSE), mean absolute percentage error (MAPE) and coefficient of determination (R^2). MAE reflects the discrepancy between the algorithm's optimal value and the theoretical optimal value. It indicates the algorithm's exploration capability and convergence accuracy. MSE measures the standard error between the predicted and true values, and as the standard error lowers, the model accuracy increases. MAPE measures the relative error between the predicted and true values as a percentage. It is useful for comparing models with different scales of data. R2 measures the model fit to the data, and as it comes closer to 1, the model fit becomes better, and vice versa. As it comes closer to 0, the fit becomes worse. The equations are as follows:

$$\text{MAE} = \frac{1}{N}\sum_{i=1}^{N}|Y_i - Z_i| \tag{18}$$

$$\text{MSE} = \frac{1}{N}\sum_{i=1}^{N}(Y_i - Z_i)^2 \tag{19}$$

$$\text{MAPE} = \frac{1}{N}\sum_{i=1}^{N}\left|\frac{Y_i - Z_i}{Y_i}\right| \times 100\% \tag{20}$$

$$R^2 = 1 - \frac{\sum_{i=1}^{n}(Y_i - Z_i)^2}{\sum_{i=1}^{n}\left(Y_i - \bar{Y}_i\right)^2} \tag{21}$$

where Y_i and Z_i represent the true and predicted values, respectively.

5.4. Model Performance Comparison Analysis

To validate the IDBO-BP model, this paper compares it with the BP, TSSA-BP, GWO-BP, IGWO-BP and DBO-BP models. The BP model is the original back-propagation neural network, and TSSA-BP, GWO-BP, IGWO-BP and DBO-BP are optimized versions of BP neural networks that incorporate the TSSA, GWO, IGWO and DBO models, respectively. Table A4 in Appendix A shows the results. For example, using the test set for illustration, IDBO-BP reduces MAE values of LCS, TRS, TME, RH, and TH models by 56%, 31%, 11%, 35% and 31%, respectively, and it reduces the MAPE by 57%, 45%, 38%, 38% and 38%, respectively, compared with the non-optimized BP neural network. For LCS, the significance values of MAE and MAPE of the test set of IDBO-PB corrected with Bonferroni are 0.002 and 0 (less than 0.05), respectively, indicating significant differences between the IDBO-BP and BP models in predicting the mechanical properties of wood. Moreover, compared with the BP, TSSA-PB, GWO-PB, IGWO-PB and DBO-PB models, the IDBO-PB model's predictions are closer to real values, indicating its superior prediction capability. Furthermore, compared with the DBO-PB model, IDBO-PB reduces the MAE values of the testing data of LCS, TRS, TME, RH, and TH models by 43%, 12%, 6%, 7% and 18%, respectively; it reduces the MSE by 78%, 21%, 10%, 8% and 26%, respectively; and it reduces the MAPE by 46%, 10%, 8%, 6% and 21%, respectively. This further verifies the effectiveness of the improved strategy in this paper.

Table A5 in Appendix A shows the rank means and overall rankings of the Friedman tests for the six models based on different evaluation metrics on different parameters. In Friedman's test, the rank mean reflects the solution quality. The rank mean is the average of the solution rank obtained by each algorithm among all the algorithms. As the rank increases, the solution quality increases, and the algorithm performance improves, i.e., it comes closer to the objective function's optimal value. Table A5 shows that IDBO-BP has the highest ranking for all five parameters, outperforming the other models. Figure 9 illustrates the distribution of the six models on their MAE for the LCS test set. The significance value of the null hypothesis for the six models with the same distribution of solutions for the MAE is 0.006, so the null hypothesis is rejected, i.e., there is a significant difference in the solution quality of the six models for MAE, and IDBO-BP has the best performance based on the rank mean.

Figure 10a–e compares the prediction results for five mechanical properties of wood using the IDBO-BP, DBO-BP and BP neural networks with the actual values. The results show that the optimized BP neural networks with the DBO or IDBO models have predictions closer to the true values, indicating that DBO or IDBO improves the BP neural network prediction accuracy. Moreover, the benchmark functions show that the IDBO model performs better than the DBO model in convergence accuracy, stability and exploration capability. This is mainly due to several factors: (1) The IDBO algorithm initializes its dung beetle population using PWLCM chaotic mapping, which enhances population diversity and initial population solution quality. (2) The IDBO algorithm uses an adaptive parameter adjustment strategy with a nonlinear decreasing producer ratio model, which improves searchability in the early and middle stages of the algorithm and increases search range and efficiency. (3) The IDBO algorithm optimizes location updates for small dung beetles by applying a foraging search strategy based on dimensional learning, which balances exploration and exploitation abilities in late iterations.

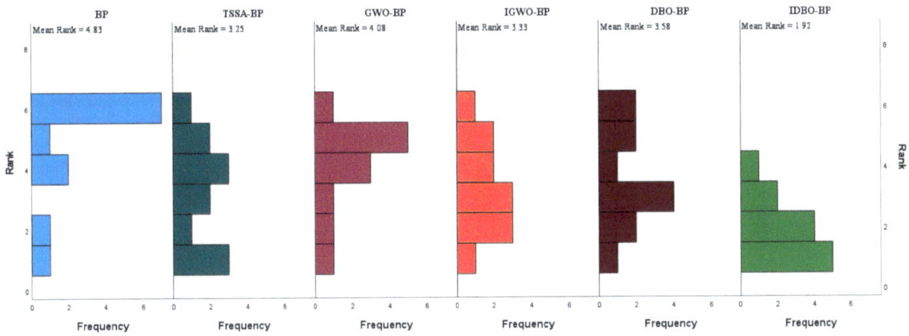

Figure 9. Friedman's two-way analysis of variances by ranks for related samples (test sets of LCS).

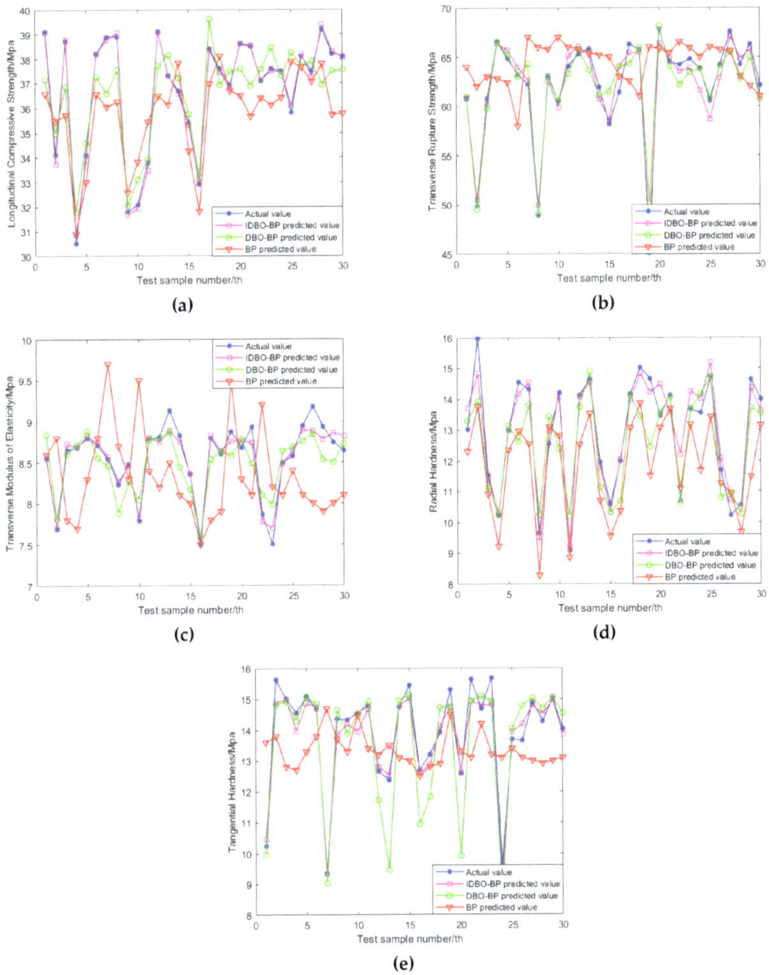

Figure 10. Comparison of results for various models based on predictions and actual values: (**a**) longitudinal compressive strength; (**b**) transverse rupture strength; (**c**) transverse modulus of elasticity; (**d**) radial hardness; (**e**) tangential hardness.

In summary, this paper proposes an IDBO algorithm that demonstrates significant improvements in search performance compared to other algorithms, thereby verifying the effectiveness of the enhancement strategy. Additionally, the results demonstrate that the presented IDBO-BP model exhibits outstanding performance in predicting wood mechanical properties.

6. Conclusions

- This article proposes the IDBO algorithm to address the limitations of the DBO algorithm. PWLCM mapping is employed to initialize the population and preserve versatility. An adaptive parameter adjustment strategy is introduced to enhance search range and efficiency. Additionally, a DLF strategy is implemented to equilibrium for exploration and exploitation search capabilities, increasing the likelihood of escaping local optima and improving later searchability. The performance of IDBO is evaluated against four basic meta-heuristic algorithms, including DBO, for 14 benchmark functions. The algorithms are ranked by applying the Wilcoxon signed-rank and Friedman tests. The outcomes demonstrate that IDBO outperforms other algorithms in finding solutions for both low- and high-dimensional functions with a single mode or multiple modes, which verifies the effectiveness of the improvement strategy, and it is highly competitive with other meta-heuristics.
- In this paper, five prediction models are separately developed using the IDBO-BP model to predict the LCS, TRS, TME, RH and TH of larch wood after heat treatment with temperature, duration and relative humidity as input variables. The outcomes indicate that the MAE, MSE and MAPE values of the IDBO-BP model are considerably diminished compared with the primitive BP neural network model. The results show that optimizing neural networks model with IDBO significantly improves the prediction accuracy of wood mechanical properties. In addition to comparing the original BP neural network model, this paper also compares it with the TSSA-BP, GWO-BP, IGWO-BP and DBO-BP models. The results denote that the forecast outcomes of the IDBO-BP model are closer to the true values, indicating significant optimization and improved prediction ability.
- This paper compares the optimal prediction models with different parameters and their corresponding topologies and activation functions, and it shows in Table A3 that the same model with different parameters does not necessarily have the same optimal topology. For the LCS, TRS, TME, RH and TH of heat-treated larch wood predicted in this paper, the five most accurate topologies of IDBO-BP that minimize the error are 3-2-1, 3-4-6-1, 3-4-1, 3-5-1 and 3-4-1, respectively.
- The Friedman test can only reflect the quality of the solution, not the diversity of the solution. Therefore, some algorithms may have significant differences in the diversity of solutions, but not in the quality of solutions. The Friedman test is also less robust in some extreme cases; for example, if an algorithm obtains an exceptionally good or bad solution, it may influence the rank and rank mean of other algorithms, thus obscuring the differences between other algorithms. This is illustrated in Table A5. The original BP model ranks first in MSE for both the training and test sets, which may be more susceptible to outlier data because MSE magnifies the prediction error. However, Table A5 also shows that, although the original BP model performs well for MSE, its overall ranking for both the test and training sets is inferior to that of the other five models.

Author Contributions: Conceptualization, R.Z.; methodology, R.Z.; software, R.Z.; validation, R.Z.; formal analysis, R.Z.; investigation, R.Z.; resources, R.Z.; data curation, R.Z.; writing—original draft preparation, R.Z.; writing—review and editing, R.Z.; visualization, R.Z.; project administration, R.Z.; funding acquisition, Y.Z. All authors have read and agreed to the published version of the manuscript.

Funding: This research was funded by the Fundamental Research Funds for the Central Universities, grant number 2572020BL01, and the Natural Science Foundation of Heilongjiang Province, grant number LH2020C050.

Data Availability Statement: From this paper, the data are openly available in a public repository that issues datasets with DOIs. The data that support the findings of this study are openly available in Bioresources at http://doi.org/10.15376/biores.10.3.5758-5776, reference number [35]. The data presented in this study are available in the article.

Conflicts of Interest: The authors declare no conflict of interest.

Appendix A

Table A1. Wood treatment conditions and corresponding mechanical properties.

Test Temperature/°C	Test Time/h	Test Humidity/%	Longitudinal Compressive Strength/MPa	Transverse Rupture Strength/MPa	Transverse Modulus of Elasticity/GPa	Radial Hardness/MPa	Tangential Hardness/MPa
120	0.5	0	41.9	67.4	9.093	14.12	15.56
120	0.5	40	39.8	65.3	9.038	13.02	14.69
120	0.5	60	39.7	69.7	9.1	14.67	15.08
120	0.5	100	39.5	67.2	8.845	14.65	15.45
120	1	0	39.5	67.8	8.649	13.98	14.36
120	1	40	39.5	66.4	8.752	12.98	15.59
120	1	60	39.4	67.8	9.245	13.78	15.32
120	1	100	39.2	63.1	7.895	14.55	14.23
120	2	0	39.2	66.9	9.074	13.33	14.23
120	2	40	39.1	68.2	8.945	12.55	14.58
120	2	60	39.1	65.2	8.854	13.25	14.89
120	2	100	39.1	63.2	8.933	13.36	14.56
120	3	0	39.1	66.5	8.9	13.56	14.78
120	3	40	39.1	67.6	8.963	13.45	14.45
120	3	60	38.9	66.6	8.745	13.01	14.69
120	3	100	38.9	64.2	8.745	12.45	14.78
140	0.5	0	38.9	66.7	8.978	14.69	15.56
140	0.5	40	38.9	67.5	8.845	13.06	15.02
140	0.5	60	38.7	66.8	9.155	14.02	14.23
140	0.5	100	38.7	65.3	8.877	15.02	15.01
140	1	0	38.6	66.5	9.179	14.16	15.68
140	1	40	38.6	64.5	9.137	13.05	15.01
140	1	60	38.5	67.2	9.024	13.49	15.17
140	1	100	38.5	63.1	8.823	13.45	15.48
140	2	0	38.4	66.3	8.823	13.54	14.69
140	2	40	38.4	65.7	8.852	14.69	14.58
140	2	60	38.2	67.1	8.799	13.99	14.74
140	2	100	38.2	62.7	8.9	14.28	15.63
140	3	0	38.2	65.4	8.811	14.39	14.23
140	3	40	38.2	64.6	8.934	13.23	14.56
140	3	60	38.1	65.5	8.654	14.23	13.65
140	3	100	38.1	62.1	8.798	13.56	14.02
160	0.5	0	38.1	66.3	8.788	14.89	14.99
160	0.5	40	38	66.9	9.011	14.87	14.36
160	0.5	60	37.9	66.3	8.745	14.58	14.78
160	0.5	100	37.8	65.8	8.712	14.69	15.69
160	1	0	37.8	62.4	8.679	13.42	14.56
160	1	40	37.6	61.4	8.645	14.09	15.3
160	1	60	37.6	62.2	8.798	14.69	15.9
160	1	100	37.6	62.8	8.679	13.58	15.63

Table A1. *Cont.*

Test Temperature/°C	Test Time/h	Test Humidity/%	Longitudinal Compressive Strength/MPa	Transverse Rupture Strength/MPa	Transverse Modulus of Elastic-ity/GPa	Radial Hard-ness/MPa	Tangential Hard-ness/MPa
160	2	0	37.6	62.2	8.727	14.63	13.92
160	2	40	37.6	62.1	8.557	14.02	14.17
160	2	60	37.5	63.1	8.687	15.17	14.28
160	2	100	37.5	60.9	8.611	14.65	15.09
160	3	0	37.5	61.9	8.611	13.65	14.36
160	3	40	37.4	61.5	8.534	13.47	14.56
160	3	60	37.3	60.8	8.601	13.58	13.89
160	3	100	37.2	60.5	8.552	13.69	14.36
180	0.5	0	37.2	65.9	8.601	15.21	14.03
180	0.5	40	37.1	65.3	8.689	15.98	14.56
180	0.5	60	37.1	66.1	8.645	16.01	13.97
180	0.5	100	36.9	65.7	8.599	14.32	14.33
180	1	0	36.9	65.4	8.623	15.09	13.79
180	1	40	36.9	64.9	8.645	14.98	14.25
180	1	60	36.7	66.3	8.579	15.45	14.08
180	1	100	36.7	64.8	8.545	14.33	13.64
180	2	0	36.6	65.1	8.574	14.65	13.69
180	2	40	36.5	65.8	8.6	14.13	13.59
180	2	60	36.5	64.5	8.532	13.99	14.49
180	2	100	36.1	64.2	8.544	15.1	13.54
180	3	0	36	64.1	8.6	14.21	14.06
180	3	40	35.9	64.2	8.541	13.99	14.21
180	3	60	35.8	64.8	8.456	14.58	13.98
180	3	100	35.8	63.8	8.499	14.99	13.69
200	0.5	0	35.8	62.1	8.483	12	13.6
200	0.5	40	35.5	60.6	8.475	11.96	12.99
200	0.5	60	35.4	59.9	8.399	11.45	13.21
200	1	0	35.4	61.9	8.422	11.69	12.98
200	1	40	35.1	60.8	8.489	11.46	12.64
200	1	60	34.6	61.2	8.321	11.54	12.35
200	2	0	34.5	61.2	8.369	11.99	13.02
200	2	40	34.5	60.8	8.354	11.15	12.69
200	2	60	34.2	60.5	8.211	10.65	12.49
200	3	0	34.1	60.9	8.249	10.68	12.73
200	3	40	34.1	59.8	8.231	11.05	12.57
200	3	60	34.1	58.2	8.011	10.22	12.37
210	0.5	0	33.9	50.1	7.856	10.23	10.98
210	0.5	40	33.8	50.8	7.789	10.59	9.98
210	0.5	60	33.2	49.9	7.865	10.55	10.23
210	1	0	32.9	50.6	7.765	10.21	10.65
210	1	40	32.9	49.8	7.712	9.98	10.21
210	1	60	32.8	48.9	7.498	10.01	10.65
210	2	0	32.5	49.1	7.689	9.98	9.64
210	2	40	32.1	49.5	7.712	9.65	9.35
210	2	60	31.8	49.6	7.623	10.03	9.67
210	3	0	31.5	47.8	7.5	9.21	8.91
210	3	40	30.8	46.5	7.412	9.1	8.21
210	3	60	30.5	45.1	7.321	9.03	8.99

Table A2. Order by Friedman test in dimensions D = 30, 50 and 100.

Alg.	D	F1	F2	F3	F4	F5	F6	F7	F8	F9	F10	F11	F12	F13	F14	Avg. Rank	Overall Rank
WOA	30	3	3	5	4.67	3.67	2.5	3	3	1.83	2.33	3	3	4.33	3.33	3.26	3
	50	3	2.33	5	4.67	3.67	2.5	3	3	2.17	3.67	3	3	4.33	3.33	3.33	3
	100	3	3	5	4.67	3.67	2.5	3	3	2.17	3.67	3	3	4.33	3.33	3.38	4
GWO	30	4	4	3	3	3.33	2.5	4	4	3.17	4	4	3.33	2.67	3.33	3.45	4
	50	4	4	3	3	3.33	2.5	4	4	3.83	2.67	4	2.67	2.67	3.67	3.38	4
	100	4	4	3	3	3.33	2.5	4	4	2.83	2.67	4	2.67	2.67	3.33	3.29	3
PSO	30	5	5	4	4.33	5	5	5	5	5	5	5	5	2.67	5	4.71	5
	50	5	5	4	4.33	5	5	5	5	4.67	5	5	5	2.67	5	4.69	5
	100	5	5	4	4.33	5	5	5	5	5	5	5	5	2.67	5	4.71	5
DBO	30	2	2	2	2	1.67	2.5	2	2	3.17	2.67	1.67	2.67	2.67	2.33	2.24	2
	50	2	2.67	2	2	1.67	2.5	2	2	2.83	2.67	1.67	3.33	2.67	1.67	2.26	2
	100	2	2	2	2	1.67	2.5	2	2	3.5	2.67	1.67	3.33	2.67	2.33	2.31	2
IDBO	30	1	1	1	1	1.33	2.5	1	1	1.83	1	1.33	1	2.67	1	**1.33**	**1**
	50	1	1	1	1	1.33	2.5	1	1	1.5	1	1.33	1	2.67	1.33	**1.33**	**1**
	100	1	1	1	1	1.33	2.5	1	1	1.5	1	1.33	1	2.67	1	**1.31**	**1**

Table A3. Optimal prediction models with different parameters and the corresponding topologies and activation functions.

Parms	Model	Neuron Configuration	Topology	Hidden and Output Activations	Train				Test			
					MAE	MSE	MAPE	R^2	MAE	MSE	MAPE	R^2
LCS	BP	2	3-2-1	LOGSIG-PURELIN	0.2206	0.0020	0.9335%	0.9770	0.1945	0.0011	0.5332%	0.9868
	GWO-BP	3	3-3-1	LOGSIG-PURELIN	0.1956	0.1054	0.5263%	0.9785	0.1372	0.0724	0.4919%	0.9940
	TSSA-BP	4	3-4-1	LOGSIG-TANSIG	0.1592	0.1282	0.4342%	0.9759	0.1481	0.0642	0.4270%	0.9887
	DBO-BP	(3, 7)	3-3-7-1	LOGSIG-PURELIN	0.1412	0.1052	0.3786%	0.9830	0.1503	0.0566	0.4260%	0.9879
	IGWO-BP	(2, 2)	3-2-2-1	PURELIN-LOGSIG	0.1652	0.1200	0.4449%	0.9805	0.1291	0.0345	0.3591%	0.9907
	IDBO-BP	2	3-2-1	LOGSIG-PURELIN	0.1315	0.0996	0.3513%	0.9834	0.0856	0.0122	0.2286%	0.9978
TRS	BP	2	3-2-1	LOGSIG-PURELIN	1.2157	0.0379	2.4030%	0.9363	1.1994	0.0753	2.4487%	0.9275
	TSSA-BP	2	3-2-1	LOGSIG-PURELIN	1.2191	2.2895	1.9534%	0.9236	1.2609	2.2655	2.0291%	0.9430
	GWO-BP	(3, 7)	3-3-7-1	LOGSIG-PURELIN	1.1097	1.8158	1.7920%	0.9433	0.9837	1.9907	1.5331%	0.9272
	IGWO-BP	3	3-3-1	LOGSIG-TANSIG	1.1020	2.2158	1.8002%	0.9310	0.9519	1.5641	1.5302%	0.9557
	DBO-BP	7	3-7-1	POSLIN-PURELIN	1.0349	1.7647	1.6610%	0.9510	0.9345	1.4363	1.4881%	0.9484
	IDBO-BP	(4, 6)	3-4-6-1	TANSIG-PURELIN	0.9032	1.4304	1.4750%	0.9601	0.8218	1.1362	1.3392%	0.9683
TME	BP	3	3-3-1	LOGSIG-PURELIN	0.1710	0.0006	2.2422%	0.8260	0.0928	0.0005	1.4955%	0.8929
	GWO-BP	(3, 7)	3-3-7-1	LOGSIG-PURELIN	0.0925	0.0306	1.0968%	0.8638	0.1010	0.0164	1.1576%	0.9002
	TSSA-BP	(2, 2)	3-2-2-1	LOGSIG-PURELIN	0.1207	0.0500	2.0647%	0.7992	0.0977	0.0158	1.1322%	0.7225
	IGWO-BP	3	3-3-1	LOGSIG-PURELIN	0.1144	0.0324	1.3611%	0.8424	0.0923	0.0155	1.0797%	0.9018
	DBO-BP	(3, 7)	3-3-7-1	LOGSIG-PURELIN	0.1024	0.0273	1.2126%	0.8658	0.0879	0.0127	1.0182%	0.9053
	IDBO-BP	4	3-4-1	POSLIN-PURELIN	0.0849	0.0266	1.0132%	0.8743	0.0824	0.0115	0.9340%	0.9156
RH	BP	(5, 6)	3-5-6-1	POSLIN-PURELIN	0.4763	0.0049	3.5198%	0.9249	0.4963	0.0078	3.8911%	0.8986
	TSSA-BP	(3, 7)	3-3-7-1	LOGSIG-PURELIN	0.4205	0.3300	4.1028%	0.8901	0.3919	0.3644	3.8402%	0.9023
	GWO-BP	5	3-5-1	POSLIN-PURELIN	0.4009	0.2375	2.9805%	0.9233	0.5008	0.3834	3.8098%	0.8785
	IGWO-BP	(4, 6)	3-4-6-1	TANSIG-PURELIN	0.4256	0.2896	3.2035%	0.9107	0.4095	0.2953	3.0822%	0.8975
	DBO-BP	2	3-2-1	TANSIG-PURELIN	0.3928	0.2466	2.9423%	0.8985	0.3478	0.2053	2.5616%	0.9285
	IDBO-BP	5	3-5-1	POSLIN-PURELIN	0.3353	0.2122	2.5198%	0.9372	0.3236	0.1889	2.4020%	0.9458
TH	BP	(7, 5)	3-7-5-1	LOGSIG-LOGSIG	0.3850	0.0024	2.8660%	0.9412	0.4308	0.0073	3.4057%	0.8996
	GWO-BP	2	3-2-1	LOGSIG-PURELIN	0.2857	0.2283	2.8154%	0.9615	0.3720	0.2421	3.0862%	0.8625
	TSSA-BP	(4, 6)	3-4-6-1	TANSIG-PURELIN	0.3743	0.2175	2.8032%	0.9354	0.3839	0.2201	2.9423%	0.8896
	IGWO-BP	4	3-4-1	TANSIG-PURELIN	0.3156	0.1671	2.3664%	0.9566	0.3824	0.2373	2.9295%	0.8899
	DBO-BP	(7, 5)	3-7-5-1	LOGSIG-LOGSIG	0.3335	0.1700	2.4652%	0.9532	0.3604	0.2092	2.6622%	0.8996
	IDBO-BP	4	3-4-1	TANSIG-PURELIN	0.2611	0.1249	1.9515%	0.9676	0.2962	0.1544	2.1062%	0.9399

Table A4. Pairwise comparisons of the prediction performance of different models with IDBO-BP.

Parms	Sample1-Sample 2	Train								Test							
		MAE	Adj. Sig.[a]	MSE	Adj. Sig.[a]	MAPE	Adj. Sig.[a]	R^2	Adj. Sig.[a]	MAE	Adj. Sig.[a]	MSE	Adj. Sig.[a]	MAPE	Adj. Sig.[a]	R^2	Adj. Sig.[a]
LCS	IDBO-BP-DBO-BP	6.85%	0.743	5.32%	1	7.22%	0.003	−0.05%	0.051	43.07%	0.436	78.44%	0.743	46.35%	0.007	−1.00%	0.743
	IDBO-BP-IGWO-BP	20.40%	1	17.05%	1	21.05%	0.011	−0.30%	0.269	33.71%	0.954	64.60%	0.954	36.35%	0.007	−0.71%	0.954
	IDBO-BP-GWO-BP	32.76%	0.572	5.52%	1	33.25%	0.001	−0.50%	1	37.65%	0.068	83.14%	0.132	53.53%	0.001	−0.38%	0.011
	IDBO-BP-TSSA-BP	17.38%	1	22.32%	1	19.10%	0.096	−0.77%	0.945	42.21%	1	80.98%	1	46.47%	0.016	−0.92%	0.572
	IDBO-BP-BP	40.38%	0.005	−4858.66%	0.068	62.37%	0	−0.66%	0.103	56.00%	0.002	−1016.77%	0.329	57.13%	0	−1.11%	0.002
TRS	IDBO-BP-DBO-BP	12.73%	0.44	18.94%	1	11.20%	1	−0.96%	0.556	12.06%	0.428	20.89%	1	10.01%	1	−2.10%	0.185
	IDBO-BP-IGWO-BP	18.04%	0.269	35.45%	1	18.06%	1	−3.13%	0.269	13.67%	0.945	27.36%	1	12.48%	1	−1.32%	0.638
	IDBO-BP-GWO-BP	18.61%	0.945	21.22%	1	17.69%	1	−1.79%	1	16.46%	0.061	42.92%	1	12.65%	1	−4.43%	0.02
	IDBO-BP-TSSA-BP	25.92%	0.035	37.52%	1	24.49%	1	−3.96%	0.945	34.83%	0.035	49.85%	1	34.00%	0.572	−2.68%	1
	IDBO-BP-BP	25.71%	0.001	−3676.83%	0.001	38.62%	0.096	−2.55%	0.103	31.48%	0.006	−1409.50%	0	45.31%	0.002	−4.41%	0.061
TME	IDBO-BP-DBO-BP	17.08%	0.794	2.37%	1	16.44%	0.164	−0.98%	0.393	6.20%	0.132	9.71%	0.572	8.27%	0.005	−1.13%	1
	IDBO-BP-IGWO-BP	25.75%	0.269	17.97%	1	25.56%	0.42	−3.79%	0.269	10.66%	0.572	25.86%	1	13.49%	0.954	−1.53%	1
	IDBO-BP-GWO-BP	8.13%	0.945	13.07%	1	7.62%	0.42	−1.22%	1	18.36%	0.002	29.85%	0.034	19.31%	0.181	−1.71%	0.007
	IDBO-BP-TSSA-BP	29.62%	0.035	46.76%	1	50.93%	0.035	−9.40%	0.945	15.60%	0.096	27.14%	0.246	17.50%	0.181	−26.72%	0.181
	IDBO-BP-BP	50.33%	0.001	−4103.20%	0.023	54.81%	0.001	−5.86%	0.103	11.15%	0.181	−2041.22%	0.436	37.55%	0	−2.54%	0.068
RH	IDBO-BP-DBO-BP	14.63%	0.361	13.98%	1	14.36%	1	−4.30%	0.185	6.97%	1.000	8.00%	1	6.23%	1	−1.86%	0.119
	IDBO-BP-IGWO-BP	21.21%	0.269	26.72%	1	21.34%	1	−2.90%	0.269	20.99%	1	36.03%	1	22.07%	1	−5.38%	0.638
	IDBO-BP-GWO-BP	16.36%	0.945	10.65%	1	15.46%	1	−1.50%	1	35.40%	1	50.74%	1	36.95%	1	−7.67%	0.02
	IDBO-BP-TSSA-BP	20.25%	0.035	35.70%	0.572	38.58%	1	−5.29%	0.945	17.44%	0.096	48.18%	0.954	37.45%	0.572	−4.82%	1
	IDBO-BP-BP	29.60%	0.001	−4200.16%	0.016	28.41%	0.048	−1.32%	0.103	34.80%	1.000	−2308.90%	0.005	38.27%	0.011	−5.26%	0.061
TH	IDBO-BP-DBO-BP	21.71%	0.7	26.50%	1	20.84%	0.087	−1.50%	0.556	17.83%	0.366	26.17%	1	20.88%	0.151	−4.48%	0.113
	IDBO-BP-IGWO-BP	17.27%	0.269	25.22%	1	17.53%	0.42	−1.14%	0.269	22.56%	0.945	34.92%	1	28.10%	1	−5.62%	0.638
	IDBO-BP-GWO-BP	8.62%	0.945	45.27%	1	30.68%	0.42	−0.63%	1	20.39%	0.061	36.20%	1	31.75%	0.035	−8.97%	0.02
	IDBO-BP-TSSA-BP	30.25%	0.035	42.55%	1	30.38%	0.035	−3.44%	0.945	22.87%	0.035	29.84%	1	28.42%	0.061	−5.64%	1
	IDBO-BP-BP	32.18%	0.001	−5117.57%	1	31.91%	0.001	−2.80%	0.103	31.25%	0.006	−2013.85%	1	38.16%	0	−4.48%	0.061

Each row compares sample 1 with sample 2 and calculates the percentage by which sample 1 reduces the error based on sample 2 (a negative sign indicates the percentage by which R^2 increases). Asymptotic significances (2-sided tests) are displayed. The significance level is 0.050. [a] Significance values have been adjusted with the Bonferroni correction for multiple tests.

Table A5. Model ranking by parameters and evaluation metrics with Friedman test.

Parms	Model	Train				Test				Avg. Rank		Overall Rank	
		MAE	MSE	MAPE	R^2	MAE	MSE	MAPE	R^2	Include MSE	Exclude MSE	Include MSE	Exclude MSE
LCS	BP	5	1	5.42	2.17	4.83	1	5.08	2.25	2.24	2.65	5	6
	TSSA-BP	2.67	4	3.08	3.92	3.25	4.08	3.5	3.58	1.64	0.83	2	2
	GWO-BP	3.83	4.5	4.08	3.33	4.08	4.75	4.08	2.58	2.43	1.69	6	5
	IGWO-BP	3.5	4.42	3.58	3.33	3.33	4.17	3.67	3.75	1.95	1.17	3	3
	DBO-BP	3.75	3.92	3.83	3.67	3.58	4.25	3.67	3.67	1.96	1.25	4	4
	IDBO-BP	2.25	3.17	1	4.58	1.92	2.75	1	5.17	0.29	−0.60	1	1
TRS	BP	3.25	1	5.17	3.25	4	1	5.08	2.92	1.67	1.89	4	6
	TSSA-BP	4.08	4.33	3.5	3	3.92	4.25	3.75	3.33	2.19	1.49	6	5
	GWO-BP	4.17	4.25	3.83	3.17	3.58	4.08	3.42	3.58	2.07	1.38	5	4
	IGWO-BP	2.83	3.75	2.67	4.33	3.5	4	3.25	3	1.58	0.82	2	2
	DBO-BP	3.17	3.58	2.75	3.67	3.5	3.92	3.33	3.42	1.65	0.94	3	3
	IDBO-BP	3.5	4.08	3.08	3.58	2.5	3.75	2.17	4.75	1.34	0.49	1	1
TME	BP	4.08	1	4.75	3.83	4.67	1	5.08	2.83	1.74	1.99	3	6
	TSSA-BP	3.58	3.75	3.17	3.67	3.83	4.5	3.5	3.08	1.95	1.22	4	3
	GWO-BP	3.17	4.67	3.67	2.5	3.67	5	4.33	2.33	2.46	1.67	6	5
	IGWO-BP	3.25	4	3	3.75	3.33	3.58	3	4	1.55	0.81	2	2
	DBO-BP	3.75	4.17	3.5	3.25	3.75	4.25	3.5	3.75	1.99	1.25	5	4
	IDBO-BP	3.17	3.42	2.92	4	1.75	2.67	1.58	5	0.81	0.07	1	1
RH	BP	4.58	1	5.08	3.42	5.17	1	5.25	3.75	1.86	2.15	4	6
	TSSA-BP	3.17	5.08	3.75	2.33	3	5.17	4.25	2.08	2.50	1.63	6	5
	GWO-BP	3.5	3.83	3.08	3.42	2.92	3.5	3	3.58	1.60	0.92	3	3
	IGWO-BP	3.5	4.08	3.25	3.58	3.92	4.25	3.42	3.67	1.90	1.14	5	4
	DBO-BP	3.25	3.5	3	4	2.92	3.33	2.42	3.83	1.32	0.63	2	2
	IDBO-BP	3	3.5	2.83	4.25	3.08	3.75	2.67	4.08	1.31	0.54	1	1
TH	BP	4.08	1	4.75	2.08	4.17	1	4.33	2.25	1.88	2.17	4	6
	TSSA-BP	4.17	4.5	3.92	3	3.83	4.08	3.08	3.33	2.16	1.45	5	4
	GWO-BP	3.33	3.83	3	3.67	3.92	4.17	3.75	3.58	1.84	1.13	3	3
	IGWO-BP	3.83	4.42	3.58	3.08	4.08	4.5	4	3.17	2.27	1.54	6	5
	DBO-BP	3.42	3.67	3	4	2.58	3.42	2.42	3.83	1.34	0.60	2	2
	IDBO-BP	3.17	3.58	2.75	3.47	2.42	3.43	2.45	3.99	1.29	0.56	1	1

For comparison purposes, R^2 is taken as a negative value when calculating the average and overall rankings.

References

1. Goli, G.; Negro, F.; Emmerich, L.; Militz, H. Thermal and chemical modification of wood—A combined approach for exclusive, high-demanding performance products. *Wood Mater. Sci. Eng.* **2022**, *18*, 58–66. [CrossRef]
2. Bekhta, P. Effect of heat treatment on some physical and mechanical properties of birch plywood. *Eur. J. Wood Wood Prod.* **2020**, *78*, 683–691. [CrossRef]
3. Esteves, B.; Ferreira, H.; Viana, H.; Ferreira, J.; Domingos, I.; Cruz-Lopes, L.; Jones, D.; Nunes, L. Termite Resistance, Chemical and Mechanical Characterization of *Paulownia tomentosa* Wood before and after Heat Treatment. *Forests* **2021**, *12*, 1114. [CrossRef]
4. Tjeerdsma, B.F.; Militz, H. Chemical changes in hydrothermal treated wood: FTIR analysis of combined hydrothermal and dry heat-treated wood. *Eur. J. Wood Wood Prod.* **2005**, *63*, 102–111. [CrossRef]
5. Kaymakci, A.; Bayram, B. Evaluation of heat treatment parameters' effect on some physical and mechanical properties of poplar wood with multi-criteria decision making techniques. *Bioresources* **2021**, *16*, 4693–4703. [CrossRef]
6. Suri, I.F.; Purusatama, B.D.; Kim, J.H.; Yang, G.U.; Prasetia, D.; Kwon, G.J.; Hidayat, W.; Lee, S.H.; Febrianto, F.; Kim, N.H. Comparison of physical and mechanical properties of Paulownia tomentosa and Pinus koraiensis wood heat-treated in oil and air. *Eur. J. Wood Wood Prod.* **2022**, *80*, 1389–1399. [CrossRef]
7. Esteves, B.M.; Pereira, H.M. Wood modification by heat treatment: A review. *Bioresources* **2008**, *4*, 370–404. [CrossRef]
8. Korkut, D.S.; Guller, B. The effects of heat treatment on physical properties and surface roughness of red-bud maple (Acer trautvetteri Medw.) wood. *Bioresour. Technol.* **2008**, *99*, 2846–2851. [CrossRef]
9. Içel, B.; Guler, G.; Isleyen, O.; Beram, A.; Mutlubas, M. Effects of Industrial Heat Treatment on the Properties of Spruce and Pine Woods. *Bioresources* **2015**, *10*, 5159–5173. [CrossRef]
10. Xue, J.; Xu, W.; Zhou, J.; Mao, W.; Wu, S. Effects of High-Temperature Heat Treatment Modification by Impregnation on Physical and Mechanical Properties of Poplar. *Materials* **2022**, *15*, 7334. [CrossRef]
11. Boonstra, M.J.; Tjeerdsma, B. Chemical analysis of heat treated softwoods. *Eur. J. Wood Wood Prod.* **2006**, *64*, 204–211. [CrossRef]
12. Hill, C.A.S. *Wood Modification*; John Wiley & Sons, Ltd.: Hoboken, NJ, USA, 2006. [CrossRef]
13. Kohonen, T. The self-organizing map. *Proc. IEEE* **1990**, *78*, 1464–1480. [CrossRef]
14. Onyiagha, C. From neuronal stochasticity to intelligent resource management of packet data networks. In Proceedings of the Fifth International Conference on Artificial Neural Networks, Venue, UK, 7–9 July 1997. [CrossRef]

15. Stergios, A.; Anthony, K.; Elli, R.; Dimitris, B. Predicting the properties corrugated base papers using multiple linear regressiom amd artificial neural networks. *Drewno* **2016**, *59*, 198. [CrossRef]
16. You, G.; Wang, B.; Li, J.; Chen, A.; Sun, J. The prediction of MOE of bamboo-wood composites by ANN models based on the non-destructive vibration testing. *J. Build. Eng.* **2022**, *59*, 105078. [CrossRef]
17. Chen, Y.; Wang, W.; Li, N. Prediction of the equilibrium moisture content and specific gravity of thermally modified wood via an Aquila optimization algorithm back-propagation neural network model. *Bioresources* **2022**, *17*, 4816–4836. [CrossRef]
18. Abualigah, L.; Yousri, D.; Elaziz, M.A.; Ewees, A.A.; Al-Qaness, M.A.; Gandomi, A.H. Aquila Optimizer: A novel meta-heuristic optimization algorithm. *Comput. Ind. Eng.* **2021**, *157*, 107250. [CrossRef]
19. Wang, Y.; Wang, W.; Chen, Y. Carnivorous Plant Algorithm and BP to Predict Optimum Bonding Strength of Heat-Treated Woods. *Forests* **2022**, *14*, 51. [CrossRef]
20. Ong, K.M.; Ong, P.; Sia, C.K. A carnivorous plant algorithm for solving global optimization problems. *Appl. Soft Comput.* **2020**, *98*, 106833. [CrossRef]
21. Li, N.; Wang, W. Prediction of Mechanical Properties of Thermally Modified Wood Based on TSSA-BP Model. *Forests* **2022**, *13*, 160. [CrossRef]
22. Xue, J.; Shen, B. A novel swarm intelligence optimization approach: Sparrow search algorithm. *Syst. Sci. Control. Eng.* **2020**, *8*, 22–34. [CrossRef]
23. Ma, W.; Wang, W.; Cao, Y. Mechanical Properties of Wood Prediction Based on the NAGGWO-BP Neural Network. *Forests* **2022**, *13*, 1870. [CrossRef]
24. Mirjalili, S.; Mirjalili, S.M.; Lewis, A. Grey Wolf Optimizer. *Adv. Eng. Softw.* **2014**, *69*, 46–61. [CrossRef]
25. Xue, J.; Shen, B. Dung beetle optimizer: A new meta-heuristic algorithm for global optimization. *J. Supercomput.* **2022**, *79*, 7305–7336. [CrossRef]
26. Tiryaki, S.; Özşahin, Ş..; Yıldırım, I. Comparison of artificial neural network and multiple linear regression models to predict optimum bonding strength of heat treated woods. *Int. J. Adhes. Adhes.* **2014**, *55*, 29–36. [CrossRef]
27. Yan, Y.; Hongzhong, M.; Zhendong, L. An Improved Grasshopper Optimization Algorithm for Global Optimization. *Chin. J. Electron.* **2021**, *30*, 451–459. [CrossRef]
28. Wu, Q. A self-adaptive embedded chaotic particle swarm optimization for parameters selection of Wv-SVM. *Expert Syst. Appl.* **2011**, *38*, 184–192. [CrossRef]
29. Wang, Y.; Chen, S.; Wang, Y. Chaos Encryption Algorithm Based on Kent Mapping and AES Combination. In Proceedings of the 2018 International Conference on Network, Communication, Computer Engineering (NCCE 2018), Chongqing, China, 26–27 May 2018. [CrossRef]
30. Wang, X.; Jin, C. Image encryption using Game of Life permutation and PWLCM chaotic system. *Opt. Commun.* **2012**, *285*, 412–417. [CrossRef]
31. Bai, H.; Chu, Z.; Wang, D.; Bao, Y.; Qin, L.; Zheng, Y.; Li, F. Predictive control of microwave hot-air coupled drying model based on GWO-BP neural network. *Dry. Technol.* **2022**, 1–11. [CrossRef]
32. Song, X.; Zhao, M.; Yan, Q.; Xing, S. A high-efficiency adaptive artificial bee colony algorithm using two strategies for continuous optimization. *Swarm Evol. Comput.* **2019**, *50*, 100549. [CrossRef]
33. Eberhart, R.; Kennedy, J. A new optimizer using particle swarm theory. In Proceedings of the Sixth International Symposium on Micro Machine and Human Science, Nagoya, Japan, 4–6 October 1995. [CrossRef]
34. Mirjalili, S.; Lewis, A. The Whale Optimization Algorithm. *Adv. Eng. Softw.* **2016**, *95*, 51–67. [CrossRef]
35. Yang, H.; Cheng, W.; Han, G. Wood Modification at High Temperature and Pressurized Steam: A Relational Model of Mechanical Properties Based on a Neural Network. *Bioresources* **2015**, *10*, 5758–5776. [CrossRef]

forests

MDPI

Article

Phase-Change-Material-Impregnated Wood for Potential Energy-Saving Building Materials

Ahmet Can [1,2,*], Seng Hua Lee [3,4], Petar Antov [5] and Muhammad Aizat Abd Ghani [6,*]

[1] Faculty of Forestry, Bartın University, Bartın 74100, Turkey
[2] Faculty of Forestry, Bursa Technical University, Bursa 16310, Turkey
[3] Department of Wood Industry, Faculty of Applied Sciences, Universiti Teknologi MARA (UiTM) Cawangan Pahang Kampus Jengka, Bandar Tun Razak 26400, Pahang, Malaysia
[4] Institute of Tropical Forestry and Forest Products, Universiti Putra Malaysia (UPM), Serdang 43400, Selangor, Malaysia
[5] Faculty of Forest Industry, University of Forestry, 1797 Sofia, Bulgaria
[6] Faculty of Tropical Forestry, University Malaysia Sabah, Pantai UMS, Jalan Masjid, Jalan UMS, Kota Kinabalu 88400, Sabah, Malaysia
* Correspondence: acan@bartin.edu.tr (A.C.); muhammad.aizat@ums.edu.my (M.A.A.G.)

Abstract: PCMs (phase change materials) are ideal for thermal management solutions in buildings. This is because they release and store thermal energy during melting and freezing. When this material freezes, it releases a lot of energy in the form of latent heat of fusion or crystallization energy. Conversely, when the material melts, it absorbs the same amount of energy from its surroundings as it changes from a solid to a liquid state. In this study, Oriental spruce (*Picea orientalis* L.) sapwood was impregnated with three different commercial PCMs. The biological properties and the hygroscopic and thermal performance of the PCM-impregnated wood were studied. The morphology of PCM-impregnated wood was characterized through the use of scanning electron microscopy (SEM), Fourier-transform infrared spectroscopy (FT-IR), thermogravimetric analysis (TGA), and differential scanning calorimetry (DSC). PCM-impregnated wood demonstrated low performance in terms of storing and releasing heat during phase change processes, as confirmed by DSC. The results show that PCMs possess excellent thermal stability at working temperatures, and the most satisfying sample is PCM1W, with a phase change enthalpy of 40.34 J/g and a phase change temperature of 21.49 °C. This study revealed that PCMs are resistant to wood-destroying fungi. After the 96 h water absorption test, the water absorption of the wood samples decreased by 28%, and the tangential swelling decreased by 75%. In addition, it has been proven on a laboratory scale that the PCM material used is highly resistant to biological attacks. However, large-scale pilot studies are still needed.

Keywords: phase change material; spruce wood; wood impregnation

Citation: Can, A.; Lee, S.H.; Antov, P.; Abd Ghani, M.A. Phase-Change-Material-Impregnated Wood for Potential Energy-Saving Building Materials. *Forests* **2023**, *14*, 514. https://doi.org/10.3390/f14030514

Academic Editors: Morwenna Spear and Miklós Bak

Received: 30 January 2023
Revised: 10 February 2023
Accepted: 21 February 2023
Published: 6 March 2023

1. Introduction

With the increase in the world population, the development of the construction industry in a globalising world has increased. Building energy consumption, which constitutes approximately 40% of the total energy consumption, has increased rapidly [1,2]. Therefore, conserving energy and utilising daylight instead of artificial lighting will be effective in reducing energy consumption in the building sector. The most efficient way to benefit from solar energy is to store the sun's energy. Heat storage technology also aids in the efficient use of energy and is crucial in terms of energy savings. Phase change materials (PCMs) have been investigated for thermal energy storage applications because they have high thermal storage densities, and the nearly isothermal process can store energy using latent heat [3,4]. PCMs are ideal for thermal management solutions in buildings. This is due to the fact that they both store and release thermal energy during melting and freezing. When this material freezes, it releases a significant amount of energy as latent heat of fusion or crystallisation energy. When a material melts, it absorbs the same amount of energy from its surroundings as it transitions from a solid to a liquid state [5]. Owing to this property, they can be used in a variety of applications, including the following: (1) building insulation, as PCMs can be used to help regulate indoor temperatures and reduce heating and cooling costs in buildings; (2) thermal energy storage: PCMs can be used in thermal energy storage systems to store and release the excess heat generated by renewable energy sources such as solar panels.; (3) refrigeration and air conditioning, PCMs can be used to improve efficiency and reduce energy consumption in refrigeration and air conditioning systems; (4) clothing and textiles: PCMs can be incorporated into clothing and textiles to provide people working in hot or cold environments with comfortable, temperature-regulating clothing; (5) packaging, PCMs can be used in packaging to help temperature-sensitive products, such as food and pharmaceuticals, to maintain their temperature; (6) electronics, PCMs can be used to regulate temperatures and reduce the risk of overheating in electronics such as laptops and smartphones; and (7) These are just a few of the many potential applications of PCMs, and research into their use is ongoing as scientists and engineers seek new and innovative ways to leverage their unique properties [6–9]. However, the long-term use of PCMs is limited due to the fact that they create a problem of leaching from the environment when they absorb heat from the environment during phase change [2,10]. Researchers have tried various methods to overcome this problem. The most important of these methods are microencapsulation and polymer blending [11,12]. The technique in which a liquid or solid package is covered with another material is called microencapsulation. In this technique, the main part of the PCM matrix, Jams, are used as shells to microencapsulate PCMs. These are melamine–formaldehyde (MF) jams [13,14], poly (urea–urethane) [15], polyurea [16,17], urea–formaldehyde (UF) jams [18], and acrylic jams [19]. The high encapsulation costs and low thermal conductivity of the encapsulated material have limited the use of the encapsulation technique [20–22]. Another method to reduce PCM leaching is to produce shape-stabilised PCMs (SPCMs) or form-stable composites. The advantage of the SPCM method is that the material used remains solid even at temperatures higher than its melting point [20–24]. Organic and inorganic materials are used in this technique. When using an organic polymer matrix, the chemical compatibility and thermal stability of polymers should be considered [20]. Various polymer materials are used in this technique, such as low-density polyethylene [25], styrene–maleic anhydride copolymer (SMA) [26], polymethyl methacrylate [27,28], high-density polyethylene [29], polyurethane [30], polypyrrole [31], polyvinyl alcohol [32–34], and biodegradable polymers, such as cellulose, chitosan, and agarose [35].

Since ancient times, wood has been used as a building material. The superiority of wood materials is enhanced by their sustainability and renewable nature. The incorporation of PCMs into the wood lignocellulosic matrices to reduce the leaching of the material from the environment represents a thermal energy storage technology that is a flexible and reliable way to store heat [36]. PCMs added to the building construction material melt with the heating of the building envelope during the day, causing less heat entry into the building. PCMs, which freeze at night when the outdoor temperature decreases, contribute to the heating of the building by releasing heat. Thus, with PCMs used in the building material, it is possible to benefit from solar energy passively while a more homogeneous temperature distribution is achieved within the building, and heating and cooling can be realized at lower costs.

There are a few studies using wood and wood flour (WF) as components of PCMs. [36–39]. In these studies, myristyl alcohol [36], fatty acid [37], a eutectic mixture of lauric acid and myristic and capric-palmitic acid [38,39], paraffin [40,41], a eutectic mixture of capric acid (CA), and stearic acid (SA) [42] were used as PCMs. Cheng and Feng [36] developed FSPCMCs (form-stable phase change composites) used on WF as a by-product of the timber industry via delignification and impregnation with myristyl alcohol (MA). The melting and crystallization enthalpy of delignified WF myristyl alcohol phase change composites were slightly higher than the calculated values, especially at the MA loading content of 65 to

75 wt%. According to the result, the latent heat of 166.5 J/g was measured in the composite material produced in the material containing 75% PCM. Jiang et al. [37] impregnated four kinds of WFs as the support material in the fatty acid (FA)/WF composite material used in the production of a form-stable phase change material (FSPCM). The phase change transition temperature and latent heat of the FA/WF composites were found in the range from 53 °C to 40 °C and 103 J/g to 88 J/g, respectively. In another study, eutectic mixtures of lauric acid and myristic acid were used as an FSPCM [38]. WF samples with porous structures were used as form-stable materials. The latent heat value of the composite material prepared with the eutectic mixture/WF was increased by 15% to 54.2% [38]. In another study, wood plastic composites were fabricated using a co-rotating twin-screw extruder using graphite and high-density polyethylene poplar (*Populus tomentasa* Carr.) WF. The authors reported a latent heat of 26.8 J/g for the composite material obtained. They also determined that the graphite added to the composite material increased the thermal conductivity of the material and reduced the bending properties [41]. In another study, Scots pine (*Pinus sylvestris* L.) sapwood was impregnated with a eutectic mixture of capric acid (CA) and stearic acid (SA) as a PCM via a vacuum process. The physical, mechanical, and thermal properties of the wood samples were investigated. According to the results obtained, after 264 h of water removal, low water absorption (WA) and high anti-swelling efficiency (ASE) increased the bending and compression strength of the wood, and the digital scanning calorimetry results showed that a good latent heat value of approximately 94 J/g was obtained at 23.94 °C [42].

Buildings receive a high percentage of solar energy. The impregnation of wood with PCMs will increase the structural applications of wood and wood-based materials, which contain a limited thermal mass, in buildings. Thus, the thermal mass of the buildings can be increased, and a contribution can be made to reduce the environment and climate change by building energy-efficient, thermal comfort, and low-carbon emission buildings. Oriental spruce (*Picea orientalis* L.) sapwood was used in this study due to being widely available in Turkey, low cost for buildings, and suitable for cross-laminated wood glulam use. Fourier-transform infrared spectroscopy (FT-IR) and scanning electron microscopy (SEM) were performed to characterize the chemical properties and morphology of the PCM-impregnated wood. The thermal degradation stability of the PCM-impregnated wood was determined by differential scanning calorimetry (DSC) and thermogravimetry (TG/DTA) techniques. In addition, the performance of the spruce wood samples impregnated with PCMs against white rot fungus (*Trametes versicolor* (L.) Lloyd (Mad-697)) and brown rot fungus (*Coniophora puteana* (Mad-515)) was also evaluated.

2. Materials and Methods

2.1. Materials

Oriental spruce (*Picea orientalis* L.) wood samples consisting of sapwood samples with a density of 0.35 g/cm^3 were used in this study. The sample sizes used in the study were cut at 5 mm (radial, R) × 15 mm (tangential, T) × 30 mm (longitudinal, L) dimensions for decay testing. The other analyses were performed using samples of these dimensions. PCMs of normal paraffin n-C14 were purchased from Sichuan Aishpeier New Material Technology Co., Ltd. (Çengdu, China). The PCM organic phase change material is based on paraffin. The features of the products are listed in Table 1. The only difference between the phase change materials used is their melting and solidification temperatures; everything else is the same. Paraffin is a common choice for use as a PCM in building construction because it has several desirable properties. One of the most important is its high thermal energy storage capacity. This means that it can absorb and store a large amount of heat energy per unit volume, which makes it very efficient for regulating the temperature of a building. Paraffin is also relatively inexpensive and widely available, which makes it a cost-effective option for use in building construction. Additionally, it is stable and non-toxic, making it safe for use in buildings. Another advantage of using paraffin is that it is a solid at room temperature, which means it will not leak or evaporate over time, making it a long-term storage solution. Finally, paraffin is also a good thermal insulator which helps to keep the heat stored inside the material, making the building more energy efficient. All of these properties make paraffin a suitable choice for use as a PCM in building construction and energy-saving projects.

Table 1. Technical specifications of the PCMs used in this work.

Product Name	Normal Paraffin n-C14
Product properties	Liquid
Appearance	Transparent
Enthalpy	180–200 J/kg
Specific heat	3.22 kj/kgK
Coefficient of thermal conductivity	0.21 Wm/K^{-1}
Color	Colorless
Melting point (PCM1, PCM2, PCM3)	20 °C, 25 °C, 30 °C
Density	0.77 g/cm^3
Solubility	Water Insoluble (0.09 ug/L, 25 °C) Soluble in: ether, alcohol, acetone
Chemical stability	Generally stable

2.2. Impregnation of Wood Samples with PCM (PCMW)

The wood samples prepared in the decay test sample size were conditioned for 1 month at 25 °C and 65% relative humidity before impregnation. The conditioned wood samples were dried in an oven (103 °C) until they reached a constant weight (2%–3%). The wood samples that reached a constant weight state were impregnated with PCMs. The impregnation process was carried out under a vacuum (650 mmHg) for 45 min and atmospheric pressure for 4 h. The temperature was fixed at 30 °C in a vacuum and atmospheric pressure process. Then, to calculate the weight gain of the samples, they were kept at 80 °C until they reached full dry weight. The PCMs were fabricated inside of the wood after over-drying, and the preparation phase change energy storage wood was named PCMW. The weight percentage gain (WPG) achieved for PCM1, PCM2, and PCM3 was 79.33%, 24.03%, and 67.17%, respectively. After the atmospheric pressure treatment, the wood samples were conditioned for 4 weeks, and the oven dry mass was recorded prior to a leaching test at 20 °C and 65% relative humidity in the air-conditioning cabinet (NÜVE TK 252).

2.3. Characterization of PCMs and PCMW

The thermal stability of the PCM and PCMW was evaluated using a thermogravimetric analyzer (Perkin Elmer Pyris™ 1 TGA) at a scanning rate of 10 °C min^{-1} in the temperature range of 20–500 °C under a sustained stream of nitrogen atmosphere. The graphics of the TG and DTG were drawn with the Origin 2019b program. The chemical structure of PCM, PCMW and spruce wood samples were recorded using a Fourier transformation infrared spectroscope (Bruker, Karlsruhe, Germany). The spectra ranged between 400 and 4000 cm^{-1} wavelength and with a 4 cm^{-1} resolution using a Bruker ATR-FTIR spectrometer. The radial surface of the wood was used to ATR-FTIR. The measurements were made with solid samples. The spectra were baseline corrected and smoothed using the OPUS software (Bruker Optics GmbH, Ettlingen, Germany). The graphics of the results obtained were drawn with the Origin 2019b program. The thermal properties of the PCM and PCMW were measured using a differential scanning calorimeter (DSC, Perkin Elmer Pyris™ 1 DSC, temperature accuracy: ±0.2 °C, enthalpy accuracy: ±5%) under a sustained stream of argon, with the cooling and heating rate of 5 °C/min. The test temperature range was set from 0 to 40 °C. The graphics of the melting and solidifying temperature were drawn with the Origin 2019b program.

2.4. Leaching and Decay Test

The leaching test was carried out according to the conditions specified by the EN84 [43] standard for 14 days prior to the decay test. The purpose of leaching is to evaluate any loss in the effectiveness of decay resistance. The decay test was carried out according to the principles of EN 113-2 [44]. Some changes were made, such as the sample dimensions, the use of Kolle flasks, and the total test period in the standard EN 113. A brown rot fungus, namely *Coniophora puteana* (Mad-515) and a white rot fungus, namely *Trametes versicolor* (L.) Lloyd (Mad-697) were inoculated to a sterile malt extract agar medium in Petri dishes. The wood samples were placed on the growing mycelium in each Petri dish and then incubated at 20 °C and 70% relative humidity for 2 months. The weight loss was calculated on the basis of the oven-dried weight before and after the decay test.

2.5. Water Absorption Test

In the study, wood samples with a dimension of 20 mm × 20 mm × 10 mm were tested for their water absorption properties. The test was carried out by immersing the oven-dried samples in a beaker filled with distilled water. The relative water absorption and tangential swelling were calculated based on the initial oven dry weight and tangential direction of the samples. Ten replicates were used for each group.

3. Results and Discussion

3.1. Thermal Stability Analysis of PCMs-Impregnated Woods

The TG and DTG curves of the PCMs tested are given in Figures 1 and 2, while the TG and DTG curves of wood and PCM-impregnated wood samples (PCMW) are shown in Figures 3 and 4.

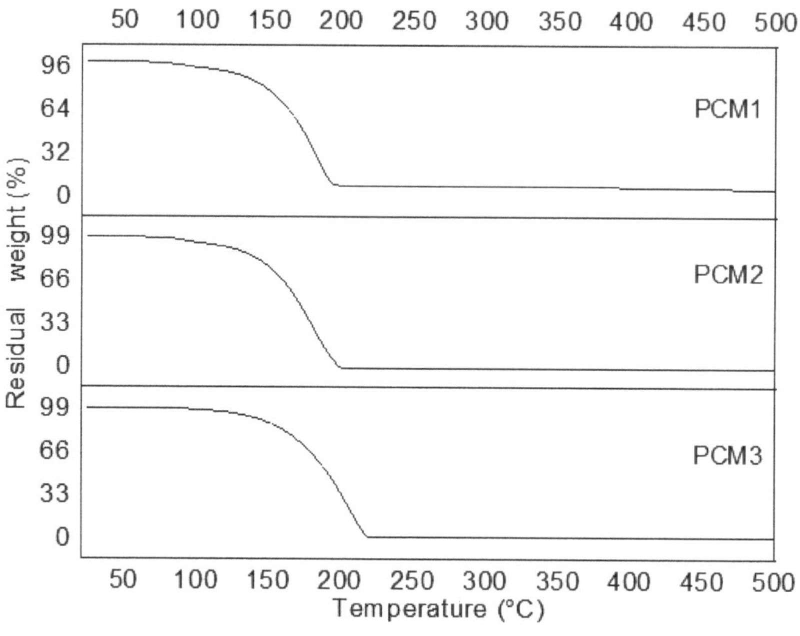

Figure 1. TGA curves of the PCM1-3.

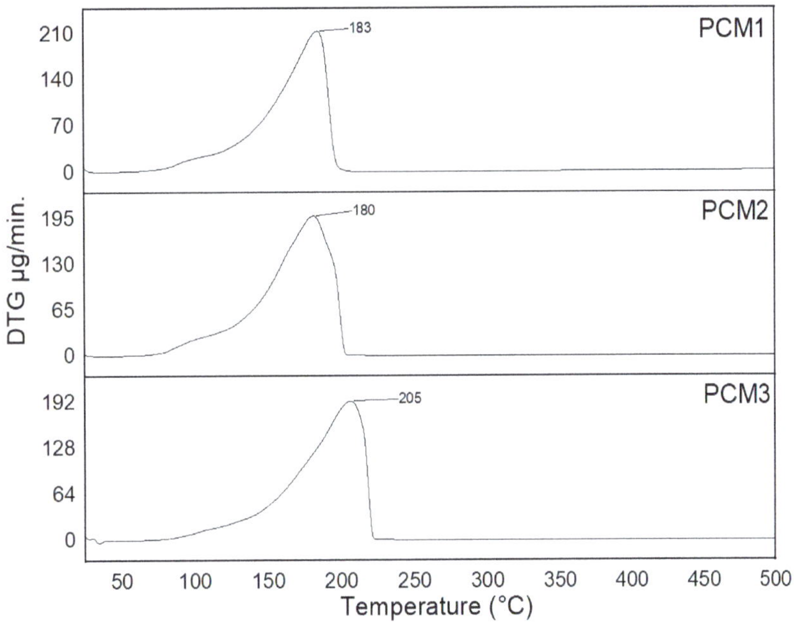

Figure 2. DTG curves of the PCM1-3.

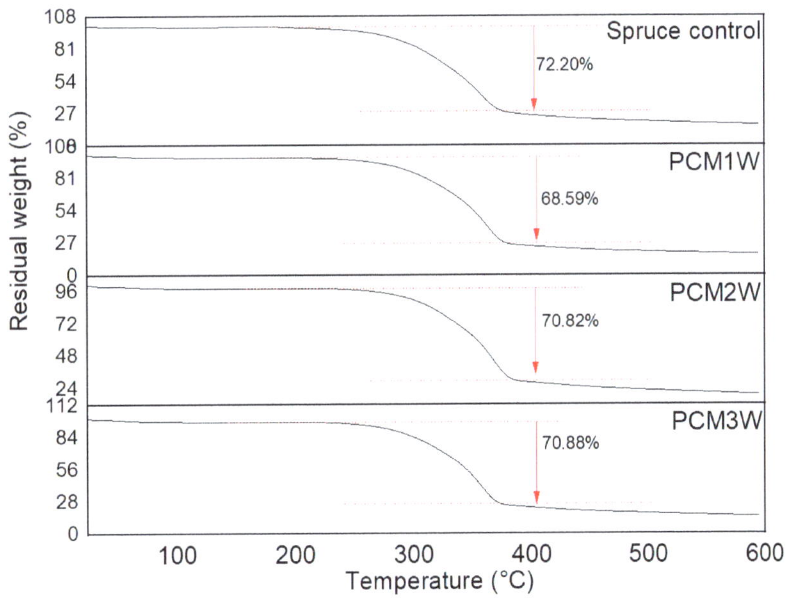

Figure 3. TGA curves of the PCMW1-3.

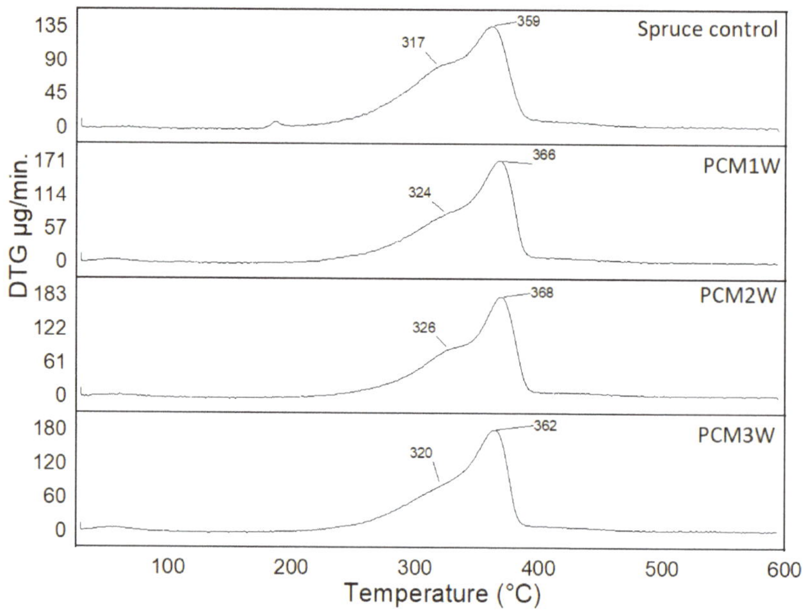

Figure 4. DTG curves of the PCMW1-3.

It appears that the PCMs in Figures 1 and 2 and Table 2 decomposed at low temperatures, specifically at 200 °C, and lost all of their weight at this temperature. This suggests that the materials were not stable at this temperature and may not be suitable for certain applications. The wood samples impregnated with PCMs were found to deteriorate at higher temperatures, in contrast to the PCMs alone, which decomposed at lower temperatures. This suggests that the addition of PCMs to the wood samples may have an impact on their thermal stability and may affect their suitability for certain applications. It would be interesting to know the exact temperature range at which the deterioration of wood samples occurs, as well as the type of PCM used for impregnating the wood. Water and volatile compounds are removed at around 100 degrees in the wood samples. Hemicellulose, cellulose, and lignin, which are the main components of wood, degrade at 100–365 °C, 270–400 °C, and 400–600 °C temperature ranges, respectively. Lignin is one of the main components of wood and is responsible for providing the material with its strength and rigidity. The temperatures mentioned, between 410 and 620 °C, are in the range of temperatures at which lignin begins to break down and degrade. At these temperatures, cellulose and hemicellulose are also broken down, but lignin is the first component to degrade [45]. Wood impregnated with PCM1 showed the slowest degradation in this range of temperatures, followed by samples impregnated with PCM2 and PCM3. Untreated wood showed the fastest degradation rate, but it revealed the least degraded sample, cumulatively. The impregnation of PCMs helps to stabilise the composite sample at 324 °C, 326 °C, and 320 °C for PCM1, PCM2, and PCM3, respectively. While the highest weight loss occurred at 366 °C (PCM1W), 368 °C (PCM2W, and 362 °C (PCM3W) in wood samples impregnated with PCMs, the highest weight loss was achieved in the spruce wood sample at 359 °C with the rate 72.20%, 68.59%, 70.82%, and 70.88%, respectively. The PCM1-, PCM2-, and PCM3-impregnated wood left a higher residue of 17.31%, 19.40%, and 15.24% at 600 °C, respectively, compared to the wood with 17.40%. This result indicates that PCMs-impregnated wood had thermal stability at extreme temperatures compared to the control wood. In the literature study, it was reported that the residual rate of the Scot's pine wood samples impregnated with capric acid was the same as the control samples, but the thermal stability increased [46].

Table 2. TGA data for PCMs, spruce wood and PCM-impregnated wood samples.

	Degradation Interval (°C)	% Mass Loss
PCM1	95–224	100
PCM2	95–203	100
PCM3	128–201	100
Spruce wood	180–400	72.20
PCM1W	200–390	68.59
PCM2W	200–390	70.82
PCM3W	200–390	70.88

3.2. FT-IR Analysis of PCMs-Impregnated Wood Samples

The FT-IR spectra of the wood, PCM, and impregnated wood are shown in Figures 5 and 6. The FTIR spectrum of spruce wood is shown in Figure 5. The 3342 cm^{-1} peaks show strong main OH stretching, and 2800–3000 cm^{-1} shows C–H stretching in the methyl and methylene groups. The other characteristics peak at 1735 cm^{-1}, acetyl groups in xylan, and other non-conjugated carbonyls), C=O Stretch in xylan (hemicelluloses), the peak at 1508 cm^{-1} shows the aromatic skeletal vibration of lignin, 1427 cm^{-1} Aromatic skeletal vibration in lignin with C-H deformation 1369 and 897 cm^{-1} are related to carbohydrates. The peak at 1265 cm^{-1} indicates a C-O stretching lignin ring in the G units and a 1026 cm^{-1} C-O vibration in cellulose and hemicellulose.

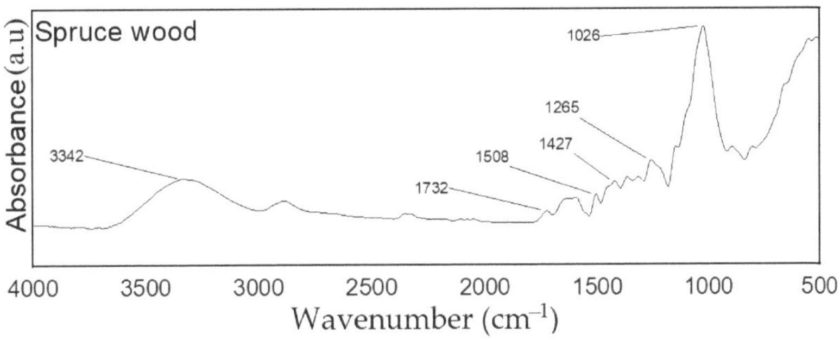

Figure 5. FTIR-ATR spectra of spruce wood (Control).

The FT-IR spectra of paraffin-based PCMs and PCM-impregnated wood (PCMW) are shown in Figure 6. Since the PCMs are paraffin-based, they showed paraffin properties. In the infrared (IR) spectrum of a PCM, the peak at 2912–2921 cm^{-1} and 2852–2847 cm^{-1} are typically associated with the symmetric stretching vibrations of the -CH_3 and -CH_2 groups, respectively. These peaks are specific to organic compounds and are a result of the bond between the carbon and hydrogen atoms in these groups. The IR spectrum of a PCM can provide valuable information about its chemical structure and properties, which can be useful for identifying and characterizing different types of PCMs. The peak at 2953–2957 cm^{-1} represents the hydrogen-bonding vibration of methyl (CH_3-). The asymmetric deformation vibration peaks of the paraffin methyl group (CH_3-) are around 1466–1470 cm^{-1}, and the symmetrical deformation vibration of the paraffin methyl group (CH3-) are around 1369–1377 cm^{-1}.

Some changes in the chemical structure of spruce wood occurred upon impregnation with PCMs. Impregnation is the process of introducing a liquid into a solid; in this case, the PCM is introduced into the spruce wood samples. Depending on the type of PCM and the impregnation method used, chemical reactions may occur between the PCM and the wood, which can lead to changes in the chemical composition of the wood. Additionally, the presence of the PCM within the wood may affect the physical properties of the wood, such as its dimensional stability. In general, the chemical structure of PCMs is also seen in wood samples. In particular, peaks of 2852 cm^{-1}, 2922 cm^{-1}, and 2958 cm^{-1} were also observed in wood samples. In addition, significant increases in the 1465 and 1261 cm^{-1} peaks in the PCM3W samples indicate asymmetric deformation vibration of the paraffin methyl group (CH3-) and C-O stretch lignin ring in guaiacyl units, respectively. In the PCM2W samples, an increase was observed at the 1259 cm^{-1} peak, which means the CO and OH groups in the hemicellulose, cellulose, and lignin; in PCM1W samples, on the other hand, a peak occurred at 1465 cm^{-1}, which means C-H deformation of lignin and carbohydrates. In addition, the 720 cm^{-1} peak, which is one of the PCMs characteristic peaks, appeared at 721 cm^{-1}, 798 cm^{-1}, and 804 cm^{-1} in wood samples (PCM1W, PCM2W, PCM3W), respectively.

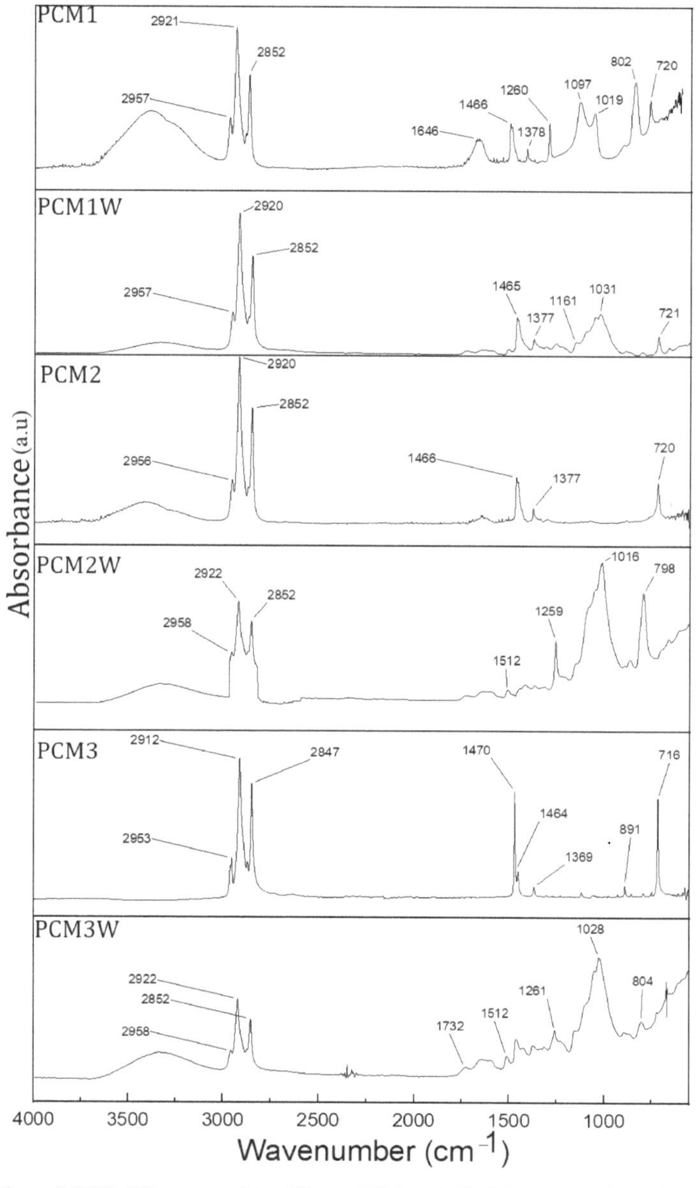

Figure 6. FTIR-ATR spectra of the different PCMs and PCMs impregnated wood (PCMW).

3.3. Thermal Properties Analysis of PCMs-Impregnated Woods

The thermal properties of PCMs and PCMW, including onset temperature, peak temperature, and latent heat during melting and solidifying processes, were analyzed using DSC, and the specific experimental data are listed in Table 3. Figure 7 describe the DSC curve of PCM1-3 and Figure 8 describe the DSC curve of PCM1-3W. The melting temperatures of PCM1, PCM2, and PCM3 were 22.07 °C, 21.77 °C, and 28.30 °C, respectively, while they were determined as 21.49 °C for PCM1W, 26.63 °C for PCM3W. The results obtained might be attributed to the following: (1) confinement of PCMs in the pores of the wood after the impregnation process, (2) the hydrogen bond between the paraffin and the lignin, cellulose, and hemicellulose components of the wood, (3) and the effect of capillary forces between the PCM molecules and cell walls of the wood. The melting latent heats of the PCMs were 168.71 J/g, 158.97 J/g, and 211 J/g, respectively, while PCMs impregnated with wood were lower than PCMs, which were 40.34 J/g for PCM1W and 30.33 J/g for PCM3W. No results were obtained for the PCM2W samples (Table 3). From the DSC data in Table 3, it can be seen that the solidifying temperatures of the PCMs are 19.27 °C, 18.43 °C, and 24.07 °C, respectively, wherein PCM1W is higher than PCM1 and PCM3W is lower than PCM3. In the literature study, it was found that the impregnation ratio of the CA-PA mixture in the DW reached 61.2%. At this impregnation rate, 94.4 J/g of latent heat was obtained [39]. Amini et al. [46] reported that a retention value of 267 kg/m^3 was obtained for the Scotch pine wood samples impregnated with an 80% concentration of capric acid. A latent heat of 70.5 J/g was obtained at this retention level. In a study by Mathis et al. [9], red oak (*Quercus rubra* L.) and sugar maple (*Acer saccharum* Marsh.) were impregnated in a reactor with a microencapsulated phase change material. A latent heat storage of 7.6 J/g was measured for the impregnated red oak samples, and 4.5 J/g was measured for the impregnated sugar maple samples.

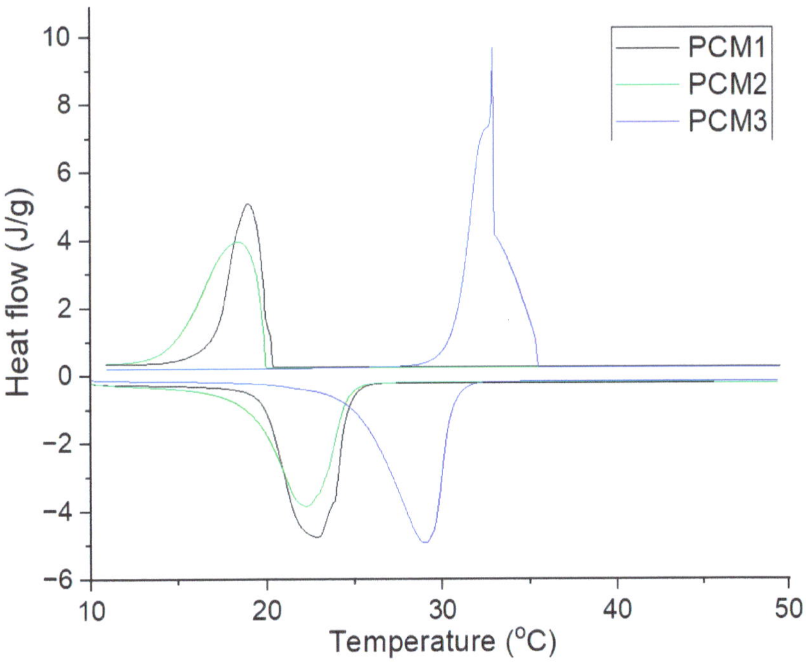

Figure 7. Melting and solidifying DSC curves of the PCMs.

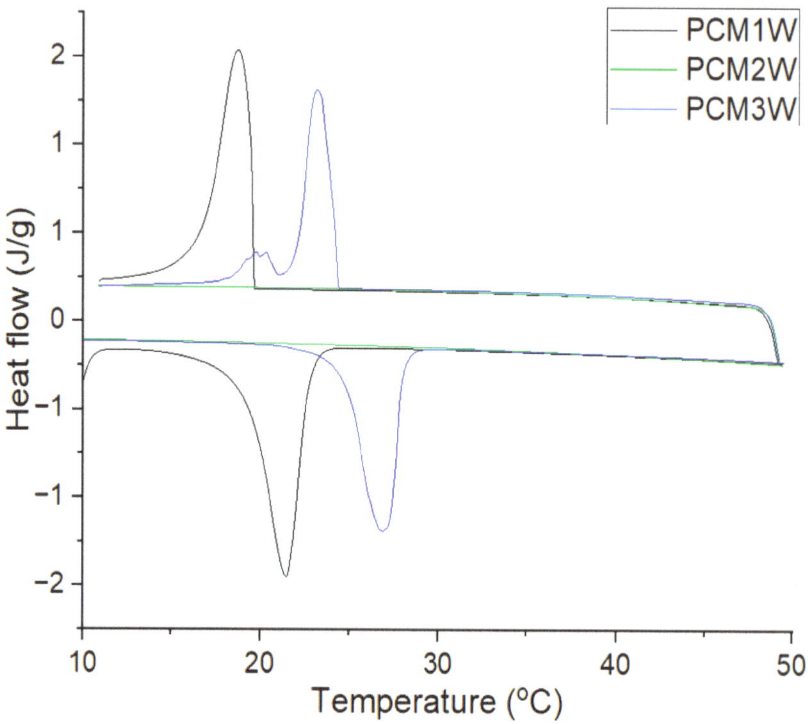

Figure 8. Melting and solidifying DSC curves of the PCM-impregnated wood (PCMW).

The energy storage efficiency can be calculated by the following formula:

$$E = [(\Delta HPCMW_m + \Delta HPCMW_s)/(\Delta HPCM_m + \Delta HPCM_s)] \times 100 \quad (1)$$

where $\Delta HPCMW_m$ and $\Delta HPCMW_s$ are the latent heat of the PCM-impregnated wood during the melting and solidifying process, $\Delta HPCM_m$ and $\Delta HPCM_s$ are the latent heat of the PCM during the melting and solidifying process, and E refers to the energy storage efficiency of PCMW. According to this Equation (1), the energy storage efficiency of the PCM1W is 23.56%, and PCM3W is 11.75%.

Table 3. DSC data of the PCM and PCM impregnated wood.

Samples	Melting			Solidifying		
	Onset Temperature (°C)	Peak Temperature (°C)	Latent Heat (J/g)	Onset Temperature (°C)	Peak Temperature (°C)	Latent Heat (J/g)
PCM1	20.63	22.07	168.71	20.27	19.27	179.09
PCM1W	19.19	21.49	40.34	19.65	18.41	41.63
PCM2	19.72	21.77	158.97	19.92	18.43	170.16
PCM2W	-	-	-	-	-	-
PCM3	25.75	28.30	211.26	25.09	24.07	210.15
PCM3W	24.77	26.63	30.33	25.04	24.18	19.19

3.4. Mass Loss of PCMs-Impregnated Woods

When the untreated control sample was exposed to white rot fungus *Trametes versicolor* (L.) Lloyd (Mad-697) and brown rot fungus *Coniophora puteana* (Mad-515) for 8 weeks of incubation, the highest mass loss (22.80% and 22.56%, respectively) was recorded. As the decay test standard (EN 113-2, 2020) stated, the minimum mass loss of the control samples should not exceed 20% mass loss. Therefore, these values corresponded to the "nondurable" class of durability ratings.

Conversely, minimum weight losses of <10% were noticed in the specimens impregnated with PCM1. The modified samples were usually more extremely attacked by the white rot fungus *T. versicolor* (from 2% to 4.7%) than by the brown rot fungus *C. puteana* (from 3% to 8.2%).

In studies on paraffin-based materials, it has been reported that wood attacked by wood-rotting fungi might degrade more slowly when impregnated with waxes. Wood-rotting fungi are a common source of decay and deterioration in wooden structures, and they can significantly reduce the strength and durability of wood. According to research, impregnating wood with waxes can slow the rate of degradation caused by wood-rotting fungi. This is because the waxes create a barrier that prevents the fungi from penetrating the wood and breaking down the cellulose and lignin. Additionally, some waxes also contain fungicides that can kill or inhibit the growth of fungi. This technique is known as "wax impregnation"; it is one of the methods used to protect the wood from decay and extend the service life of wooden structures. Paraffin wax, when applied to wood, forms a barrier that prevents the penetration of both fungal enzymes and fungal mycelium into the cell walls and wood structural components. Filling the wood cell walls and lumens with paraffin wax creates a physical barrier that makes it difficult for fungal enzymes to break down the cellulose and lignin in the wood. Furthermore, the wax barrier prevents fungal mycelium, the vegetative growth of the fungus, from penetrating the wood, preventing the fungus from spreading and colonising the wood. This helps to protect the wood and extend its service life by slowing fungal degradation. The wax penetration can be conducted by using a pressure method, by melting the wax and immersing the wood in the wax, or by using a vacuum method. Specific changes in the molecular structure of wood that cause an increase in its resistance to decay occur as a result of parallel PCM impregnation. The inhibition of fungal growth is also a result of polysaccharide dihydroxylation, resulting in a lower moisture content of the wood [47–49]. In this study, while an 80% final humidity was obtained in the control samples, the final humidity was obtained between 40% and 50% in the test samples. According to Liu et al. [50], the highest mass loss (64%) occurred when the untreated post-decay sample was exposed to white rot fungus for 12 weeks of incubation, indicating that *Trametes versicolor* was active. The Chinese standard states that a mass loss higher than 45% demonstrates the effectiveness of the fungal tests. The mass losses of the three paraffin wax emulsion-treated groups were reduced to 62%, 61%, and 58%. It has been reported that the paraffin wax emulsion process used at low concentrations only slows and does not completely prevent fungal attacks. According to Humar et al. [51] paraffin wax inhibits the growth of fungi by reducing the moisture content. The durability of the wood improved when treated with paraffin wax. Other reports found that the paraffin wax treatment can slow the growth of basidiomycetes due to the hydrophobicity of the paraffin wax [48,51]. However, in this study, it seems that the paraffin wax material included in the capsule provided a high level of protection.

3.5. Hygroscopic Properties of PCMs-Impregnated Woods

Figure 9 shows the water absorption and tangential swelling of spruce wood impregnated with the three PCMs from 20 min to 96 h soaking time.

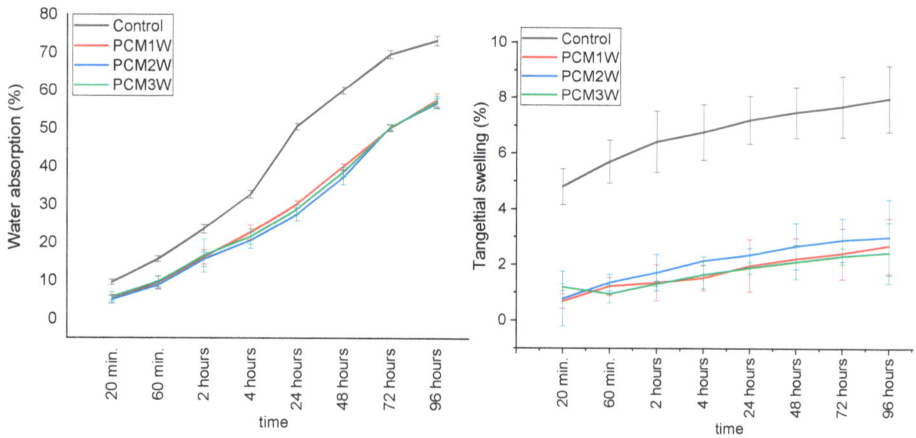

Figure 9. Water absorption (%) and tangential swelling (%) of wood impregnated with PCMs.

In this study, all of the PCM-impregnated samples exhibited significantly lower water absorption and tangential swelling than that of the control samples. The 96 h water absorption test results show that the control samples had a higher water content (73.4%) compared to the test samples (57.69%, 56.8%, and 57.24%). Similarly, while the tangential swelling of the control samples was 7.98%, the tangential swelling rates of the test samples were 2.7%, 3.0% and 2.45%, respectively. For water absorption, the effects were less prominent during the first 2 h of soaking, but the absorption inhibition effects of the PCM-treated wood started to show significant differences as the soaking period progressed. As for tangential swelling, PCM impregnation bestowed significantly better dimensional stability to the wood, and the effect can be observed immediately even after merely 20 min soaking period. The PCMs used in this study are not water-soluble and do not leach easily. When applied to wood, they are deposited into the wood cells and lumen, which helps to limit moisture absorption and protect the wood from decay. The interaction between the wood and paraffin can also prevent hydrogen bonds between the wood and water molecules from forming. Because paraffin is a hydrocarbon, its molecules can interact with the wood cells and form a barrier, making it more difficult for water molecules to penetrate and be absorbed [52]. There was no difference between the PCMs used. The wood samples impregnated with all three PCMs absorbed the same amount of water. Longer soaking periods can result in higher water absorption in wood. When the wood is soaked in water for a longer period, it allows more time for moisture to penetrate into the wood. As the water is absorbed, the wood cells expand, causing the wood to increase in weight and volume. Water absorption refers to this increase in weight and volume, and the longer the wood is soaked, the higher the water absorption value. Water absorption can cause the wood to swell and change dimensions, which can affect the physical and mechanical properties of the wood [46].

4. Conclusions

In this study, important properties of spruce wood samples impregnated with PCMs were revealed. Wood samples with three different PCMs were impregnated using the vacuum pressure process. Wood samples impregnated with PCMs showed positive properties for use as thermal-regulating components in buildings.

- The DSC results revealed melting enthalpy values of up to 41.63 J/g to PCM1, which are beneficial for latent heat energy storage.
- The fungal test results showed that wood samples impregnated with PCMs showed resistance to fungi, specifying that it is safe to treat wood with PCMs.
- Spruce wood samples impregnated with PCMs have been shown to have potential as thermal regulation building materials. PCMs are capable of absorbing and releasing heat energy as they change phase (i.e., solid to liquid and vice versa), which can help to minimize temperature fluctuations in a building, thereby reducing energy consumption for heating and cooling. This is particularly useful in reducing temperature changes during the day and night.
- Potential future plans for PCMW include the following:
- The integration of PCMW into building materials, such as wall insulation and roof tiles, for passive thermal energy storage.
- Investigation of the potential for using PCMW in combination with thermal management systems, such as heat pumps and refrigeration systems, to improve overall energy efficiency.

Author Contributions: Writing—review and editing, A.C., S.H.L. and P.A.; Writing—original draft, A.C.; Visualization, A.C.; Validation, P.A.; Resources, A.C.; Methodology, A.C.; Investigation, A.C.; Formal analysis, A.C. and S.H.L.; Data curation, A.C.; Conceptualization, P.A. and M.A.A.G.; Funding acquisition, M.A.A.G. All authors have read and agreed to the published version of the manuscript.

Funding: The Article Processing Charge (APC) was covered by the Universiti Malaysia Sabah.

Data Availability Statement: No data were used for the research described in the article.

Acknowledgments: The authors would like to thank Bursa Technical University, Faculty of Forestry Dean's Office for the use of the FTIR device.

Conflicts of Interest: The authors declare no conflict of interest.

References

1. Cao, X.; Dai, X.; Liu, J. Building energy-consumption status worldwide and the state-of-the-art technologies for zero-energy buildings during the past decade. *Energy Build.* **2016**, *128*, 198–213. [CrossRef]
2. Qiu, F.; Song, S.; Li, D.; Liu, Y.; Wang, Y.; Dong, L. Experimental investigation on improvement of latent heat and thermal conductivity of shape-stable phase-change materials using modified fly ash. *J. Clean. Prod.* **2020**, *246*, 118952. [CrossRef]
3. Sharma, A.; Tyagi, V.V.; Chen, C.R.; Buddhi, D. Review on thermal energy storage with phase change materials and applications Renew. *Sustain. Energy Rev.* **2009**, *13*, 318–345. [CrossRef]
4. Pielichowska, K.; Pielichowski, K. Phase change materials for thermal energy storage. *Prog. Mater. Sci.* **2014**, *65*, 67–123. [CrossRef]
5. Soares, N.; Costa, J.J.; Gaspar, A.R.; Santos, P. Review of passive PCM latent heat thermal energy storage systems towards buildings' energy efficiency. *Energy Build.* **2013**, *59*, 82–103. [CrossRef]
6. Fragnito, A.; Bianco, N.; Iasiello, M.; Mauro, G.M.; Mongibello, L. Experimental and numerical analysis of a phase change material-based shell-and-tube heat exchanger for cold thermal energy storage. *J. Energy Storage* **2002**, *56*, 105975. [CrossRef]
7. Pekdogan, T.; Tokuç, A.; Ezan, M.A.; Başaran, T. Experimental investigation on heat transfer and air flow behavior of latent heat storage unit in a facade integrated ventilation system. *J. Energy Storage* **2021**, *44*, 103367. [CrossRef]
8. Ben Khedher, N. Numerical Study of the Thermal Behavior of a Composite Phase Change Material (PCM) Room. *Eng. Technol. Appl. Sci. Res.* **2018**, *8*, 2663–2667. [CrossRef]
9. Mathis, D.; Blanchet, P.; Landry, V.; Lagière, P. Impregnation of wood with microencapsulated bio-based phase change materials for high thermal mass engineered wood flooring. *Appl. Sci.* **2018**, *8*, 2696. [CrossRef]
10. Montanari, C.; Li, Y.; Chen, H.; Yan, M.; Berglund, L.A. Transparent wood for thermal energy storage and reversible optical transmittance. *ACS Appl. Mater. Interfaces* **2019**, *11*, 20465–20472. [CrossRef]
11. Li, X.; Chen, H.; Liu, L.; Lu, Z.; Sanjayan, J.G.; Duan, W.H. Development of granular expanded perlite/paraffin phase change material composites and prevention of leakage. *Sol. Energy* **2016**, *137*, 179–188. [CrossRef]
12. Xia, R.; Zhang, W.; Yang, Y.; Zhao, J.; Liu, Y.; Guo, H. Transparent wood with phase change heat storage as novel green energy storage composites for building energy conservation. *J. Clean. Prod.* **2021**, *296*, 126598. [CrossRef]
13. Chen, Z.; Wang, J.; Yu, F.; Zhang, Z.; Gao, X. Preparation and properties of graphene oxide-modified poly(melamine-formaldehyde) microcapsules containing phase change material n-dodecanol for thermal energy storage. *J. Mater. Chem. A* **2015**, *3*, 11624–11630. [CrossRef]
14. Su, J.-F.; Wang, L.-X.; Ren, L. Preparation and characterization of double-MF shell microPCMs used in building materials. *J. Appl. Polym. Sci.* **2005**, *97*, 1755–1762. [CrossRef]

15. Yoo, Y.; Martinez, C.; Youngblood, J.P. Synthesis and Characterization of Microencapsulated Phase Change Materials wit Poly(urea urethane) Shells Containing Cellulose Nanocrystals. *ACS Appl. Mater. Interfaces* **2017**, *9*, 31763–31776. [CrossRef]
16. Park, S.; Lee, Y.; Kim, Y.S.; Lee, H.M.; Kim, J.H.; Cheong, I.W.; Koh, W.-G. Magnetic nanoparticle-embedded PCM nanocapsules based on para_n core and polyurea shell. *Colloids Surf. A Physicochem. Eng. Asp.* **2014**, *450*, 46–51. [CrossRef]
17. Zhang, H.; Wang, X. Synthesis and properties of microencapsulated n-octadecane with polyurea shells containing di_erent soft segments for heat energy storage and thermal regulation. *Sol. Energy Mater. Sol. Cells* **2009**, *93*, 1366–1376. [CrossRef]
18. Su, J.; Wang, L.; Ren, L. Fabrication and thermal properties of microPCMs: Used melamine-formaldehyde resin as shell material. *J. Appl. Polym. Sci.* **2006**, *101*, 1522–1528.
19. Kaygusuz, K.; Alkan, C.; Sarı, A.; Uzun, O. Encapsulated Fatty Acids in an Acrylic Resin as Shape-stabilized Phase Change Materials for Latent Heat Thermal Energy Storage. *Energy Sources Part. A Recover. Util. Environ. E* **2008**, *30*, 1050–1059. [CrossRef]
20. Umair, M.M.; Zhang, Y.; Iqbal, K.; Zhang, S.; Tang, B. Novel strategies and supporting materials applied to shape-stabilize organic phase change materials for thermal energy storage—A review. *Appl. Energy* **2019**, *235*, 846–873. [CrossRef]
21. Khadiran, T.; Hussein, M.Z.; Zainal, Z.; Rusli, R. Encapsulation techniques for organic phase change materials as thermal energy storage medium: A review. *Sol. Energy Mater. Sol. Cells* **2015**, *143*, 78–98. [CrossRef]
22. Zhang, Y.; Xiu, J.; Tang, B.; Lu, R.; Zhang, S.-F. Novel semi-interpenetrating network structural phase change composites with high phase change enthalpy. *AIChE J.* **2017**, *64*, 688–696. [CrossRef]
23. Regin, A.F.; Solanki, S.; Saini, J. Heat transfer characteristics of thermal energy storage system using PCM capsules: A review. *Renew. Sustain. Energy Rev.* **2008**, *12*, 2438–2458. [CrossRef]
24. Wang, C.; Feng, L.; Li, W.; Zheng, J.; Tian, W.; Li, X. Shape-stabilized phase change materials based on polyethylene glycol/porous carbon composite: The influence of the pore structure of the carbon materials. *Sol. Energy Mater. Sol. Cells* **2012**, *105*, 21–26. [CrossRef]
25. Krupa, I.; Miková, G.; Luyt, A.S. Phase change materials based on low-density polyethylene/paraffin wax blends. *Eur. Polym. J.* **2007**, *43*, 4695–4705. [CrossRef]
26. Sánchez, L.; Lacasa, E.; Carmona, M.; Rodríguez, J.; Sánchez, P.; Silva, M.L.S.; Rodríguez, J. Applying an Experimental Design to Improve the Characteristics of Microcapsules Containing Phase Change Materials for Fabric Uses. *Ind. Eng. Chem. Res.* **2008**, *47*, 9783–9790. [CrossRef]
27. Alkan, C.; Sarı, A. Fatty acid/poly(methyl methacrylate) (PMMA) blends as form-stable phase change materials for latent heat thermal energy storage. *Sol. Energy* **2008**, *82*, 118–124. [CrossRef]
28. Wang, L.; Meng, D. Fatty acid eutectic/polymethyl methacrylate composite as form-stable phase change material for thermal energy storage. *Appl. Energy* **2010**, *87*, 2660–2665. [CrossRef]
29. Cai, Y.; Wei, Q.; Huang, F.; Lin, S.; Chen, F.; Gao, W. Thermal stability, latent heat and flame retardant properties of the thermal energy storage phase change materials based on para_n/high density polyethylene composites. *Renew. Energy* **2009**, *34*, 2117–2123. [CrossRef]
30. Wang, X.; Yu, X.; Tian, C.; Wang, J. Preparation and characterization of form-stable para_n/polyurethane composites as phase change materials for thermal energy storage. *Energy Convers. Manag.* **2014**, *77*, 13–21.
31. Silakhori, M.; Metselaar, H.S.C.; Mahlia, T.M.I.; Fauzi, H.; Baradaran, S.; Naghavi, M.S. Palmitic acid/polypyrrole composites as form-stable phase change materials for thermal energy storage. *Energy Convers. Manag.* **2014**, *80*, 491–497. [CrossRef]
32. Sari, A.; Akcay, M.; Soylak, M. Polymer–stearic acid blends aform formrm-stablese change material for thermal energy storage. *Energy Sources* **2005**, *27*, 1535–1546.
33. Sari, A.; Kaygusuz, K. Poly (vinyl alcohol)/fatty acid blends for thermal energy storage. *Energy Sources* **2007**, *29*, 873–883. [CrossRef]
34. Li, Z.; He, W.; Xu, J.; Jiang, M. Preparation and characterization of in situ grafted/crosslinked polyethylene glycol/polyvinyl alcohol composite thermal regulating fiber. *Sol. Energy Mater. Sol. Cells* **2015**, *140*, 193–201. [CrossRef]
35. Sentürk, S.B.; Kahraman, D.; Alkan, C.; Gokce, I. Biodegradable PEG/cellulose, PEG/agarose and PEG/chitosan blends as shape stabilized phase change materials for latent heat energy storage. *Carbohydr. Polym.* **2011**, *84*, 141–144. [CrossRef]
36. Cheng, L.; Feng, J. Form-stable phase change materials based on delignified wood flour for thermal management of buildings. *Compos. Part A* **2020**, *129*, 105690. [CrossRef]
37. Liang, J.; Zhimeng, L.; Ye, Y.; Yanjun, W.; Jingxin, L.; Changlin, Z. Fabrication and characterization of fatty acid/wood-flour composites as novel form-stable phase change materials for thermal energy storage. *Energy Build.* **2018**, *171*, 88–99. [CrossRef]
38. Ma, L.; Guo, C.; Ou, R.; Sun, L.; Wang, Q.; Li, L. Preparation and characterization of modified porous wood flour/lauric-myristic acid eutectic mixture as a form-stable phase change material. *Energy Fuel.* **2018**, *32*, 5453–5461. [CrossRef]
39. Ma, L.; Wang, Q.; Li, L. Delignified wood/capric acid-palmitic acid mixture stable-form phase change material for thermal storage. *Sol. Energy Mater. Sol. Cells* **2019**, *194*, 215–221. [CrossRef]
40. Barreneche, C.; Vecstaudza, J.; Bajare, D.; Fernandez, A. PCM/wood composite to store thermal energy in passive building envelopes. *Proc. IOP Conf. Ser. Mater. Sci. Eng.* **2017**, *251*, 012111. [CrossRef]
41. Guo, X.; Zhang, L.; Cao, J.; Peng, Y. Paraffin/wood flour/high-density polyethylene composites for thermal energy storage material in buildings: A morphology, thermal performance, and mechanical property study. *Polym Compos.* **2018**, *39*, E1643–E1652. [CrossRef]
42. Temiz, A.; Hekimoğlu, G.; Köse Demirel, G.; Sarı, A.; Mohamad Amini, M.H. Phase change material impregnated wood for passive thermal management of timber buildings. *Int. J. Energy Res.* **2020**, *44*, 10495–10505. [CrossRef]
43. *EN 84*; Wood Preservatives—Accelerated Ageing of Treated Wood Prior to Biological Testing—Leaching Procedure. European Committee for Standardization: Brussels, Belgium, 1997.

44. *EN 113-2*; Durability of Wood and Wood-Based Products—Test Method against Wood Destroying Basidiomycetes—Part 2: Assessment of Inherent or Enhanced Durability. European Committee for Standardization: Brussels, Belgium, 2020.
45. Barzegar, R.; Yozgatligil, A.; Olgun, H.; Atimtay, A.T. TGA and kinetic study of different torrefaction conditions of wood biomass under air and oxy-fuel combustion atmospheres. *J. Energy Inst.* **2020**, *93*, 889–898. [CrossRef]
46. Amini, M.H.M.; Temiz, A.; Hekimoğlu, G.; Demirel, G.K.; Sarı, A. Properties of Scots pine wood impregnated with capric acid for potential energy saving building material. *Holzforschung* **2022**, *76*, 744–753. [CrossRef]
47. Archer, K.; Leebow, S. *Primary Wood Processing: Principles and Practice*; Chapter 9—J.C.F. Walker, Wood Preservation; Springer: Dordrecht, The Netherlands, 2006; pp. 297–338. ISBN 978-1-4020-4393-2.
48. Lesar, B.; Humar, M. Use of wax emulsion for improvement of wood durability and sorption properties. *Eur. J. Wood Wood Prod.* **2011**, *69*, 231–238. [CrossRef]
49. Reinprecht, L.; Repák, M. The impact of paraffin-thermal modification of beech wood on its biological, physical and mechanical properties. *Forests* **2019**, *10*, 1102. [CrossRef]
50. Liu, M.; Zhong, H.; Ma, E.; Liu, R. Resistance to fungal decay of paraffin wax emulsion/copper azole compound system treated wood. *Int. Biodeterior. Biodegrad.* **2018**, *129*, 61–66. [CrossRef]
51. Humar, M.; Krzisnik, D.; Lesar, B.; Thaler, N.; Ugovsek, A.; Zupancic, K.; Zlahtic, M. Thermal modification of wax-impregnated wood to enhance its physical, mechanical, and biological properties. *Holzforschung* **2016**, *70*, 411–419. [CrossRef]
52. Chau, T.; Ma, E.; Yang, T.; Cao, J. Moisture Sorption and Hygroexpansion of Paraffin Wax Emulsion–Treated Southern Pine (*Pinus* spp.) under Dynamic Conditions. *For. Prod. J.* **2017**, *67*, 463–470. [CrossRef]

MDPI AG
Grosspeteranlage 5
4052 Basel
Switzerland
Tel.: +41 61 683 77 34

Forests Editorial Office
E-mail: forests@mdpi.com
www.mdpi.com/journal/forests

www.ingramcontent.com/pod-product-compliance
Lightning Source LLC
Chambersburg PA
CBHW041346210926
51850CB00013B/3010